Organolithium Compounds/ Solvated Electrons

Some Aspects of the Chemistry of Polylithiated Aliphatic Hydrocarbons

A. Maercker and M. Theis

Heteroatom Directed Aromatic Lithiation Reactions for the Synthesis of Condensed Heterocyclic Compounds

N. S. Narasimhan and R. S. Mali

Electrochemistry of Solvated Electrons

N. M. Alpatova, L. I. Krishtalik and Yu. V. Pleskov

AKADEMIE-VERLAG BERLIN

Topics in Current Chemistry · Vol. 138

Organolithium Compounds/Solvated Electrons

Organolithium Compounds/ Solvated Electrons

With 14 Figures and 17 Tables

AKADEMIE-VERLAG BERLIN
1987

Die Originalausgabe erscheint im Springer-Verlag
Berlin · Heidelberg · New York · London · Paris · Tokyo
in der Schriftenreihe "Topics in Current Chemistry",
Volume 138

Vertriebsrechte für alle sozialistischen Länder:
Akademie-Verlag Berlin
Vertriebsrechte für alle Staaten mit Ausnahme der sozialistischen Länder:
Springer-Verlag Berlin · Heidelberg · New York · London · Paris · Tokyo

ISBN 3-540-16931-8 Springer-Verlag Berlin · Heidelberg · New York
ISBN 0-387-16931-8 Springer-Verlag New York · Heidelberg · Berlin

ISBN 3-05-500344-6 Akademie-Verlag Berlin

Erschienen im Akademie-Verlag Berlin, DDR-1086 Berlin, Leipziger Straße 3—4
Lizenznummer: 202 · 100/508/87
Printed in the German Democratic Republic
Gesamtherstellung: VEB Druckerei „Thomas Müntzer", 5820 Bad Langensalza
LSV 1275
Bestellnummer: 763 718 0 (3072/138)
13400

Editorial Board

Table of Contents

Some Aspects of the Chemistry of Polylithiated Aliphatic Hydrocarbons
A. Maercker, M. Theis 1

Heteroatom Directed Aromatic Lithiation Reactions for the Synthesis of Condensed Heterocyclic Compounds
N. S. Narasimhan, R. S. Mali. 63

Electrochemistry of Solvated Electrons
N. M. Alpatova, L. I. Krishtalik, Y. V. Pleskov 149

Author Index Volumes 101–138 221

Some Aspects of the Chemistry of Polylithiated Aliphatic Hydrocarbons

Adalbert Maercker and Manfred Theis

Institut für Organische Chemie der Universität Siegen,
Adolf-Reichwein-Straße, D-5900 Siegen, FRG

Table of Contents

1 Introduction 2

2 Theoretical Results 2
2.1 Lithiated Methanes 2
2.2 Lithiated Ethanes 5
2.3 Lithiated Ethenes 9
2.4 Dilithioacetylene 11
2.5 1,3-Dilithiopropane 11
2.6 Other Lithiated C_3-Hydrocarbons 12
2.7 Lithiated C_4-Hydrocarbons 14
2.8 Hypervalent and Related Species 20

3 Syntheses 21
3.1 Halogen-Metal Exchange Relations 21
3.2 Pyrolysis Reactions 22
3.3 Mercury-Lithium Exchange Reactions 25
3.4 Transmetalation Reactions with other Metals 29
3.5 Reductive Metalations 31
3.6 Metalation of Acidic Hydrocarbons 36
3.7 Reactions with Lithium Vapor 39
3.8 Fragmentations 40

4 Properties 41
4.1 Reactions 41
4.2 Structures 47

5 Acknowledgement 57

6 References 57

1 Introduction

Certain acidic hydrocarbons are known to undergo polymetalation by alkyllithium compounds, often with N,N,N',N'-tetramethylethylenediamine (TMEDA) as a catalyst, to give polylithiated organic compounds. Those propargylic, allylic, and benzylic polymetalations already have been reviewed by West [1] and Klein [2,3], the leaders of two most active research groups in this field. The present review article is mainly dealing with polylithiated aliphatic hydrocarbons which cannot be obtained by simple metalation reactions. Often special techniques have been used in order to reach this goal, e.g. reactions with lithium vapor, a method developed by Lagow and his coworkers [4,5]. Many interesting polylithium organic compounds still have to be synthesized, although their remarkable structures — sometimes with anti-van't Hoff geometries — have already been proposed by theoretical calculations especially by Schleyer, Pople, and coworkers [6,7] but also by Streitwieser [8] and others.

2 Theoretical Results

For compounds not available by synthetic methods theoretical calculations provide the only information concerning structural details. Moreover polylithium organic compounds have attracted the interest of theoretical chemists because the geometries in general do not follow classical structural considerations, i.e. replacement of hydrogen in a hydrocarbon by lithium almost always results in some fundamental change in geometry. As in monolithium compounds lithium prefers bridging positions and in certain polylithiated hydrocarbons even double bridging within the molecule is possible. Thereby the lithium atoms by using their free p orbitals reach a higher coordination number which is favorable energetically. On the other hand, factors other than those involving lithium p orbitals, e.g. electrostatic interactions, might also be responsible for bridging and sometimes provide a simpler explanation than multi center covalent bonding.

2.1 Lithiated Methanes

The lithiated methanes have been investigated calculationally already ten years ago by Schleyer, Pople et al. [9,10]. The most exciting result is the decrease of the energy differ-

Table 1. Calculated Planar-Tetrahedral Energy Differences (kcal/mol) for Lithiated Methanes (RHF) [9]

Molecule[a]	STO-3G	4-31G
1	240	168
2	52	42
3 (*trans*)	54	47
3 (*cis*)	17	10
3 (*cis*, triplet)	10	3
4	10	7
5	22	16

[a] Singlet state, if not marked otherwise

ences between the tetrahedral and the (*cis*-)planar structures with increasing lithiation (Table 1).

1a

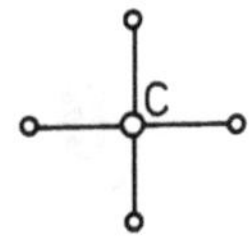

1b

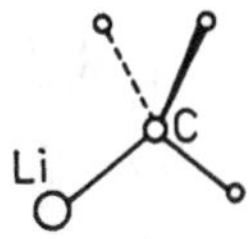

2a

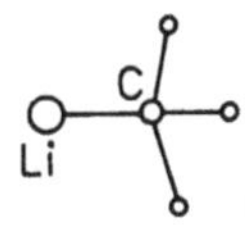

2b

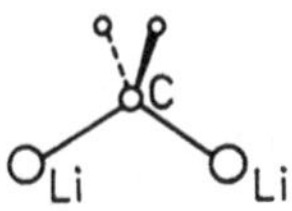

3a

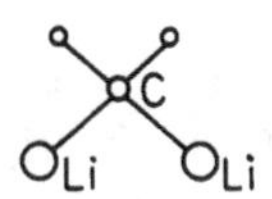

3b (cis)

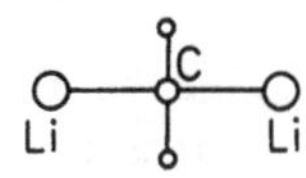

3b (trans)

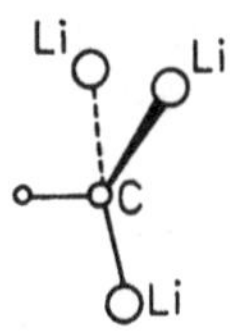

4a

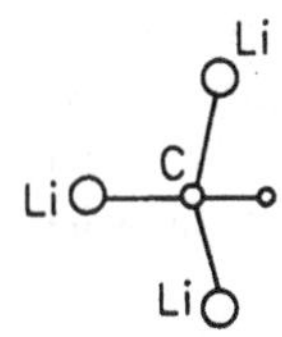

4b

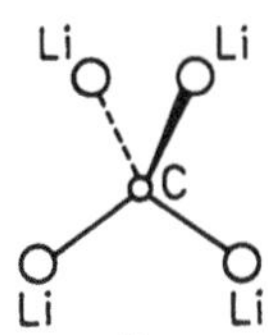

5a

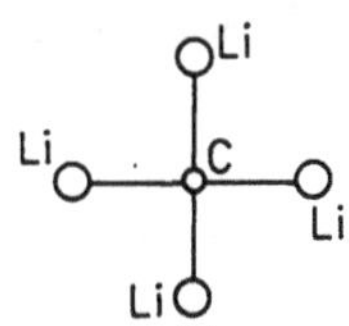

5b

The stabilization of the planar carbon by lithium was explained by its simultaneous σ-donor and π-acceptor ability which can be shown by the following resonance structures of the *cis*-planar dilithiomethane *3b* (*cis*):

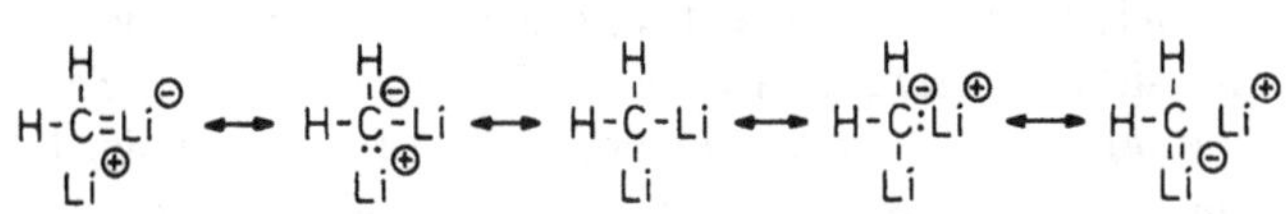

Another consequence of the donor/acceptor properties of the lithium atom is the dramatic stabilization by 37 kcal/mol (155 kJ/mol) of the *cis*-planar dilithiomethane *3b* (*cis*) compared with the *trans*-planar structure *3b* (*trans*). In contrast to *3b* (*trans*) the lithium atoms in *3b* (*cis*) can interact electronically to form together with the p orbital at the carbon atom a (4n + 2)-Hückel system (n = 0) isoelectronically to the cyclopropenyl cation *6*:

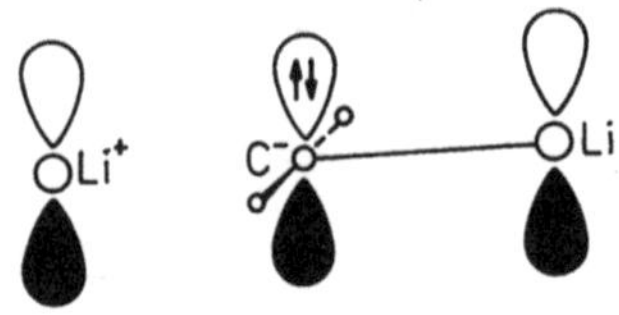

3b(trans)

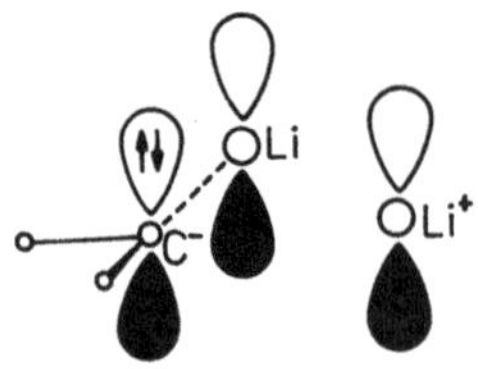

3b(cis)

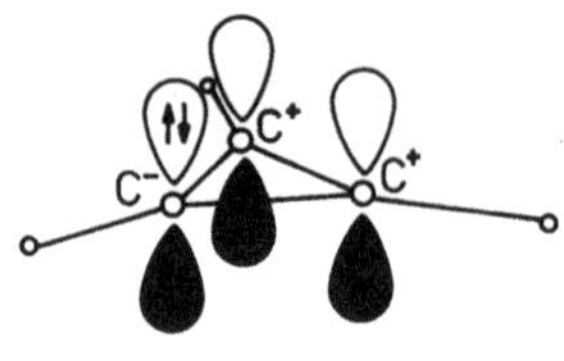

6

It is remarkable that the lowest inversion barrier (2.9 kcal/mol, 12.1 kJ/mol) of all calculated species was found for *3b* (*cis*) in its triplet state. This was confirmed by more accurate calculations considering large scale configuration interaction and cluster correction delivering even lower values of 1.7 and 1.8 kcal/mol (7.1 and 7.5 kJ/mol) respectively [11]. The original result that the triplet states of dilithiomethane *3* might be energetically favorable, however, could not be confirmed (Table 2). But anyhow, the energies were found to lie so close together that in *molecular* dilithiomethane *3* four nearly degenerate states have to be discussed.

Table 2. Calculated Energy Differences (kcal/mol) and Dipole Moments (Debye) for Dilithiomethane Conformers of Different Multiplicity

Conformer (Multiplicity)	STO-3G [9]	4-31G [9]	Large Scale CI [11]	Cluster Correction [11]	μ [11]
3a (singlet)	0	0	0.0	0.0	5.42
3a (triplet)	−12	−16	−0.8	2.1	−0.76
3b (*cis*, singlet)	17	10	7.4	8.3	4.85
3b (*cis*, triplet)	−2	−13	0.9	3.9	−1.22

Most interestingly the dipole moments of the triplet states are not only very much smaller but also show reversed polarization: $C^{\oplus}Li^{\ominus}$. Hand in hand a dramatic diminuation of the angle ∢LiCLi in dilithiomethane *3* is found on going from the singlet to the triplet state [9, 11]. This is due to the population of a MO in which some Li—Li σ bond is maintained. Therefore at least for the triplet state another three resonance structures have to be added:

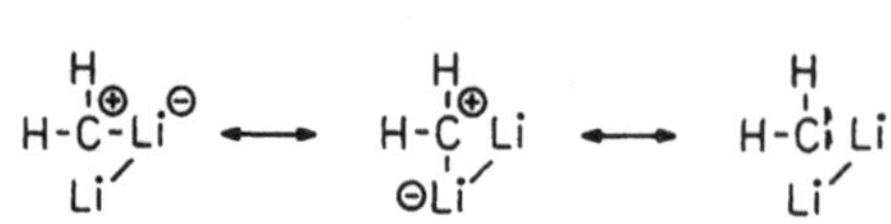

The structure and bonding in dilithiomethane *3* was reexamined recently in detail by Streitwieser [12] (electron density analysis) and Harrison [13] (Multi-Configuration SCF calculations), essentially confirming the previous results.

Dimers of dilithiomethane *3* have also been calculated by Jemmis, Schleyer, and Pople [14]. The head to head dimer *3c* (D_{2d} symmetry) of planar monomers *3b* (*cis*) the four lithium atoms bridging two perpendicular CH_2 units was found to be the most stable.

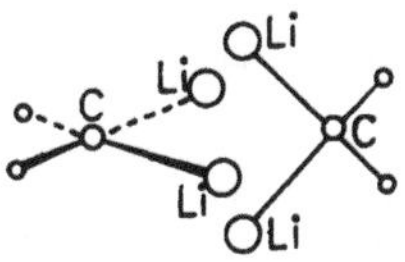

3c

The dimerization energy relative to the energy of two tetrahedral monomers *3a*, 37 kcal/mol (155 kJ/mol) (4-31G/5-21G(Li)), is considerable and should be even larger with higher association. Possible trimer and polymer structures have also been discussed [14].

2.2 Lithiated Ethanes

The potential energy surface of 1,2-dilithioethane *7* was examined by Schleyer, Gleiter, Pople, and coworkers [15] at several levels of ab initio theory. The global energy minimum was found to be the partially bridged trans conformation *7b* (C_{2h} symmetry), although the symmetrically trans doubly bridged structure *7a* (D_{2h} symmetry) is only 1.9 kcal/mol (7.95 kJ/mol) higher in energy.

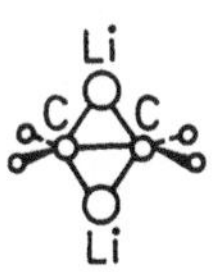

7a

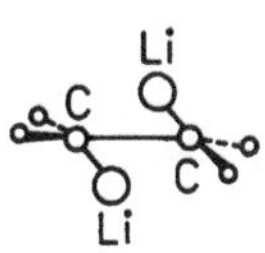

7b

Triplet states this time are all lying significantly higher in energy and need not be considered in this connection. The energy gained by partial bridging in *7b*, 16.4 kcal/mol (68.6 kJ/mol) (4-31G), can be assessed by comparing the energy of fully optimized *7b* with that of the standard geometry (180° dihedral angle) used by Radom et al. [16] in calculations of the rotational potential surface of 1,2-dilithioethane *7*.

The stability of *7b* toward possible dissociation models (Eqs. 1–3) was also examined [15].

7b →  + Li_2 −1.5 kcal/mol   (1)

7b → 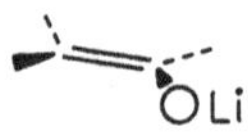 + LiH   29.3 kcal/mol   (2)

8

7b →  + H_2 32.7 kcal/mol   (3)

9a

Monomeric 1,2-dilithioethane 7 should thus be marginally unstable thermodynamically toward dissociation into ethylene and Li_2, but stable toward the loss of lithium hydride or of hydrogen. Equation 2, however, is not in agreement with experimental results of Rautenstrauch [17] and Bogdanović [18] who found that 1,2-dilithioethane 7 — if it is formed at all by the addition of lithium to ethylene — looses lithium hydride spontaneously. These findings prompted a reconsideration of the nature of 1,2-dilithioethane 7 and its tendency to eliminate lithium hydride [19]. Most interestingly it was found that the vinyllithium-lithium hydride complex *10* is 29.8 kcal/mol (124.7 kJ/mol) (3-21G//3-21G) more stable than *7b*, 51.0 kcal/mol (213.4 kJ/mol) (3-21G//3-21G) more stable than the separated fragments (Eqs. 4 and 5)! Since the driving force for the formation of the LiH complex *10* is so large, elimination of lithium hydride from *7b* is expected to be quite rapid.

7b → 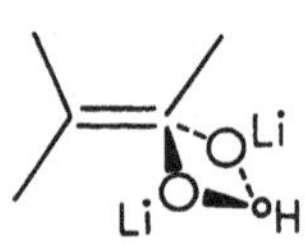   −29.8 kcal/mol   (4)

10

8 + LiH ⟶ 10   −51.0 kcal/mol   (5)

This does not mean, however, that 1,2-dilithioethane 7 cannot exist at all and that the claimed syntheses by Kuus [20,21] must be in error. In dilithioethane aggregates, stabilization due to lithium-lithium interactions might result in species which are stable toward loss of lithium hydride.

Since high-level ab initio calculations on dimers or higher oligomers of 1,2-dilithioethane 7 are impracticable, Schleyer [15] used the semiempirical MNDO method of Dewar and Thiel [22] to examine several dimer geometries. The best energy was obtained for structure *7c*, the dimer of a *cis* doubly bridged form corresponding to a distorted lithium tetrahedron to which perpendicular H_2CCH_2 units are bound on opposite edges. The huge dimerization energy found, 101 kcal/mol (422 kJ/mol), may be overestinated by MNDO [15] but aggregates of dilithioethane are certainly expected to be much more stable than the monomer.

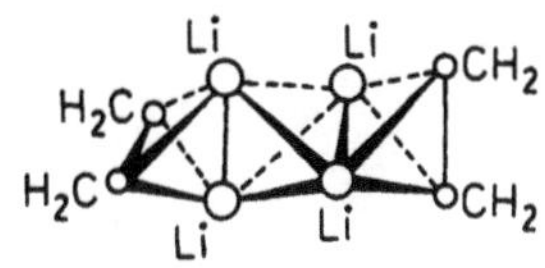

7c

This was shown especially well with 1,1-dilithioethane *11*, which we were able to synthesize although — according to ab initio calculations performed in cooperation with Schleyer — it is even 17.6 kcal/mol (73.6 kJ/mol) (3-21G) less stable than 1,2-dilithioethane *7* [23].

$$\underset{11}{CH_3CHLi_2} \longrightarrow \underset{7}{LiCH_2CH_2Li} \quad -17.6 \text{ kcal/mol}$$

$$11 \longrightarrow 10 \quad -47.5 \text{ kcal/mol}$$

The structure *11a* optimized using the 3-21G basis set is "classical" and exhibits no special Li ... H or Li ... Li interactions. On the other hand the planar conformer *11b* is only 1.1 kcal/mol (4.6 kJ/mol) less stable and displays a strong Li ... H interaction.

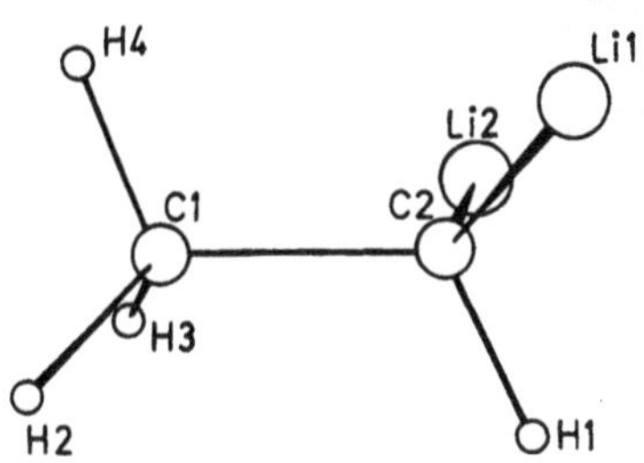

11a

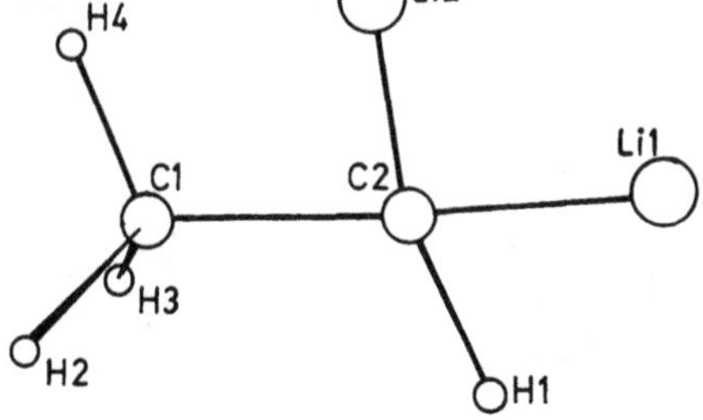

11b

According to recent calculations by Houk and Rondan [24] *11c*, 10.2 kcal/mol (42.7 kJ/mol) higher in energy than *11a*, is the transition state for the lithium hydride elimination which was found to take place within 8 h at room temperature [23].

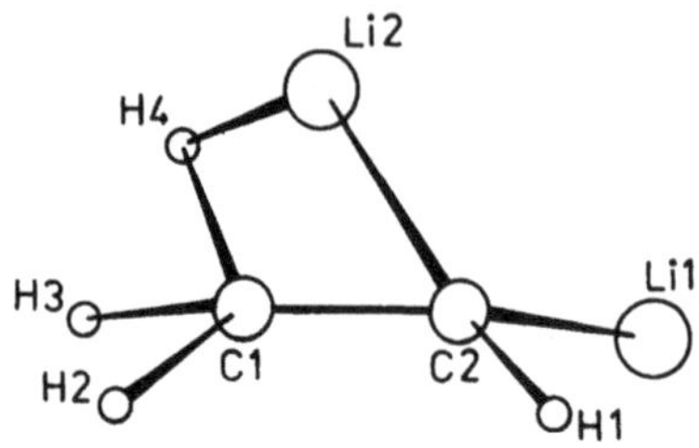

11c

An activation energy of only 10.2 kcal/mol (42.7 kJ/mol) for decomposition of course would not allow to isolate *11* at room temperature. Therefore stabilization of *11* by aggregation again has to be taken into account. In order to get some insight, the dimer *11d* was calculated using the MNDO method and found to be stabilized by 55 kcal/mol (230 kJ/mol). [25]

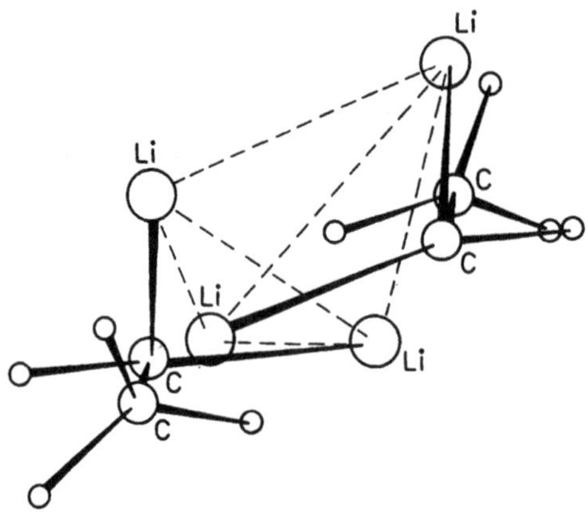

11d

In a preliminary report Schleyer, Thiel, and coworkers [26] also describe some calculations on structural isomers of perlithioethane *12*. Of the seven calculated structures the low energy isomers, *12a* (C_{2h} symmetry), *12b* (D_{4h} symmetry), and *12c* (C_{2v} symmetry), are candidates for the global energy minimum in the gas phase. In the solid state, however, C_2Li_6 should aggregate and might exhibit consequent structural modification.

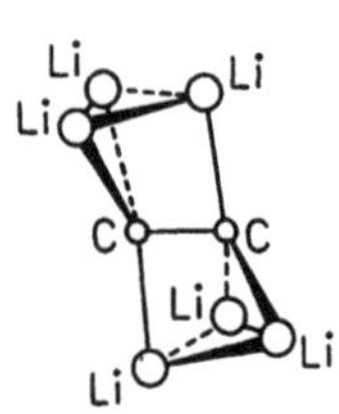

12a

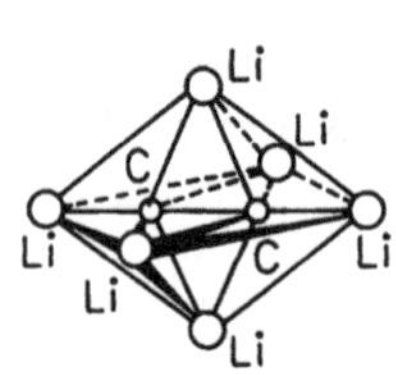

12b

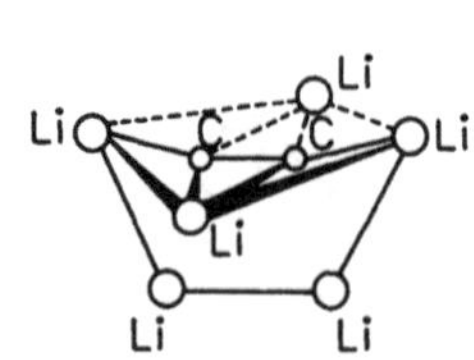

12c

2.3 Lithiated Ethenes

1,1-Dilithioethylene *13* was calculated by several groups [27–30] the highest level being used by Laidig and Schaefer [29]. Most interestingly this molecule was found to prefer a triplet ground state with perpendicular geometry *13b* showing nearly free rotation about the double bond. The planar triplet with the relative energy of 1.4 kcal/mol (5.9 kJ/mol) is followed energetically by the twisted singlet (10.5 kcal/mol, 44 kJ/mol) and the planar singlet (12.5 kcal/mol, 52.3 kJ/mol).

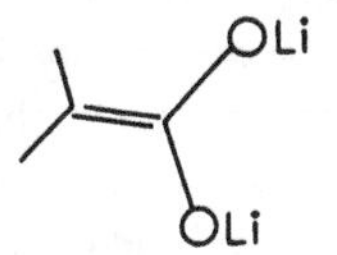

13a

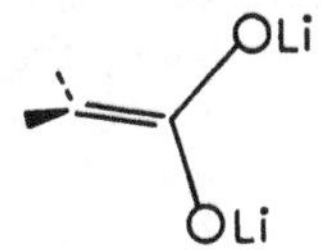

13b

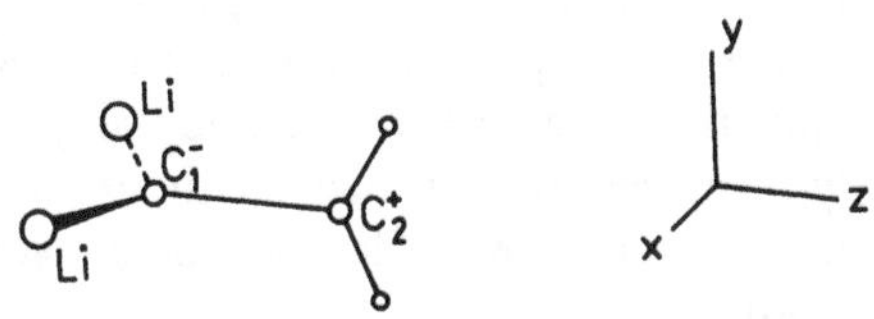

13c

The high stability of the twisted conformers *13b* is due to the special combination of σ-donor and π-acceptor character of lithium, the operation of which can be understood by considering the formal zwitterionic structure *13c*, the anionic part of which in the singlet state is stabilized by delocalization of the two p_y electrons in a cyclopropenium-type aromatic system — similar to that found in *cis*-planar dilithiomethane *3b* (*cis*) [9].

The same factors stabilize the twisted triplet which can be visualized as having transferred one of the p_y electrons on Cl in *13c*, with spin inversion, to a σ orbital which bonds the two lithium atoms.

14a *9a*

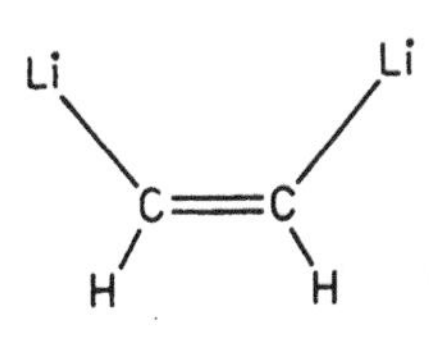

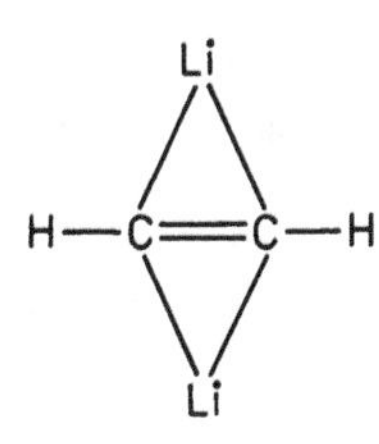

14b *9b*

The cationic center is stabilized by very strong hyperconjugation of the p_x orbital on C2 with the two coplanar C—Li bonds and thus balances the transfer of electrons from Cl to the lithium atoms. Due to this back-donation into the formally empty p_x orbital 1,1-dilithioethylene is only slightly polarized and the C—C bond — although twisted — actually is a double bond as represented in *13b*.

In contrast to 1,1-dilithioethylene *13*, triplet 1,2-dilithioethylenes are lying significantly higher in energy than the singlet states and need not be considered in this case [30]. The *trans*-isomer was found to prefer a distorted planar, partially bridged geometry *9a* (C_{2h} symmetry). The more symmetrical bridged structure *9b* (D_{2h} symmetry) is much higher in energy. On the other hand, the nonplanar doubly bridged *cis*-isomer *14a* (C_{2v} symmetry) was found to be slightly more stable than *9a*, and substantially lower in energy than the "classical" *cis* geometry *14b* (C_{2v} symmetry) [30].

In light of our successful synthesis of 1,2-dilithioethylene and the observation that the *cis*-isomer *14a* seems to rearrange into the more stable *trans*-isomer *9a* [31] all $C_2H_2Li_2$ geometries have now been reoptimized with the polarized split valence 6-31G* basis set and the relative energies have been recalculated at the correlated MP2/6-31G*//6-31G* level [32]. While these higher level relative energies for *9a*, *9b*, *14a*, and *14b* differ only modestly from the earlier values [30], the *trans*-isomer *9a* is now found to be 0.9 kcal/mol (3.8 kJ/mol) more stable than the *cis*-isomer *14a*, in agreement with the experimental findings [31].

Schleyer [32] has also proposed a mechanism for the isomerization *14a* → *9a*. The most favorable pathway — according to these calculations — does not involve rotation around the carbon—carbon double bond, but rather in-plane inversion of one of the CH groups. The barrier for the *cis-trans*-isomerization *14a* → *9a* was predicted to be about 22 kcal/mol (92 kJ/mol) for a hypothetical gas phase process.

Due to calculations [30] *trans*-1,2-dilithioethylen *9a* should be a very stable lithium-organic compound — even more stable than methyllithium and vinyllithium:

$$\underset{2}{CH_3Li} + \underset{8}{CH_2{=}CHLi} \longrightarrow \underset{9a}{LiCH{=}CHLi} + \underset{1}{CH_4} \qquad -8.6\ \text{kcal/mol}$$

$$2\,\underset{8}{CH_2{=}CHLi} \longrightarrow \underset{9a}{LiCH{=}CHLi} + CH_2{=}CH_2 \qquad -2.6\ \text{kcal/mol}$$

This, however, is not in agreement with the unsuccessful attempts by Seyferth and Vick [33] to prepare *9* by tin-lithium exchange reactions using *n*-butyllithium.

Moreover, elimination of hydrogen, lithium hydride, or lithium all should be unfavorable reactions [30]:

$$9a \longrightarrow H_2 + LiC{\equiv}CLi \qquad 4.1\ \text{kcal/mol}$$

$$9a \longrightarrow LiH + HC{\equiv}CLi \qquad 14.0\ \text{kcal/mol}$$

$$9a \longrightarrow Li_2 + HC{\equiv}CH \qquad 7.3\ \text{kcal/mol}$$

2.4 Dilithioacetylene

Molecular orbital calculations by Schleyer et al. [6, 34] have indicated that dilithioacetylene in the gas phase may prefer a bridged structure *15b* (D_{2h} symmetry) rather than the conventional linear structure *15a* ($D_{\infty h}$ symmetry).

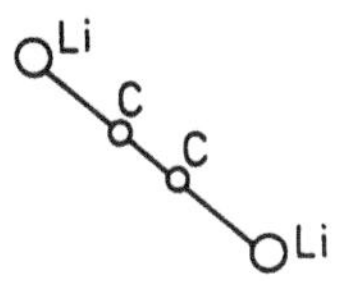

15a

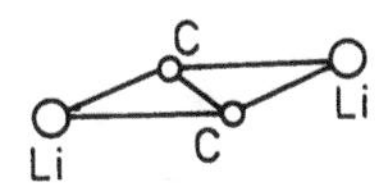

15b

However, the results for C_2Li_2 are particularly sensitive to the theoretical level employed. STO-3G, indeed, favors the doubly bridged form *15b* by 20.5 kcal/mol (85.8 kJ/mol) over the linear geometry *15a*. On the other hand, split valence basis sets favor the linear form, the preference increasing with the size of the basis set. When d type polarization functions are added, however, the bridged form becomes more stable once again by up to 9.8 kcal/mol (41 kJ/mol) (MP3/6-31G*). The triplet forms of *15a* and *15b* and other possible structures of dilithioacetylene were also calculated but found to be significantly less stable [34].

According to vibrational frequency calculations by Ritchie [35] both bridged and linear dilithioacetylenes are minima on the potential energy surface.

2.5 1,3-Dilithiopropane

The successful synthesis of 1,3-dilithiopropane *16* by Bickelhaupt and coworkers [36] prompted a thorough computational study of this interesting compound by Schleyer et al. [6, 19]. The symmetrical doubly lithium bridged structure *16a* (C_{2v} symmetry) was found to be the minimum energy isomer (3-21G basis set), 24.6 kcal/mol (103 kJ/mol) more stable than the corresponding extended W-conformation.

$$LiCH_2CH_2CH_2Li \longrightarrow CH_2{=}CH{-}CH_2Li + LiH \qquad 16.5\ \text{kcal/mol}$$

16

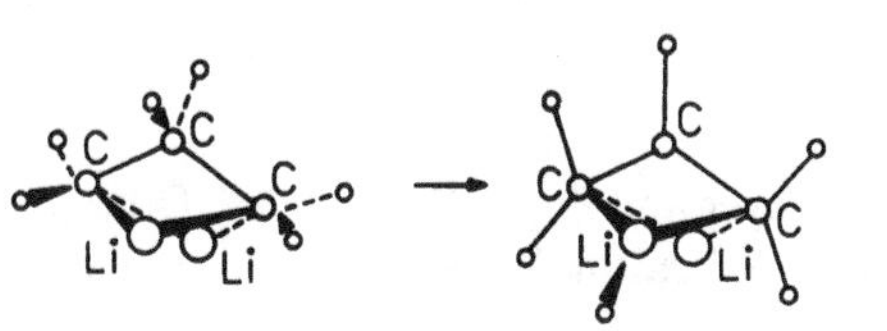

-5.4 kcal/mol

16a *17*

The elimination of lithium hydride with the formation of allyllithium observed by Bickelhaupt [36] to occur with a half reaction time of 1 h at room temperature at first could not be understood, because this reaction was calculated to be endothermic by 16.5 kcal/mol (69 kJ/mol). However, as was the case with 1,2-dilithioethane, primarily rearrangement to a very stable allyllithium-lithium hydride complex *17* might take place, which is calculated to be exothermic by 5,4 kcal/mol (22.6 kJ/mol) [6].

2.6 Other Lithiated C_3-Hydrocarbons

Klein and Medlik-Balan [37] discovered that propene can be metalated with *n*-butyllithium/TMEDA to form a 1,3-dilithiated allyl system. According to ab initio calculations by Schleyer [6,38] and Streitwieser [39] with the 3-21G basis set the symmetrically doubly lithiumbridged structure *18a* is the most stable, although *18b* with "turned off" allylic resonance is only 0.7 kcal/mol (2.9 kJ/mol) less stable.

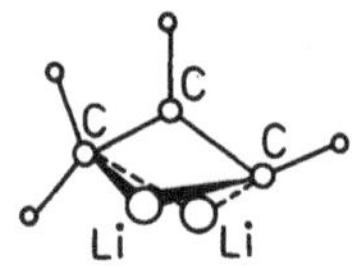

18a

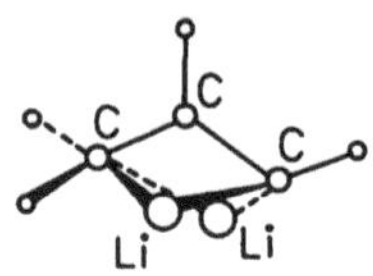

18b

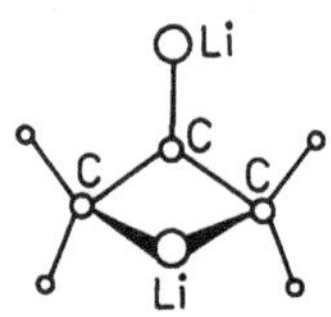

19a

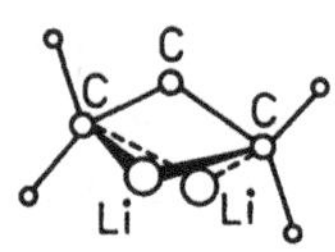

19b

On the other hand the 2,3-dilithiopropene *19a* [39] is 11.5 kcal/mol (48 kJ/mol) less stable than *18a* but 2 kcal/mol (8.4 kJ/mol) more stable than the corresponding doubly bridged structure *19b* [39]. This is consistent with the experimental observation that CH_2CLiCH_2Li species are not formed by further lithiation of allyllithium.

Also in agreement with the metalation studies of propene [37], the first step yielding allyllithium was found calculationally to be somewhat more exothermic than the second [38]:

$$CH_2{=}CHCH_3 + CH_3Li \longrightarrow [CH_2CHCH_2]Li \qquad -18.0\ \mathrm{kcal/mol}$$

$$[CH_2CHCH_2]Li + CH_3Li \longrightarrow \underset{18a}{[CH_2CHCH]Li_2} \qquad -15.4\ \mathrm{kcal/mol}$$

According to calculations by Schleyer and Pople [9] planar 1,1-Dilithiocyclopropane *20*, another $C_3H_4Li_2$ isomer, is by 7 kcal/mol (29.3 kJ/mol) more stable than the tetrahedral geometry.

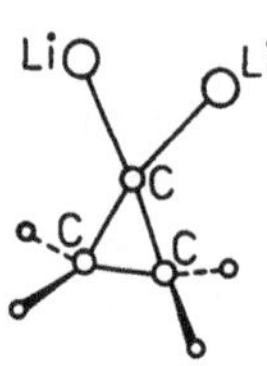

20

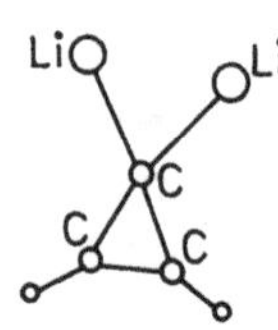

21

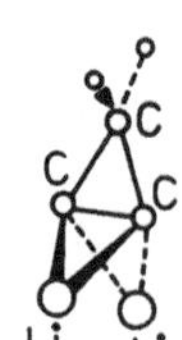

22

3,3-Dilithiocyclopropene ($C_3H_2Li_2$) *21* is even more favorable in its preference for the planar geometry [9]. The isomeric 1,2-Dilithiocyclopropene with its doubly bridged structure *22*, however, is much more stable [40].

The most stable $C_3H_2Li_2$ isomer *23* was found by optimization of 1,3-dilithiopropyne *23a* and 1,1-dilithioallene *23b*, respectively [40].

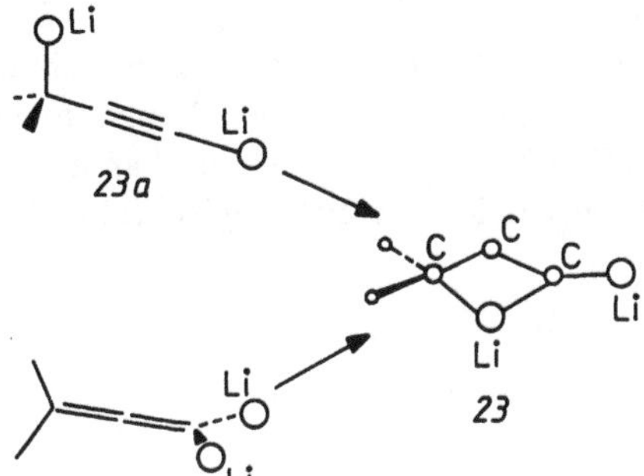

23a *23* *23b*

1,3-dilithioallene *24a* leads to the doubly bridged structure *24*, which, however, is by 12.9 kcal/mol (54 kJ/mol) less stable than *23* [40].

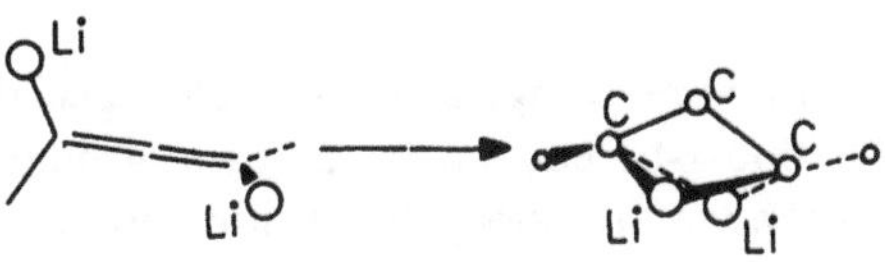

24a *24*

Similar structures were found for trilithiopropyne (C_3HLi_3) *25* and perlithiopropyne (C_3Li_4) *26* which in addition have one and two acetylide-type C—Li bonds, respectively [40,41].

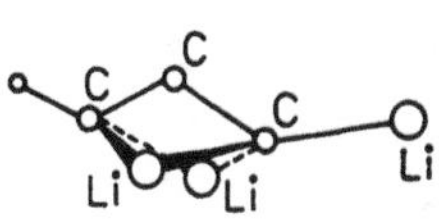

25

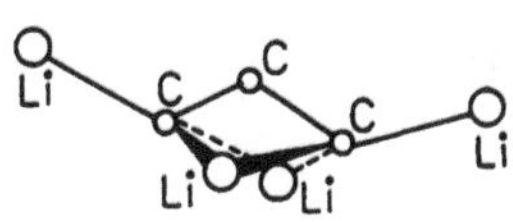

26

Finally even the exotic molecule dilithiocyclopropyne (C_3Li_2) has been calculated by Frenking [42]. Geometry optimization (3-21G) resulted in a planar geometry *27* with a planar tetracoordinated carbon atom C3 and C—C bond lengths inter-

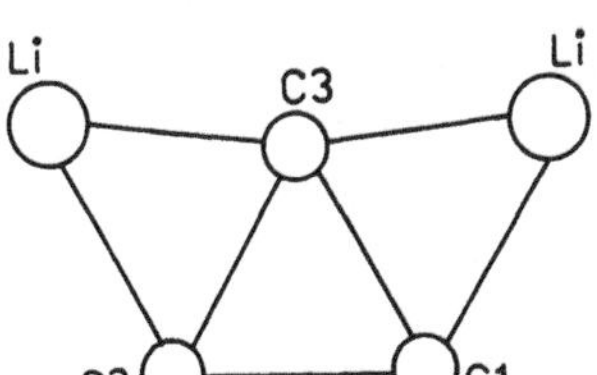

27

mediate between single and double bonds. The lithium atoms are stronger bonded to C1 and C2 than to C3 and therefore C_3Li_2 is not a derivative of cyclopropyne but of the more stable cyclopropylidene, i.e. an aromatic compound.

2.7 Lithiated C_4-Hydrocarbons

According to ab initio calculations by Schleyer et al. [19] 1,4-Dilithiobutane *28* has a similar doubly bridged structure as 1,3-dilithiopropane *16* although somewhat twisted (C_2 symmetry). *28* is 34.6 kcal/mol (144.8 kJ/mol) more stable than the extended zick-zack form of 1,4-dilithiobutane, the cyclization energy being 10 kcal/mol (41.8 kJ/mol) higher than in the case of *16*.

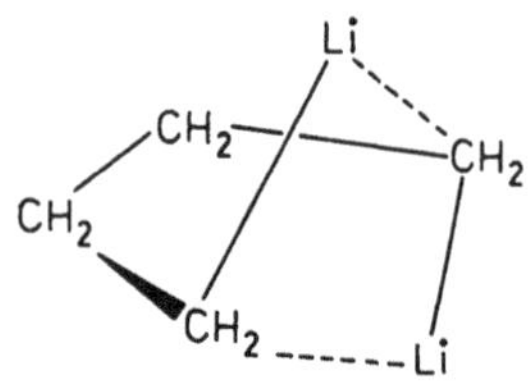

28

Due to our engagement in carbanion rearrangements we were especially interested in the homoallylic system 4,4-dilithio-1-butene *33*. In the case of homoallyllithium *29* ring-chain equilibrium strongly favors the open-chain lithiumorganic compound [43], and the introduction of four methyl groups into the α- and β-positions is necessary to shift the equilibrium *31* ⇌ *32* to the cyclopropylcarbinyl side [44].

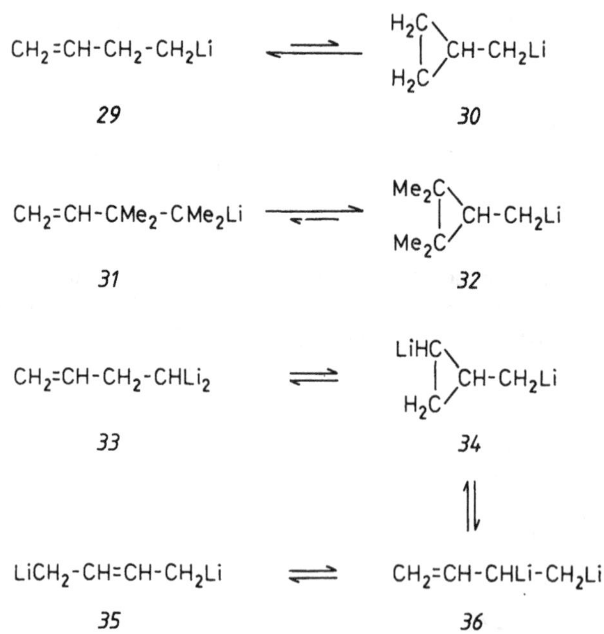

Assuming a similar reaction path starting with *33* one might expect that according to calculations of Dill, Greenberg, and Liebmann [45] the lithium substituent should stabilize the cyclopropane ring in the cyclopropylcarbinyl species *34* by 7–8 kcal/mol (29–33.5 kJ/mol). On the other hand ring-opening of *34* finally could yield the most stable 1,4-dilithio-2-butene *35*.

At first the geometries of the monolithio species *29* and *30* were optimized by MNDO calculations [25] and structures *29a* and *30a* were found to be the most stable. Interestingly the lithium atom in *29a* is placed beyond the π system of the double bond forcing the methylene hydrogens to adopt ecliptic arrangements. In *30a* which according to the calculations is 14 kcal/mol (58.6 kJ/mol) less stable than *29a*, the lithium atom is located symmetrically over the cyclopropane ring showing some electronic interaction with the ring orbitals. Due to this interaction the C2–C3 and C2–C4 bonds become longer (1.551 Å compared with 1.534 Å for C3–C4) facilitating ring opening by loosening of one of these weakened bonds.

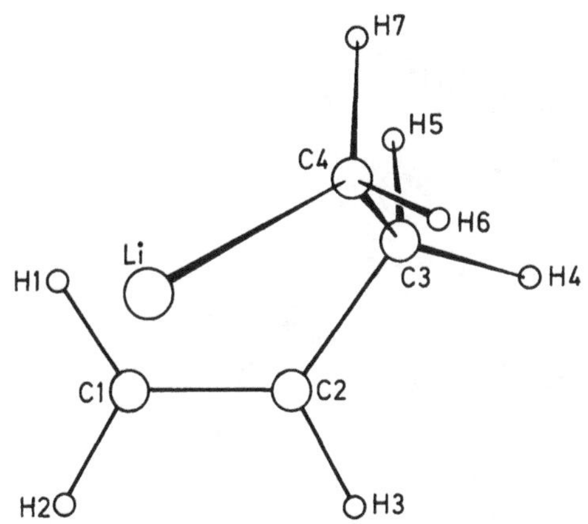

29a

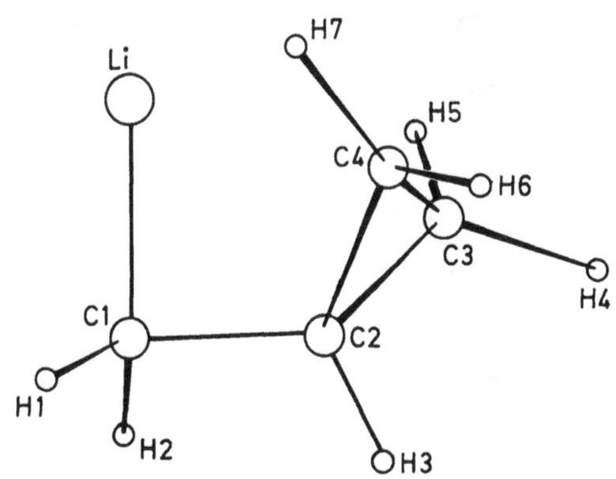

30a

For the dilithio compounds *33*, *34*, and *35* the MNDO geometries *33a*, *34a* (*trans*), *34a* (*cis*), and *35a* have been found [25]. In *33a* the interaction of the two

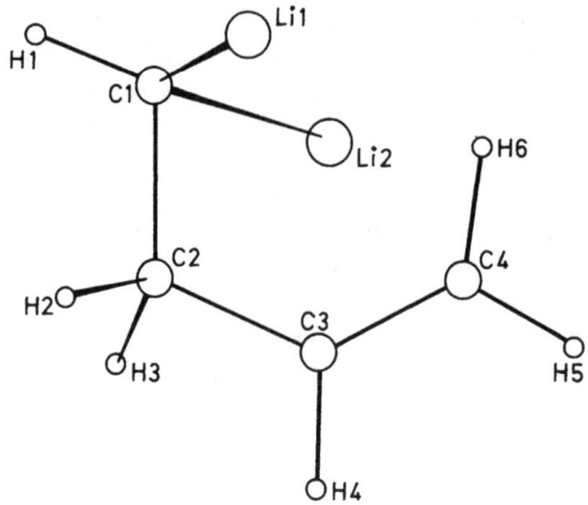

33a

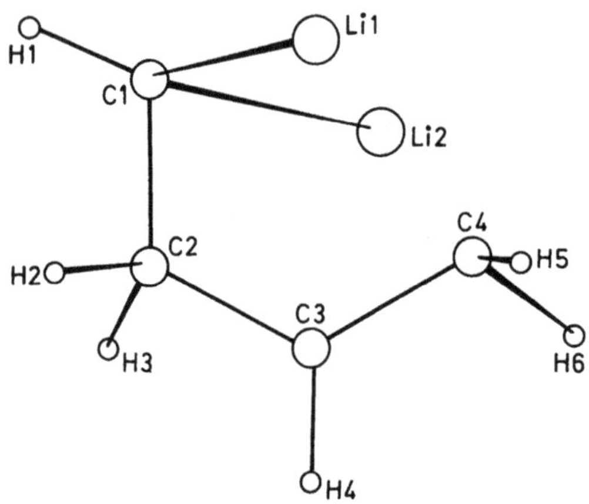

33b

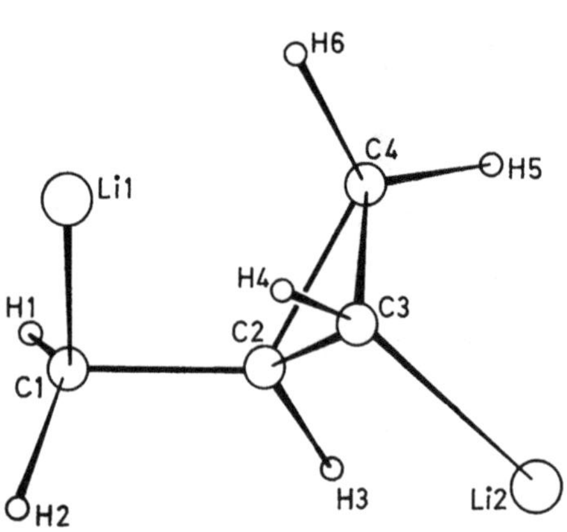

34a(trans)

lithium atoms with the π system leads to weakening of the double bond as shown by the structure *33b* with a twisted terminal methylene group (C4–H5–H6) which is only 28 kcal/mol (117 kJ/mol) less stable than *33a*, while in unsubstituted ethylene the rotational barrier is known to be 65 kcal/mol (272 kJ/mol) [46].

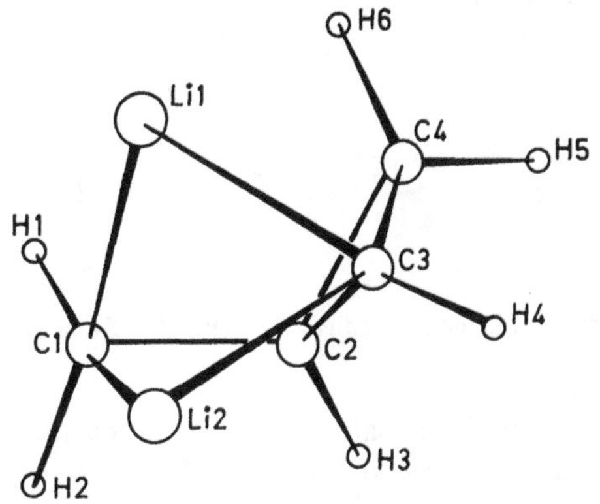

34a (cis)

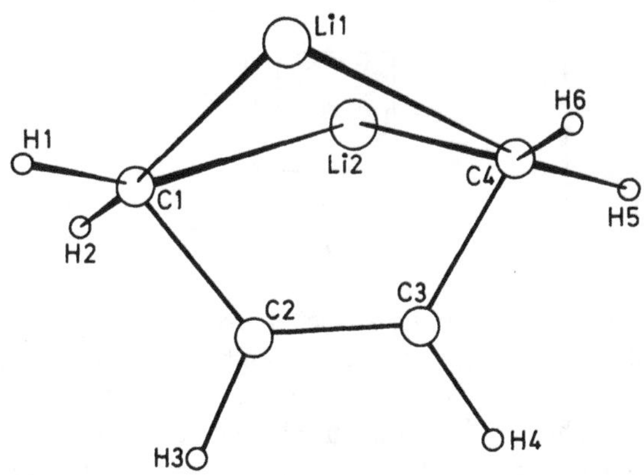

35a

For the cyclopropylcarbinyl species *34* this time two isomers have to be discussed. While the trans isomer *34a* (*trans*) is more or less "classical" comparable with *30a*, the *cis* isomer *34a* (*cis*) was found to prefer a lithium doubly bridged structure, 22 kcal/mole (92 kJ/mol) more stable than *34a* (*trans*) and even 26 kcal/mol (108.8 kJ/mol) more stable than the open-chain starting material *33a* (Fig. 1).

Another 29 kcal/mol (121.3 kJ/mol) are gained by going to the most stable end-product, 1,4-dilithio-2-butene *35*, which again was found by MNDO calculations to show double lithium bridging *35a* [25].

Schleyer [47] has recently shown by ab initio calculations (3-21G) *35* to be even more stable by 6.7 kcal/mol (28 kJ/mol) in a conformer with the terminal methylene groups in plane with the carbon skeleton. According to these calculations [47] *35* is also 17.6 kcal/mol (73.6 kJ/mol) more stable than 1,4-dilithiobutane *28*.

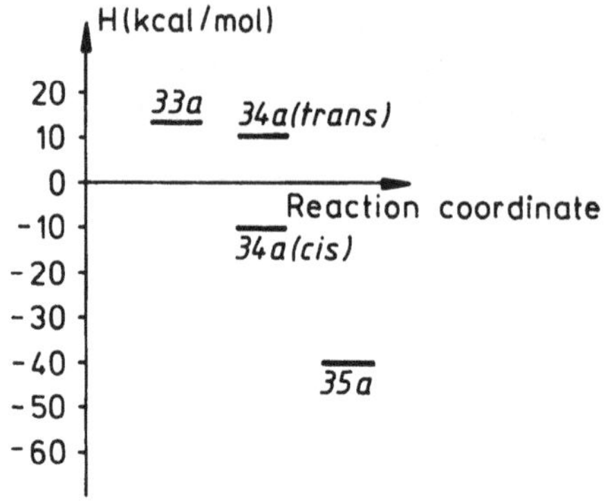

Fig. 1. MNDO Enthalpies of Formation for the Homoallylic System *33* ⇌ *34* ⇌ *35*

Another $C_4H_6Li_2$ isomer is the dilithium salt of the cross-conjugated dianion of isobutylen [48–50], the trimethylenemethane dianion *37* (D_{3h} symmetry). This 6π electron Y-shaped system is distinguished by a novel enhanced stabilization presumeably due to the so-called acyclic "Y-aromaticity", although the treatment of this question is still controversal [3,51–62].

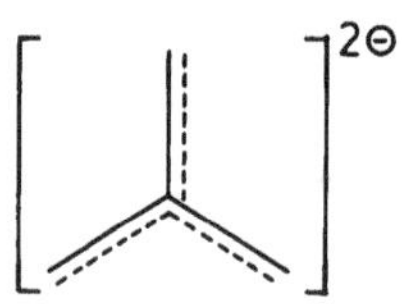

37

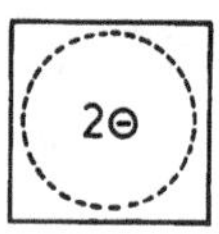

38

In contrast, cyclobutadiene dianion *38*, a Hückel aromatic compound, was shown calculationally by Schleyer [57] to be — due to coulombic repulsion — far less stable than Y-delocalized spezies with more favorable π charge distributions. This was confirmed by Boche and Thiel [63,64]: Although *38* itself could not be prepared until now, phenyl substituted derivatives show no special charge delocalization within the cyclobutadiene ring.

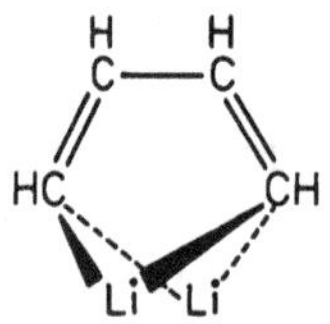

39

On the other hand, the isomeric (*Z*,*Z*)-1,4-dilithio-1,3-butadiene in its symmetrically doubly bridged form *39* (C_{2v} symmetry) is a very stable compound, 11.2 kcal/mol (46.9 kJ/mol) more stable than 1,4-dilithio-2-butene *35* [47]. While favorable electrostatic interactions in *39* will contribute greatly to stabilization, Schleyer [65] has called attention to 4π electron aromaticity, Möbius in character, which undoubtedly is also significant.

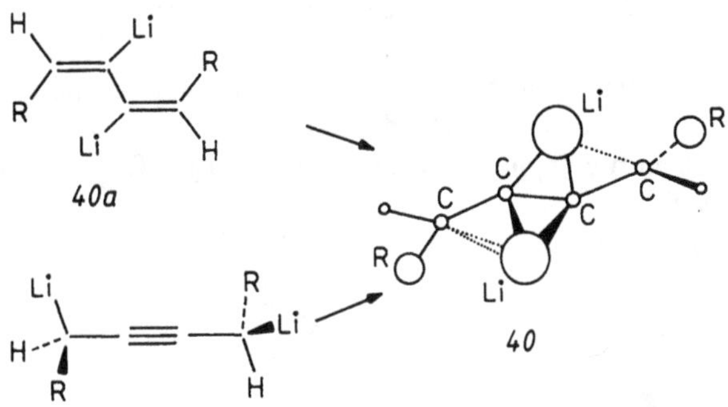

40a 40 40b

The isomeric 2,3-dilithio-1,3-butadiene was calculated independently by Seeger [66] and Schleyer [67] to prefer a lithium doubly bridged bisallylic geometry *40* (C_2 symmetry) the structure strongly dependent on the substituents R. The unsubstituted and even more the methylsubstituted compound is largely a 1,3-butadiene derivative in its gauche conformation, the lithium atoms showing only minor interaction with the 1,4 carbon atoms [66]. With replacement of the methyl by phenyl groups, however, the butadiene character is lost in favor of the 2-butyne structure [67]. Internal rotation in 1,4-dilithio-2-butyne *40b* already had been studied by Radom and coworkers [68] with the aid of ab initio calculations. They had found a pronounced preference for a conformation in which the dihedral angle LiC ... CLi is slightly greater than 90°.

On the other hand, 1,4-dilithiobutatrien ($C_4H_2Li_2$) according to MNDO calculations [69] has a dimeric *cis*-structure *41* with two different kinds of lithium atoms and a true butatriene geometry:

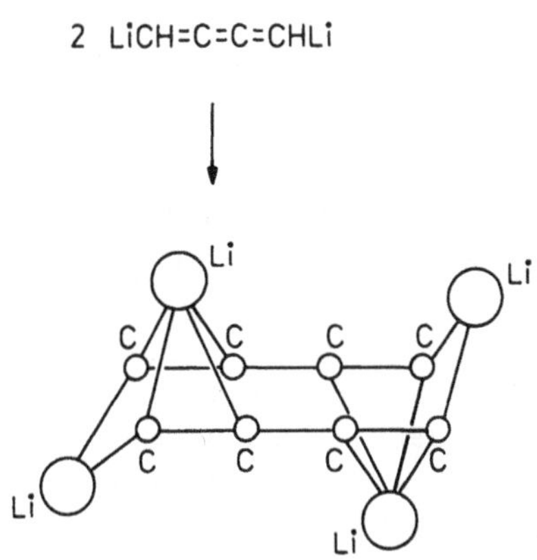

41

Finally the structure of C_4Li_4 is still controversal. According to recent ab initio SCF and RMP2 calculations by Disch and Ritchie [70,71] tetralithiotetrahedrane *42* (T_d symmetry) suggested by Schleyer [72] is not a minimum on the potential energy surface. They found a dimer of dilithioacetylene "tetralithiodiacetylene" *43* (D_{2d} symmetry) as the most stable geometrical isomer of molecular formula C_4Li_4, over 100 kcal/mol (418 kJ/mol) more stable than *42*. The synthesis of *42* therefore should be reproduced inasmuch as hydrolysis yielded acetylene only [72] and the reaction with methyl iodide to give tetramethyltetrahedrane reported by a Russian research group [73] is highly questionable.

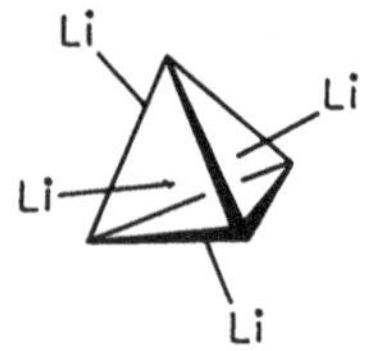

42

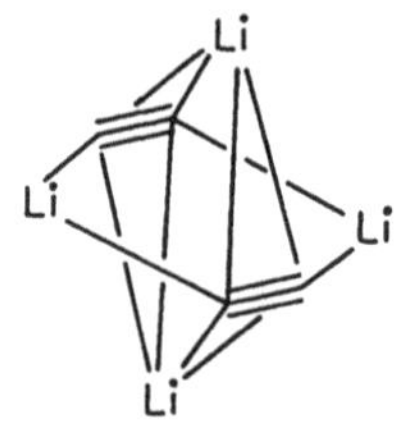

43

2.8 Hypervalent and Related Species

Schleyer, Pople and coworkers [74] have shown calculationally that polylithiated methanes with one or two additional lithium atoms should form very stable hypervalent species of high symmetry. Thus both trigonal-bipyramidal CLi_5 *44* (D_{3h} symmetry) and octahedral CLi_6 *45* (O_h symmetry) deduced from CLi_4 *5* are indicated by ab initio calculations (3-21G basis set) to be highly stable toward loss of a lithium atom from *44* or loss of Li_2 from *45*:

$$\underset{44}{CLi_5} \longrightarrow \underset{5}{CLi_4} + Li \qquad 54.1\ \text{kcal/mol}$$

$$\underset{45}{CLi_6} \longrightarrow \underset{5}{CLi_4} + Li_2 \qquad 65.2\ \text{kcal/mol}$$

The related series of molecules deduced from $CHLi_3$ *4* and CH_2Li_2 *3*, respectively, behave similarly: $CHLi_4$ *46* (C_{3v} symmetry), $CHLi_5$ *47* (C_{4v} symmetry), CH_2Li_3 *48* (C_{2v} symmetry), and CH_2Li_4 *49* (C_{2v} symmetry).

The charge on carbon, however, does not increase appreciably as more lithium atoms are added. Thus the extra electrons beyond the usual octet are not associated with carbon but contribute to Li—Li bonding building a metallic cage around the central atom. Hypermetalated molecules therefore are not hypervalent in the strict sense. Additional calculations on hyperlithiated methanes have been performed by Waite and Papadopoulos [75] as well as Weinhold et al. [76].

The corresponding hyperlithiated methanonium ions $CH_nLi_{5-n}^{\oplus}$ prefer the same geometries and are indicated by ab initio molecular orbital calculations to be also highly stabilized species [77, 78].

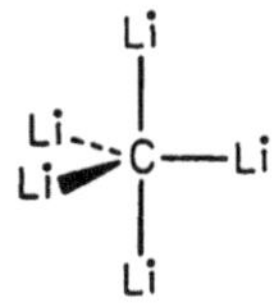

44

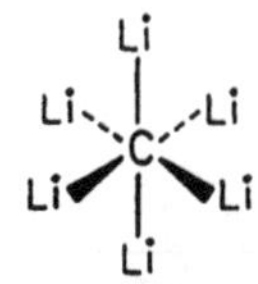

45

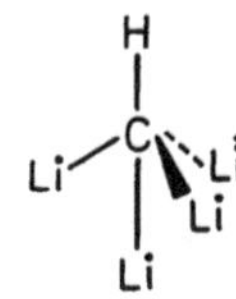

46

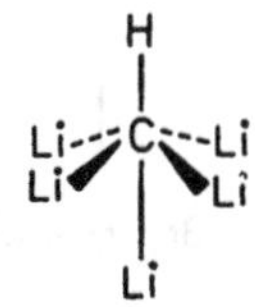

47

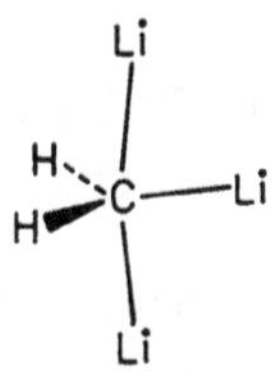

48

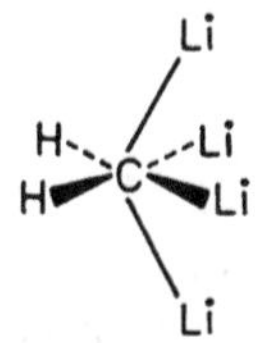

49

3 Syntheses

3.1 Halogen-Metal Exchange Reactions

The direct replacement of halogen in organic molecules by treatment with lithium metal, the most powerful method for the synthesis of alkyllithium compounds, often is only of limited value for the synthesis of polylithium organic compounds due to α, β, or γ elimination of lithium halide after the first step being faster than the second step. In their classical study of the lithiation of 1,ω-dibromoalkanes, West and Rochow [79] were able to prepare 1,4-dilithiobutane *28* and higher 1,ω-dilithioalkanes *57* ($n \geqq 4$) but not 1,2-dilithioethane *7* and 1,3-dilithiopropane *16*. Bromomethyllithium *52* could be trapped by working in the presence of chlorotrimethylsilane, when bis(trimethylsilyl)methane *51* was obtained in 6% yield via *53* and *54* in a stepwise reaction.

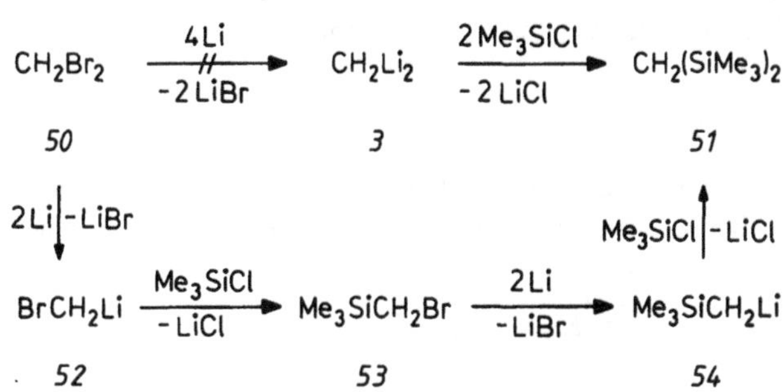

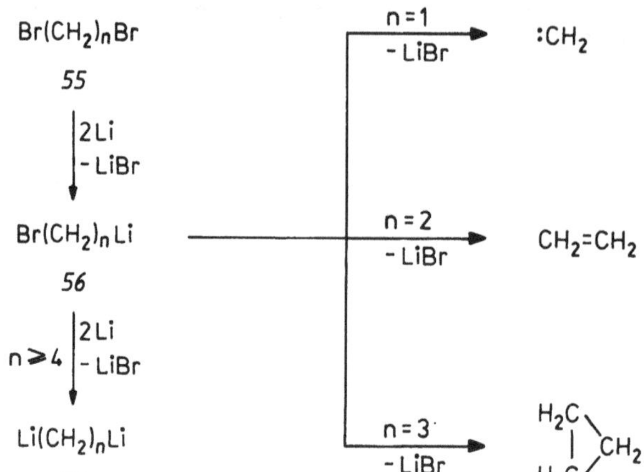

In light of these results the reported synthesis [20] of 1,2-dilithioethane *7* in 6–9% yield from 1,2-dichloro- or 1,2-dibromoethane and lithium should be reinvestigated, inasmuch as characterization was only by hydrolysis to yield some ethane.

Vinylic iodine is better replaced by lithium on treatment with *n*-butyllithium. This way (*Z,Z*)-1,4-dilithio-1,3-butadiene *39* [80] and (*Z,Z*)-1,5-dilithio-1,4-pentadiene *58* [81] have been prepared.

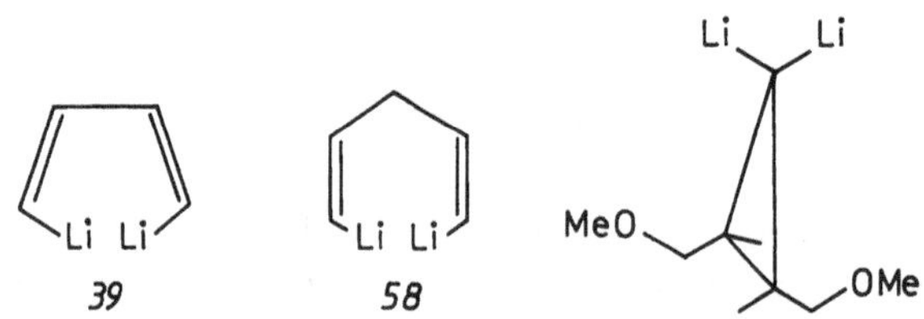

39 *58* *59*

On the other hand, the reaction of *n*-butyllithium or even *tert*-butyllithium with isocentric polyhalogenated alkanes [82,83] and alkenes [83] as well as with geminal dibromo- [83–85] and dichlorocyclopropanes [86] usually stops after the replacement of one halogen only. Reported [87] derivatives of dilithiocompounds have to be explained by involving reaction of the first-step carbenoid with the electrophile, followed by halogen-lithium exchange and subsequent quenching with a second equivalent of the electrophile as was the case with the formation of *51*. Only by intramolecular solvation Warner [88] succeeded in replacing two bromine atoms by lithium in the doubly methoxylated cyclopropane derivative *59*.

3.2 Pyrolysis Reactions

Already in 1955 Ziegler and coworkers [89] had reported that halide-free methyllithium *2* upon pyrolysis disproportionates into methane *1* and dilithiomethane *3* in excellent yields.

$$\underset{2}{2\ CH_3Li} \xrightarrow{225\,^{\circ}C} \underset{1}{CH_4} + \underset{3}{CH_2Li_2}$$

This method — inproved by Lagow [90–92] — still represents the best procedure for preparing dilithiomethane *3*, although temperature control is important because at temperatures higher than 230 °C decomposition to hydrogen (H_2), lithium hydride (LiH) and lithium carbide (C_2Li_2) takes place. The thermolability of *3*, already mentioned by Ziegler [89], was thoroughly investigated by Lagow and coworkers [90,91]. By pyrolysis at different temperatures and reaction times they found out that perlithiopropyne (C_3Li_4) is an intermediate finally loosing elementary lithium.

$$CH_2Li_2 \xrightarrow[6\,\text{min}]{350\,°C} \underset{(20\%)}{C_3Li_4} + \underset{(80\%)}{C_2Li_2}$$

$$CH_2Li_2 \xrightarrow[10\,\text{min}]{400\,°C} \underset{(99\%)}{C_2Li_2}$$

$$C_3Li_4 \xrightarrow[3\,\text{min}]{300\,°C} \underset{(40\%)}{C_3Li_4} + \underset{(60\%)}{C_2Li_2}$$

$$C_3Li_4 \xrightarrow[10\,\text{min}]{350\,°C} \underset{(100\%)}{C_2Li_2}$$

The endproduct of the thermal decomposition of tetralithiomethane *5* is also lithium carbide (C_2Li_2). This time in addition to perlithiopropyne (C_3Li_4) perlithioethylene (C_2Li_4) was found as an intermediate [91].

$$CLi_4 \xrightarrow[2\,\text{min}]{225\,°C} \underset{(60\%)}{CLi_4} + \underset{(20\%)}{C_2Li_4} + \underset{(20\%)}{C_3Li_4}$$

$$CLi_4 \xrightarrow[8\,\text{min}]{225\,°C} \underset{(20\%)}{CLi_4} + \underset{(30\%)}{C_2Li_4} + \underset{(40\%)}{C_3Li_4} + \underset{(10\%)}{C_2Li_2}$$

$$CLi_4 \xrightarrow[10\,\text{min}]{225\,°C} \underset{(100\%)}{C_2Li_2}$$

Lagow et al. [90,91] during these experiments in addition made the most important discovery, that polylithioalkanes prior to pyrolysis are stable in the gas phase for a short period of time. In this way for the first time mass spectra have been achieved of those compounds which have no observable vapor pressure below 650 °C even in the highest possible vacuum. A temperature of 1500 °C has been reached in less than three seconds by a special flash-vaporization apparatus, whereby e.g. dilithiomethane *3* could be transported over a distance of 10 cm with less than 10% decomposition.

$$CH_2Li_2 \xrightarrow[2\,s]{1500\,°C} \underset{(90\%)}{CH_2Li_2} + \underset{(2\%)}{C_3Li_4} + \underset{(8\%)}{C_2Li_2}$$

The main disadvantage of the Ziegler procedure is the fact that it cannot be used for the preparation of sensitive polylithium organic compounds especially when elimination of lithium hydride can take place. Thus Lagow and coworkers [92] have not been successful in applying this reaction technique to cyclopropyllithium but it worked excellently for the synthesis of 1,1-dilithio-2,2,3,3-tetramethylcyclopropane *62*.

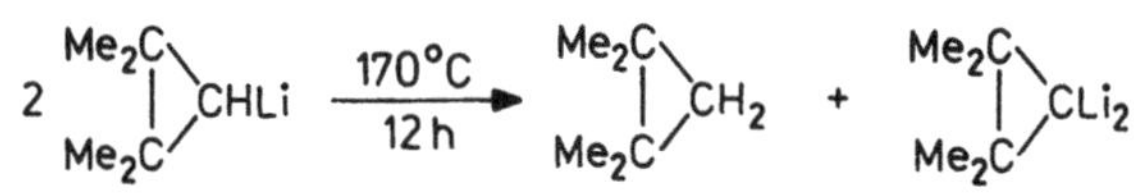

60 *61* *62*

*Inter*molecular elimination of lithium hydride may also complicate the reaction as we have found starting with halide-free benzyllithium *63* [25]. After a 4 h pyrolysis reaction at 150–170 °C the black solid was treated with D_2O at −50 °C and yielded 75% 3-benzyltoluene (d_1–d_5) via *64* as the main reaction product besides 10% 9,10-dihydroanthracen *66* (d_1–d_4) via *65* and 15% toluene (d_1–d_3). No attack at the p-position could be detected. The high deuterium content of the products is due to strong polymetalation reactions additionally taking place.

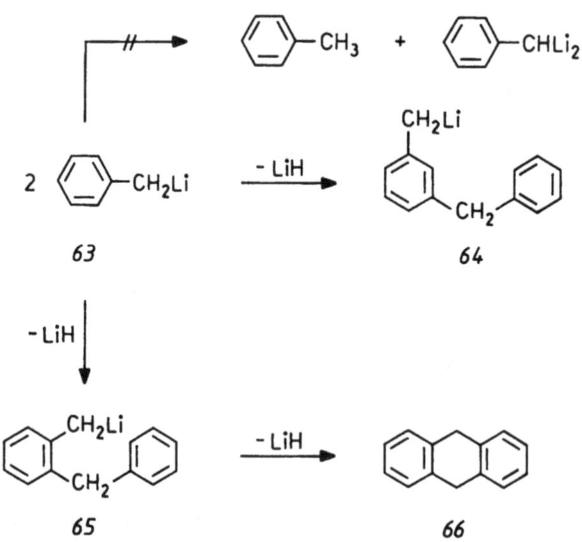

Most interestingly halide-free allylithium *69* under the same conditions showed lithium hydride elimination only *after* the Ziegler disproportionation had taken place. As the endproduct after 4 h at 150–170 °C we found perlithiopropyne (C_3Li_4) in 95%

yield [25]. The intermediate 3,3-dilithio-1-propene *18* could be detected in 7% yield as a 1:1 mixture of the two bis(trimethylsilyl)propenes *67* and *68* by working at 100–120 °C, when after 4 h 70% of the starting material *69* was still intact and only 23% of C_3Li_4 *26* had yet been formed.

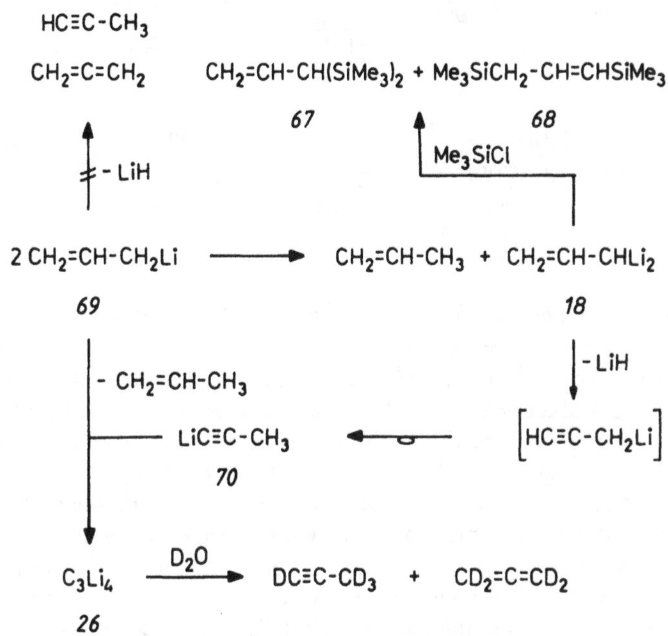

3.3 Mercury-Lithium Exchange Reactions

The procedure to replace mercury in organic compounds by lithium was developed already in 1917 by Schlenk and Holtz [93] and it is still the method of choice for the preparation of halide-free organolithium compounds. Wittig and Bickelhaupt [94] were the first who used a mercury-lithium exchange reaction for the synthesis of a dilithiocompound not available by halogen-lithium exchange, *o*-dilithiobenzene *75*.

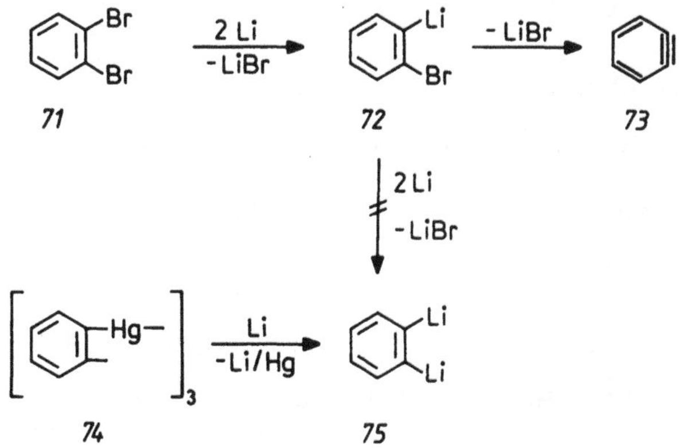

A recent example for this reaction is the successful synthesis of heterosubstituted 1,3-dilithiopropanes *77* and 1,4-dilithiobutanes *79* by Barluenga and coworkers [95,96].

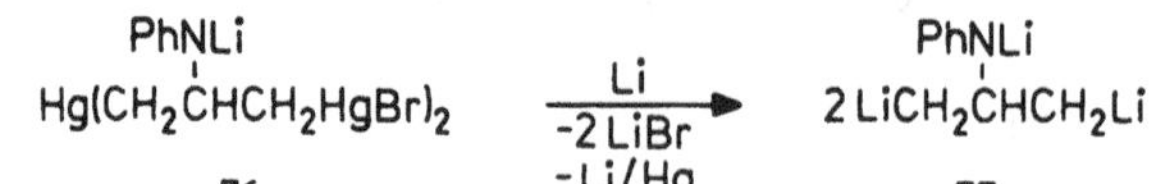

76 *77*

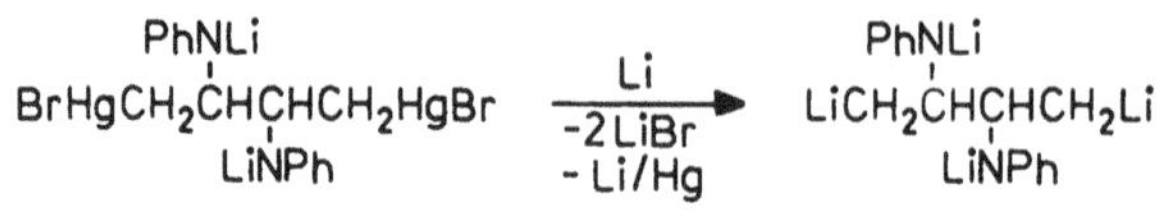

78 *79*

We used this procedure for the preparation of isocentric polylithiated hydrocarbons, i.e. compounds bearing two (geminal) or more than two Li atoms on one C atom as in CH_2Li_2 *3* [23], CLi_4 *5* [97], CH_3CHLi_2 *11* [23], $CH_2{=}CHCH_2CHLi_2$ *33* [25] and 1,1-dilithio-1-alkenes *81* [98,99]. Diethylether at room temperature turned out to be the best medium, the reaction with lithium powder (2% sodium) being complete after 2–3 hours — starting with vinyl compounds *80* already after 1 hour.

The mechanism proceeds via radicals and is obviously rather complicated passing even through dimers as intermediates. Thus during synthesis of 1,1-dilithio-1-alkenes *81* [98] from *80* by direct mercury-lithium exchange, we often obtained up to 7% of the 2,3-dilithio-1,3-butadienes *83* which might have been formed from the corresponding mercury compounds *82*. That *82* is in fact an intermediate in the

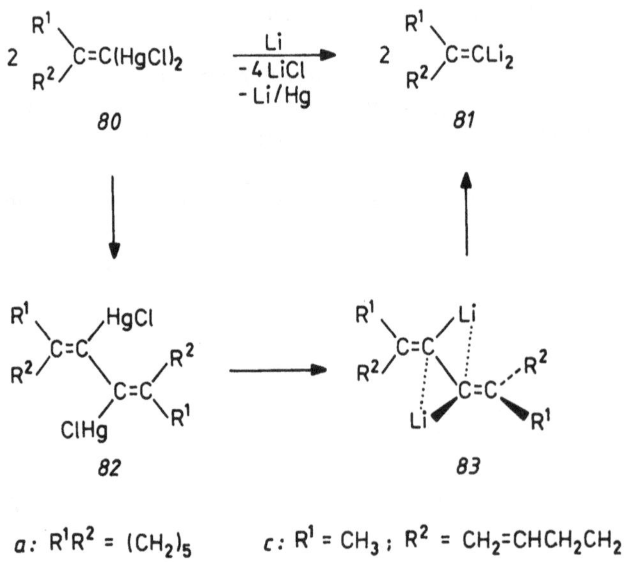

80 *81* *82* *83*

a: $R^1R^2 = (CH_2)_5$ *c:* $R^1 = CH_3$; $R^2 = CH_2{=}CHCH_2CH_2$

b: $R^1 = R^2 = CH_3$ *d:* $R^1R^2 = (CH_2)_2CH(CH_2)_2$ with t-Bu on CH

transformation of *80* to *81* could be shown by interrupting the reaction starting with *80a* after 10 minutes: *82a* was formed in 95% yield and afterwards could be cleaved to *81a* with additional lithium dust [100]) *83a* again is formed as a side product in yields up to 8%.

A similar mechanism might become operative during synthesis of 1,1-dilithioalkanes, because besides CH_2Li_2 traces of lithium carbide (C_2Li_2) have been found upon treatment of $CH_2(HgI)_2$ with lithium dust in diethyl ether [25]).

Starting with $C(HgCl)_4$, finally the dimeric products hexalithioethane (C_2Li_6) and tetralithioethylene (C_2Li_4) (ratio 4:3) have been obtained predominantly, without any doubt attributable to dimerization of radical intermediates [25]). The direct mercury-lithium exchange reaction therefore is not very suitable for the synthesis of CLi_4 *5* [97]).

$$\underset{\textit{84}}{BrHgCH_2CR_2CH_2HgBr} \xrightarrow[-2\ LiBr]{2\ t\text{-}BuLi}$$

$$\underset{\textit{85}}{t\text{-}BuHgCH_2CR_2CH_2Hg\text{-}t\text{-}Bu} \xrightarrow[-2\ t\text{-}Bu_2Hg]{2\ t\text{-}BuLi} \underset{\textit{86}}{LiCH_2CR_2CH_2Li}$$

a: R = H

b: R = CH_3

As with halogen-lithium exchange, however, replacement of mercury can also be achieved upon reaction with alkyllithium compounds instead of lithium metal yielding amalgam-free products. Bickelhaupt and coworkers [36]) were the first who used this variant for the synthesis of 1,3-dilithiopropanes *86*, not available by halogen-lithium exchange reactions [79,101]). The first step *84* → *85* with 2 equivalents of *tert*-butyllithium is rapid and was performed at 0 °C in pentane, precipitated LiBr was removed by filtration. Addition of 2 further equivalents of *tert*-butyllithium to the solution of *85* led to a slow reaction at room temperature (several hours for R=H, 1 week for R=CH_3) during which *86* precipitated as a white powder which can be freed from di-*tert*-butylmercury by washing with pentane.

This was also the method of choice for the synthesis of CLi_4 *5* from $C(HgCl)_4$ *87*, although we were not able to separate the LiCl after the first step [97,102]).

$$\underset{\textit{87}}{C(HgCl)_4} \xrightarrow[-4\ LiCl,\ -4\ t\text{-}Bu_2Hg]{8\ t\text{-}BuLi} \underset{\textit{5}}{CLi_4} \xleftarrow[-4\ EtLi,\ -4\ t\text{-}Bu_2Hg]{8\ t\text{-}BuLi} \underset{\textit{88}}{C(HgEt)_4}$$

We therefore synthesized $C(HgEt)_4$ *88* hoping finally to be able to wash out the di-*tert*-butylmercury as well as the ethyllithium, both being soluable in cyclopentane. Surprisingly, however, after stirring *88* in cyclopentane with *tert*-butyllithium for one day at room temperature we obtained a deep red-brown very light-sensitive *solution* containing a complex of CLi_4 *5* with di-*tert*-butylmercury and ethyllithium. The accompanying products therefore had to be removed by sublimation and interestingly the pure extremely pyrophoric tetralithiomethane *5* afterwards did no longer dissolve in hydrocarbon solvents.

Using this transmetalation reaction with *tert*-butyllithium we also succeeded for the first time in synthesizing *cis*- and *trans*-1,2-dilithioethylene [31]. According to preliminary experiments the *cis*-isomer *14* seems to rearrange into the more stable *trans*-isomer *9*, the results, however, have to be confirmed.

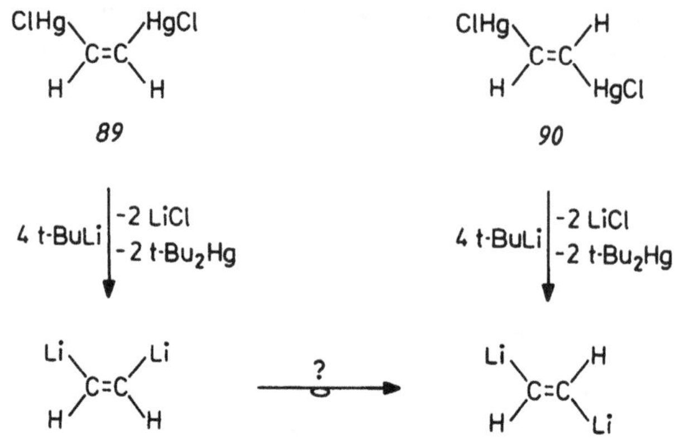

As these compounds are very unstable we had to work at —50 °C introducing a cyclopentane solution of *tert*-butyllithium into a suspension of the mercury compound in diethyl ether the reaction being complete already after one hour. This was shown by the amount of di-*tert*-butylmercury finally remaining constant as detected by gas chromatography using an internal standard. On the other hand no reaction of the mercury compounds *89* and *90* could be observed with lithium metal, neither in cyclopentane nor in diethyl ether or THF as the solvent presumeably because of their poor solubility.

Transmetalation with *tert*-butyllithium, however, is not always the better way to achieve mercury-lithium exchange reactions. Although we had no problems with the

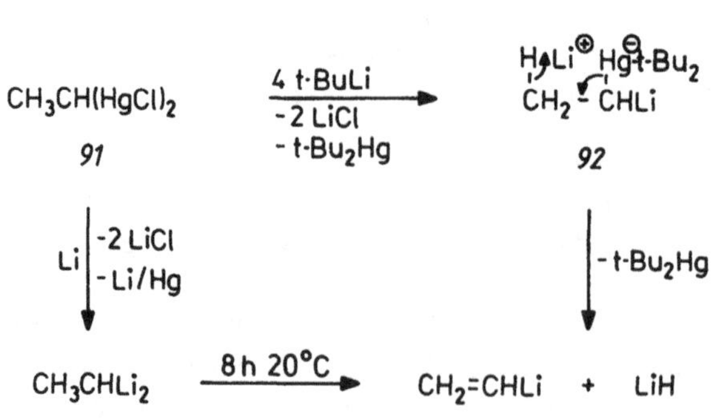

synthesis of dilithiomethane *3* [23], the corresponding reaction of 1,1-bis(chloromercurio)ethane *91* with *tert*-butyllithium — although faster than with lithium metal — yielded vinyllithium *8* instead of 1,1-dilithioethane *11*. The ate-complex intermediate *92* obviously splits off lithium hydride faster than does the geminal dilithio compound *11* [23]. It seems reasonable to assume that part of the allyllithium found by Bickelhaupt [36] during the reaction of *85a* with *tert*-butyllithium has the same origin.

While 1,1-dilithioethane *11* prepared from *91* and lithium dust in diethyl ether is rather stable decomposing at room temperature only within 8 h to afford lithium hydride and vinyllithium *8*, higher 1,1-dilithioalkanes are much less stable. Thus 1,1-dilithioheptane *94* decomposes already during its preparation — even on using lithium metal — yielding 1-lithio-1-heptene *95* at once [25].

$$\underset{93}{C_6H_{13}CH(HgCl)_2} \xrightarrow[\substack{-2\,LiCl \\ -Li/Hg}]{Li} \underset{94}{[C_6H_{13}CHLi_2]} \longrightarrow \underset{95}{C_5H_{11}CH{=}CHLi} + LiH$$

Due to intramolecular coordination (see Sect. 2.7) 4,4-dilithio-1-butene *33* is more stable and can be prepared by a direct mercury-lithium exchange reaction although lithium hydride elimination yields a conjugated system *97* [25]. Rearrangement to a cyclopropylcarbinyl species *34*, however, was not observed (see Sect. 2.7).

$$\underset{96}{CH_2{=}CH{-}CH_2{-}CH(HgCl)_2} \xrightarrow[\substack{-2\,LiCl \\ -Li/Hg}]{Li}$$

$$\underset{33}{CH_2{=}CH{-}CH_2{-}CHLi_2} \longrightarrow \underset{97}{CH_2{=}CH{-}CH{=}CHLi} + LiH$$

3.4 Transmetalation Reactions with Other Metals

Transmetalation using tin-lithium exchange is an excellent method — even better than mercury-lithium exchange — for the preparation of organolithium compounds, but it works only for compounds being more stable than *n*-butyllithium. This is also true for the synthesis of polylithiumorganic compounds. Examples are (*E*,*E*)-1,5-dilithio-1,4-pentadiene *99* [103], the (*Z*,*Z*)-1,5-dilithio-1,4-pentadiene derivative *101* [104] — the unsubstituted compound surprisingly could not be prepared this way [103] — and a compound tentatively formulated as 1,4-dilithio-2-butyne *103* [105].

Neither *trans*-1,2-dilithioethylene *9* [33] nor 1,1-dilithio-1-alkenes *81* [106], however, could be prepared by tin-lithium exchange, the reaction stops after the replacement of one stannyl group only.

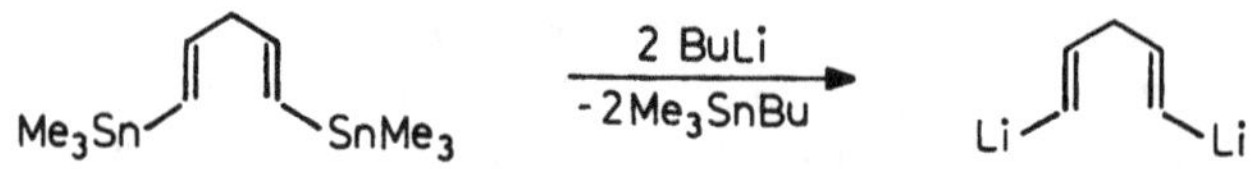

98 99

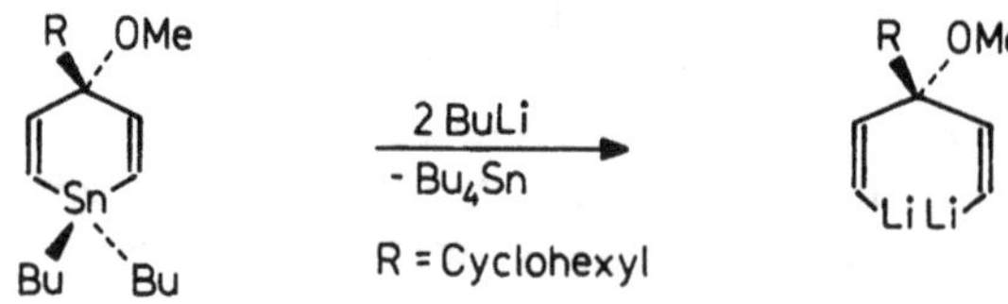

100 101

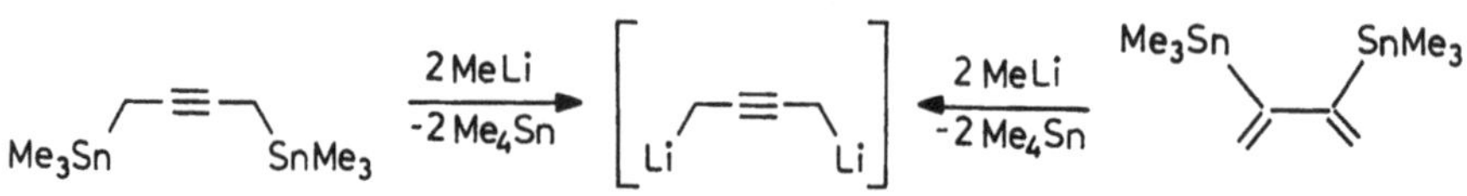

102 103 104

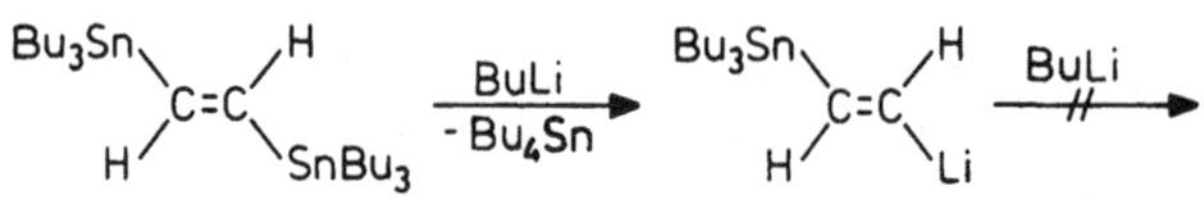

105 106

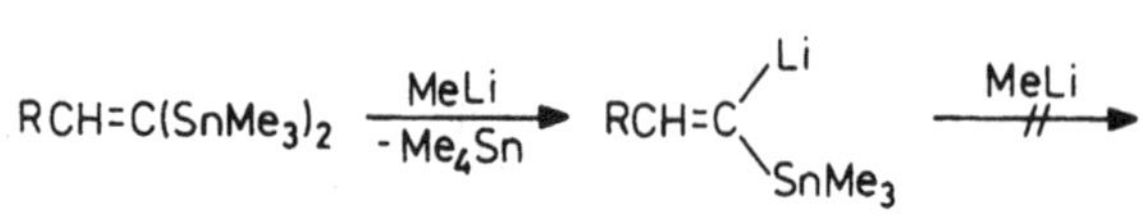

107 108

In our view this means that dilithioethylenes are less stable than *n*-butyllithium in contrast to theoretical proposals [30] concluding that *trans*-1,2-dilithioethylene *9* should be more stable than methyllithium *2* and even vinyllithium *8*.

Therefore we used a stronger base than *n*-butyllithium, *tert*-butyllithium, but in this case one has to switch over to mercury-lithium exchange (see Sect. 3.3), because it is not possible to place four *tert*-butyl groups around the tin atom due to steric crowding. Thus Kuivila and coworkers [107] recently have found that upon treatment of tetramethylstannane with *tert*-butyllithium only two methyl groups can be displaced by *tert*-butyl groups.

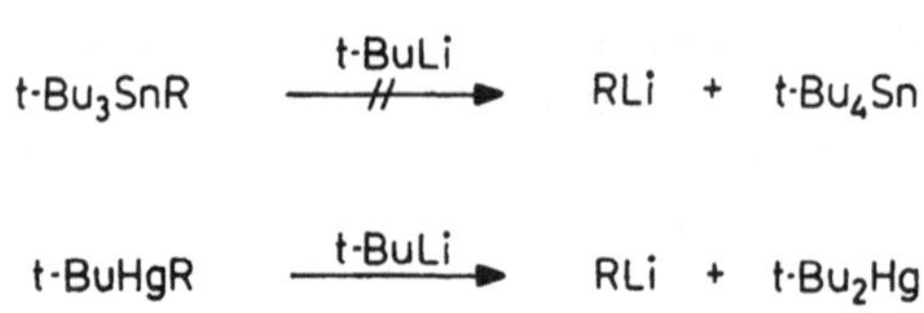

Transmetalation reactions with several other elements of the periodic table can also be used for the synthesis of lithiumorganic compounds [108, 109], but polylithiated hydrocarbons have not yet been prepared this way. As with the corresponding tinorganic compounds the reaction of *gem*-organodiboron compounds *109* with butyllithium stops after replacement of one boron group only, the endproduct is an ate-complex *110* with the remaining boron atom as the central atom [110–112].

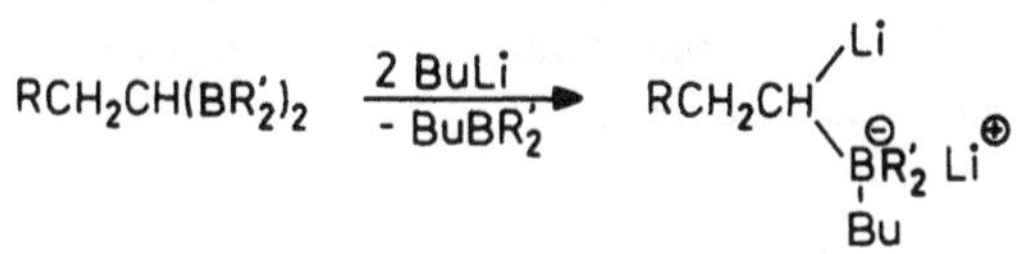

109 *110*

A trivial example for the replacement of potassium by lithium in a dianion finally has to be mentioned [113]:

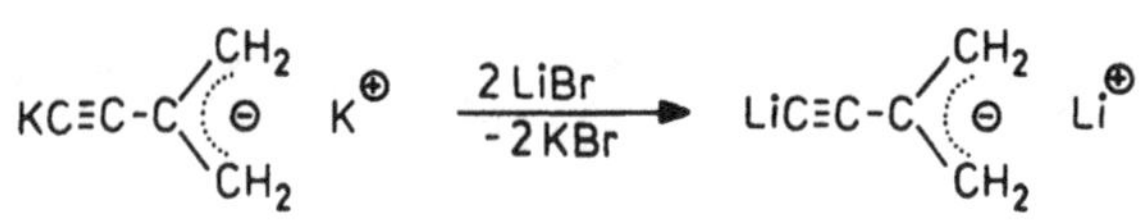

111 *112*

3.5 Reductive Metalations

Phenyl substituted ethylenes, acetylenes, and allenes are known to add lithium metal — sometimes with dimerization — to yield stable benzyl-type 1,2- or 1,4-dilithium compounds, respectively. The first observations in this connection already go back to Schlenk and Bergmann [114] in 1928.

Within the last years, however, it has been found that even nonactivated aliphatic alkenes and alkynes can be forced to react with metallic lithium — catalized or even uncatalized — whereby dilithiated hydrocarbons are formed, although sometimes only as intermediates.

Peterson and coworkers [115] were the first who showed that 1-alkenes can react with lithium dispersion under relatively mild conditions to give the corresponding 1-alkynyllithium compounds and lithium hydride as the major products. Thus 1-hexene *113* was converted into 1-hexynyllithium *116* in 65% yield within 1 h at the reflux temperature of the olefin (64 °C). It is reasonable to assume that 1,2-dilithiohexane *114* is formed first loosing lithium hydride to give 1-hexenyllithium *115*, which was detected as a side-product in 4% yield.

113 *114*

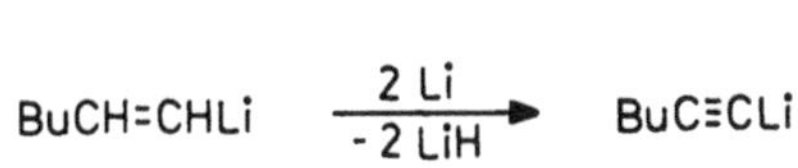

115 *116*

The reaction starting with ethylene can be stopped at the vinyllithium step by using a special catalyst as was shown by Bogdanović [18,116] and earlier by Rautenstrauch [17], the latter, however, using biphenyl and naphthalene as the catalyst, had obtained considerable amounts of 1,4-dilithiobutane *28* and other 1,ω-dilithioalkanes *57* as by-products. Starting with propene *117* the reaction interestingly can be directed to yield allyllithium *69* as the main product simply by adding $PtCl_2$ or $PdCl_2$ to the Bogdanović catalyst [18,116]. Always in addition minor amounts of the isomers *120* and *121* had been formed. On the other hand not even traces of the postulated 1,2-dilithioalkanes *118* ever could be detected.

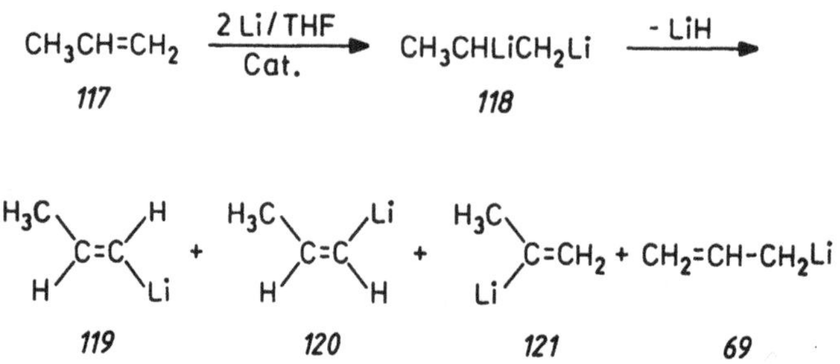

In light of these results the reported uncatalyzed synthesis [21] of 1,2-dilithioethane *7* in 27–30% yield from ethylene and lithium powder in dioxane as the solvent should be reproduced, inasmuch as characterization was only by hydrolysis to yield ethane.

In contrast we found that the addition of lithium to the central double bond of butatrienes *122* takes place very easily [100]. The resulting 2,3-dilithio-1,3-butadienes *123* which are stable towards excess lithium interestingly are cleaved to 1,1-dilithio-1-alkenes *81* and *125* in the presence of mercury(II) chloride (see Sect. 3.3).

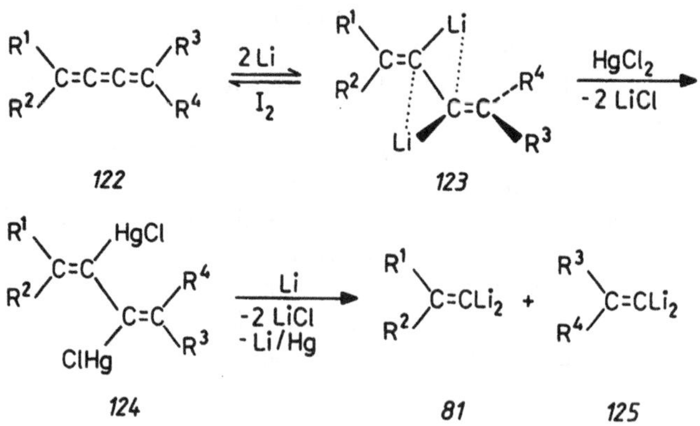

a: $R^1 = R^2 = R^3 = R^4 = CH_3$

b: $R^1 = R^3 = CH_3$; $R^2 = R^4 = t\text{-}Bu$

c: $R^1 = R^2 = CH_3$; $R^3R^4 = (CH_2)_5$

d: $R^1R^2 = R^3R^4 = (CH_2)_5$

Other cumulenes, e.g. the tetraene *126* and the pentaene *128* add lithium as well, the end products *127* and *129* obtained in 48% and 65% yield, respectively, are shown in their most stable tautomeric structures [117]. The real geometries, however, are not known.

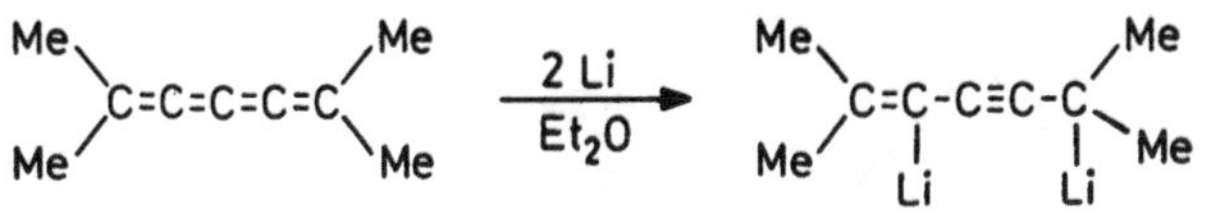

126 *127*

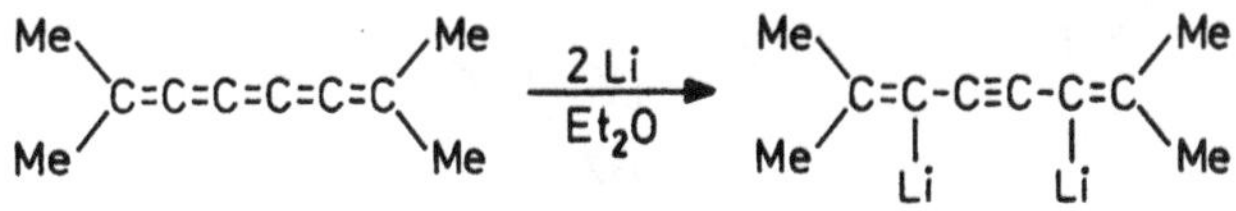

128 *129*

In special cases even a single bond can be reductively cleaved as was found by Goldstein et al. [118,119] who obtained two diastereoisomeric dimers of "dilithium semibullvalenide" *131* on treating semibullvalene *130* with lithium in THF or dimethyl ether at —78 °C.

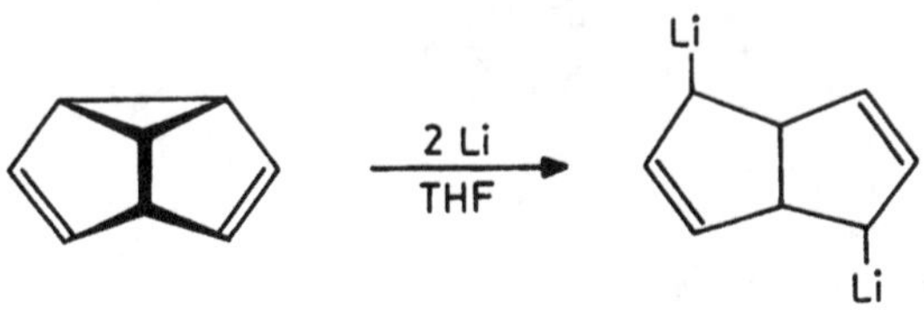

130 *131*

A *cis*-1,4-dilithiobutatriene *133* has been obtained by Schleyer et al. [69] upon conjugate addition of lithium to di-*tert*-butyl-diacetylene *132*.

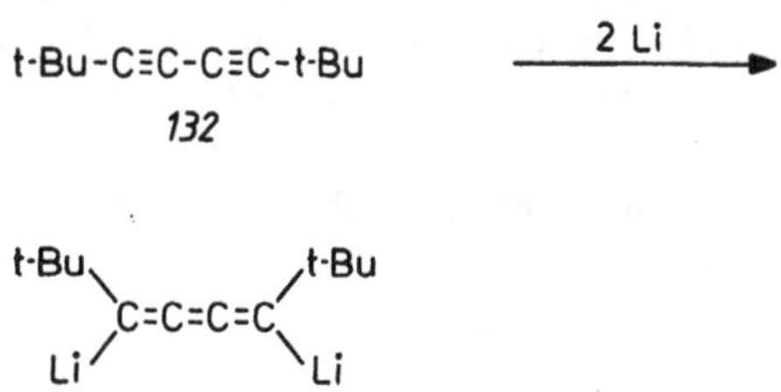

132

133

Isolated triple bonds, on the other hand, react rather slowly with lithium dust except in strained rings. Thus we succeeded in adding lithium to the triple bond of cyclooctyne *134* even at —35 °C in diethyl ether to yield *cis*-1,2-dilithiocyclooctene *136* as a yellow solution containing up to 20% of the 1,4-dilithio-1,3-butadiene derivative *137*, the product of dimerizing addition [120].

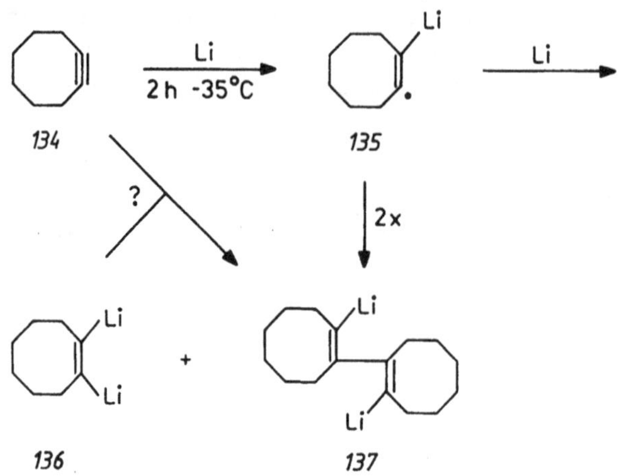

With open-chain aliphatic alkynes *138*, however, very slow *trans*-addition of lithium takes place, reaction times of up to 48 hours at room temperature were necessary [120]. This time we observed only small amounts of dimeric side products and the vicinal *trans*-dilithioalkenes *139* turned out to be insoluble in diethyl ether.

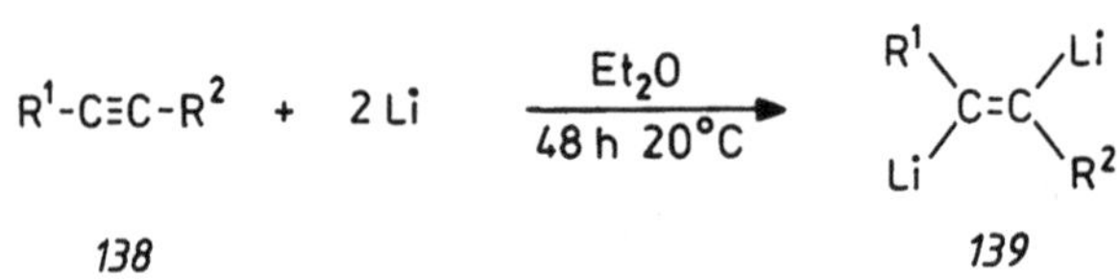

a: $R^1 = C_2H_5$; $R^2 = C_4H_9$

b: $R^1 = R^2 = C_3H_7$

c: $R^1 = R^2 = C_4H_9$

It is not known if *cis*-addition occurs first followed by *cis-trans*-isomerization. The infrared spectrum of the primary product of lithium atoms and acetylene molecules in an argon matrix at 15 K speaks for a planar structure $LiC_2H_2^{\cdot}$ with lithium bridging the π system and the hydrogen atoms situated at the opposite side in a *cis*-arrangement [121].

Geminal and/or vicinal dilithioalkenes in principle should also be available by the addition of alkyllithium compounds to lithium acetylides:

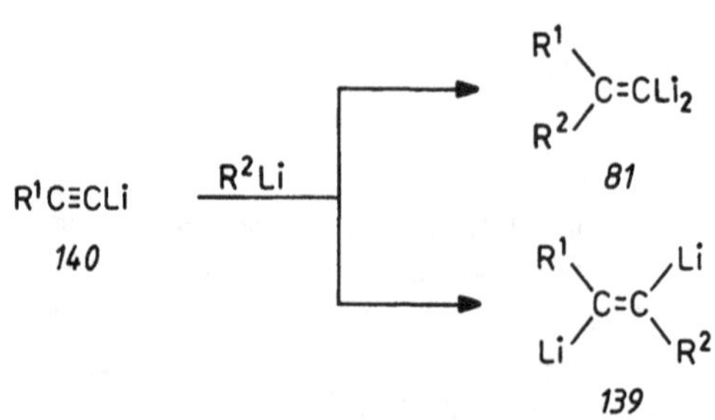

We tried to add *tert*-butyllithium to lithium *tert*-butylacetylide *141*, but the yield of 1,1-dilithio-2,2-di-*tert*-butylethylene *142*, was only 9%[117]. The presence of di-*tert*-butylmercury as a catalyst is important, presumably yielding $(t\text{-Bu})_2C{=}CLi(Hg\text{-}t\text{-Bu})$ as an intermediate[122]

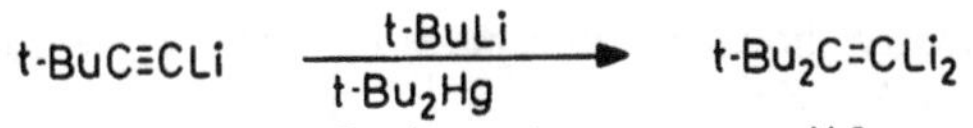

141 *142*

An intramolecular variant of this reaction is the cyclization of 1,6-dilithio-1-hexyne *144* to dilithiomethylenecyclopentane *145* which we found to take place in 28% yield on standing in diethyl ether for 10 minutes at −78 °C[117].

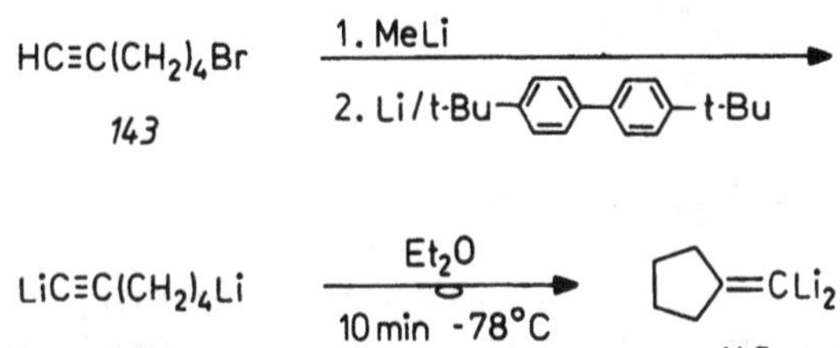

143 *144* *145*

Better intermolecular yields were obtained by switching over to propargylic alcohol[117], a reaction which is already known to work with Grignard reagents[123]. The yield was 48% using ethyllithium, but *n*-butyllithium and phenyllithium have also been added in 15% and 42% yield, respectively[117]. The reaction is catalyzed by 10% copper(I) iodide. Without this catalyst the yields are only about half of the amounts given.

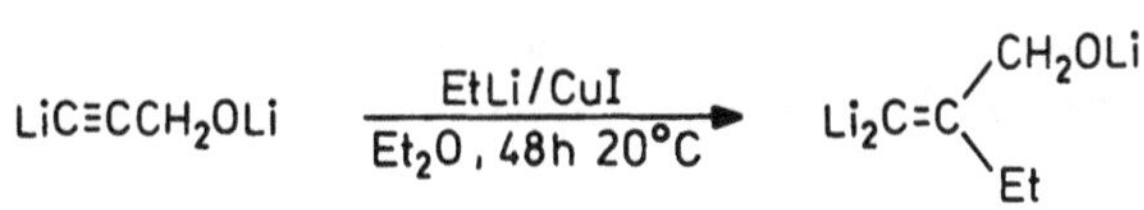

146 *147*

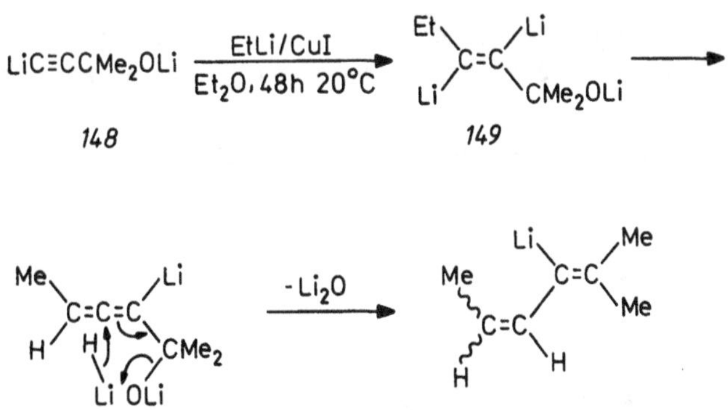

148 *149* *150* *151*

The dilithium salt of the corresponding dimethylpropargylic alcohol *148* interestingly adds ethyllithium in a different manner. The primarily formed 2,3-dilithio-2-pentenolate *149* obviously eliminates lithium hydride which attacks the allene intermediate *150* splitting off lithium oxide in an S_N2' reaction. The endproduct, an *E*/*Z*-mixture of 3-lithio-2-methyl-2,4-hexadiene *151* finally was obtained in 30% yield [117]. The last step was independently shown to take place by treating the corresponding allenic alcoholate with lithium hydride [117].

Isopropyllithium can also be added yielding 3-lithio-2,5-dimethyl-2,4-hexadiene *152* in 36% yield [117].

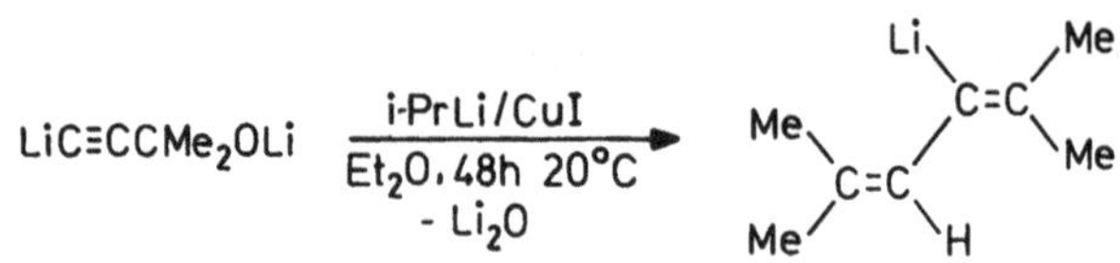

148 *152*

3.6 Metalation of Acidic Hydrocarbons

Numerous polylithiated aliphatic hydrocarbons have been prepared by polymetalation of the corresponding hydrocarbons [1–3] and only a few characteristic examples can be given.

Propyne interestingly can be perlithiated with *n*-butyllithium in hexane to yield the "lithiocarbon" C_3Li_4 *26* as a deep red-brown solution [124, 125]. After evaporation of the solvent the solid cannot be resolved in hexane as was the case with CLi_4 *5* [97].

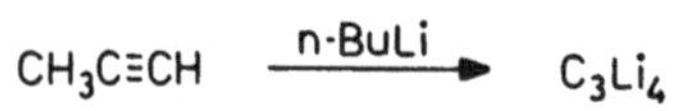

26

The isomeric allene, on the other hand, using *n*-butyllithium in hexane/THF (1:1) at −50 °C is metalated only to the dianion *24* [126].

24

Controlled dilithiation of propyne can be achieved by using two equivalents of *n*-butyllithium in hexane/ether in the presence of one equivalent of TMEDA [127].

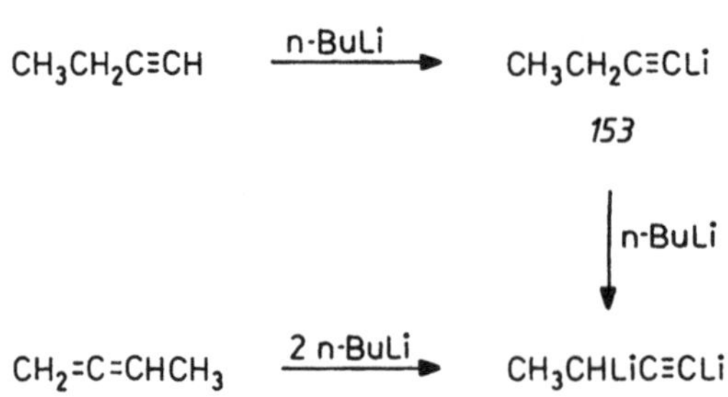

153

154

Upon boiling 1-butyne with excess *n*-butyllithium in hexane a rapid dimetalation takes place without the intermediate acetylide *153* being precipitated [128]. The same dilithiobutyne *154* was obtained starting with 1,2-butadiene.

Switching over to the stronger base *tert*-butyllithium a third lithium atom ca be introduced [125].

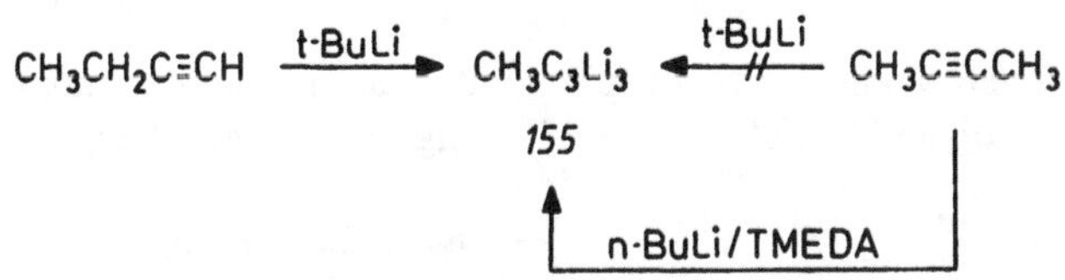

Interestingly 2-butyne cannot be deprotonated under the same conditions [125]. Only by using *n*-butyllithium in the presence of TMEDA — one of the strongest known metalating reagents — trimetalation takes place here too yielding the same product *155* as with 1-butyne [129].

3-Methyl-1-butyne, on the other hand, is dilithiated again by boiling with excess *n*-butyllithium in hexane [129].

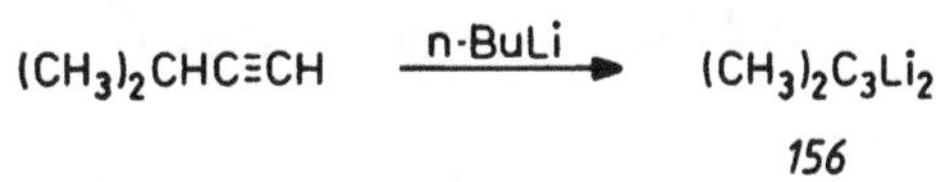

Dilithiation of 1- and 2-alkynes with *n*-butyllithium can be achieved already at room temperature by using diethyl ether as the solvent instead of hexane [130]. 3-Alkynes, however, under these conditions yield only monoanions. On the other hand up to four lithium atoms could be introduced into 1,8-cyclotetradecadiyne with *n*-butyllithium in THF or in the presence of TMEDA even in hexane as the solvent [131].

Upon treatment of 1,3-pentadiyne with excess *n*-butyllithium/TMEDA in hexane at room temperature another "lithiocarbon" C_5Li_4 *157* is obtained as was the case with propyne [132].

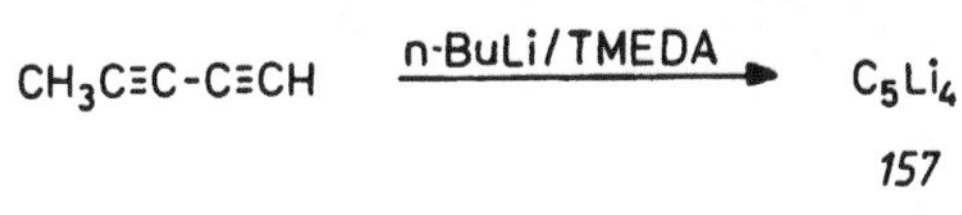

2,4-Hexadiyne under the same conditions [133] or even in the absence of TMEDA [134] yield the trilithiated product *158*.

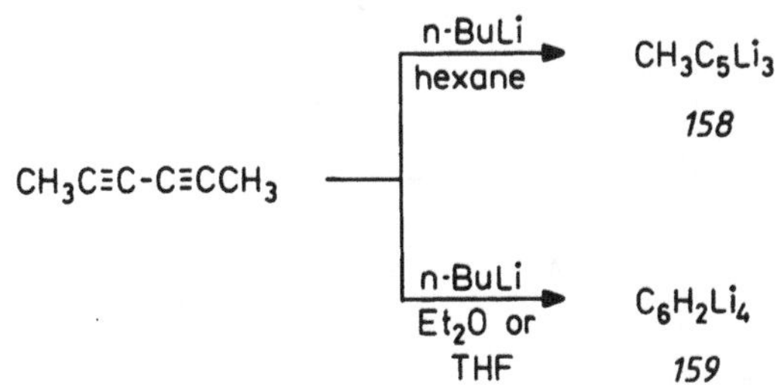

Using more polar solvents as diethyl ether or THF tetraanions were found as well, while 2,4-octadiyne again yields a trianion *160* [134].

$$C_3H_7C{\equiv}C{-}C{\equiv}CCH_3 \xrightarrow[Et_2O]{n\text{-}BuLi} \underset{160}{C_3H_7C_5Li_3}$$

In 1,6-heptadiyne finally the two triple bonds react isolated as was the case with 1,8-cyclotetradecadiyne [131] yielding the following di-, tri-, and tetraanions [134]:

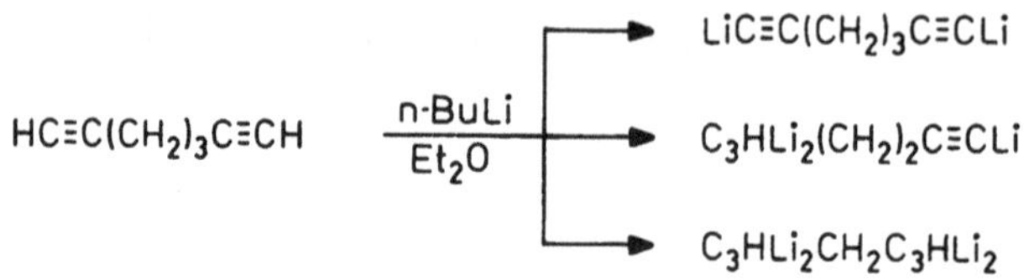

Dimerization was observed upon treatment of 3-penten-1-yne with butyllithium in a mixture of THF and hexane [135]:

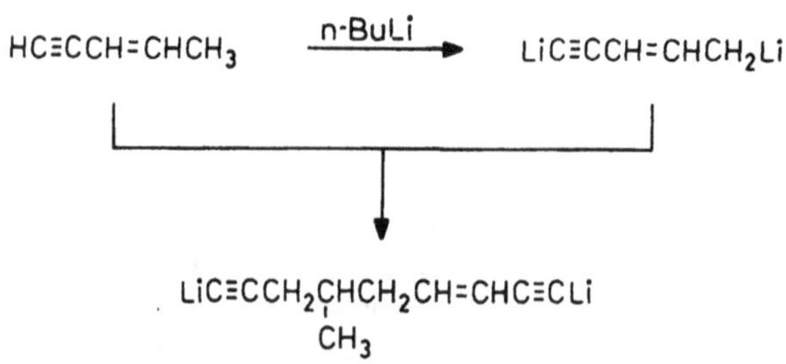

Linear and branched *olefins* can also be polymetalated provided that the corresponding polyanions are resonance stabilized. Thus propene by treatment with

18

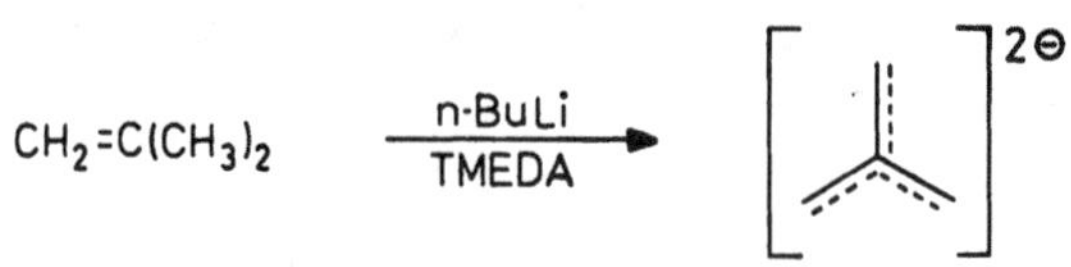

37

35

n-butyllithium/TMEDA in hexane yields a dilithiated allyl system *18* [37] and isobutylen the dilithium salt of the cross-conjugated trimethylenemethane dianion *37* [48–50]. The isomeric 2-butene can be dilithiated to give 1,4-dilithio-2-butene *35* [49].

Numerous higher alkadienes and alkatrienes — also cyclic ones — up to C_9 with isolated as well as conjugated double bonds have been polylithiated and the structure and reactivity of the products have been studied mainly by Bates and coworkers [49, 52, 137–144a], but also by Klein [50, 136] and by Mills [54–56]. These results, although very interesting, cannot be discussed here in any detail and should be taken from the original literature. The same is true for aryl and heterofunctionally substituted systems of any kind [145].

Finally it should be mentioned that strained ring compounds bearing C—H bonds with high s character can be dilithiated although the corresponding dianions are not resonance stabilized. Examples are 1,2-dilithiocyclopropene *22* [146] and certain bicyclobutane derivatives e.g. *161* [147, 148].

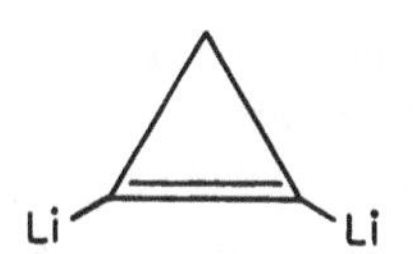

22

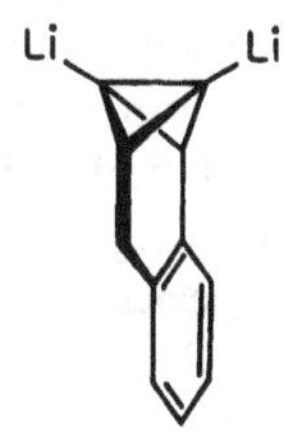

161

3.7 Reactions with Lithium Vapor

Lithium vapor is much more reactive than even the finest lithium dust. This was shown already in 1955 by Hérold [149] who obtained lithium carbide from graphite in quantitative yield.

In 1972 Lagow and coworkers [150] reported a general technique for preparing polylithium organic compounds utilizing reactions between organic substrates and lithium vapor. Lithium vapor syntheses involve cocondensing lithium atoms with a hydrocarbon or halocarbon substrate under vacuum. The reactions are carried out in a special stainless steel reactor consisting of a Knudsen cell containing the lithium metal heated in a furnace, an inlet tube, and a liquid nitrogen cold finger. The main advantage of this non-solution method besides the high reactivity of lithium vapor is the reduction of rearrangements and secondary reactions from the rapid quenching of the products onto the cold finger.

Tetralithiomethane *5* (CLi_4) and hexalithioethane *12* (C_2Li_6), the first examples of perlithiated alkanes, were prepared by reacting lithium vapor with the appropriate halocarbon between 800 and 1000 °C [150].

$$CCl_4 + 8\ Li\,(g) \xrightarrow{850\,^\circ C} CLi_4 + 4\ LiCl$$

The main side products were C_2Li_4 and C_2Li_2. Starting with partially halogenated hydrocarbons in addition part and sometimes even all of the hydrogens are replaced by lithium [151,152]. The selectivity can be improved by working at 750 °C instead of 850 °C [153,154]. This way CH_2Li_2 and CLi_4 could be synthesized in 65.7% and 40.5% yield, respectively [154]. However, even at 750 °C, the lability of the hydrogen atoms toward substitution by a lithium atom showed up significantly when chloroform was used as a substrate: $CHLi_3$ was obtained in only 15.5% yield, the main by-products being 19.1% CH_2Li_2, 20.1% CLi_4, and 39.7% C_2Li_2 [153,154]. It is further observed that the greater the number of C—Cl bonds per chloromethane, the greater the yield of C_2 molecules, most notably lithium carbide. Thus even the best CLi_4 contained 58.5% C_2Li_2 [154].

The purest C_2Li_6 *12* was obtained by using diethylmercury instead of C_2Cl_6 as the starting material, while $(C_2H_5)_4Sn$ and $(C_2H_5)_4Pb$ delivered mainly ethyllithium (C_2H_5Li) [155].

The reaction of carbon vapor with atomic lithium, on the other hand, yielded C_3Li_4 *26* as the main product [156].

Lagow and coworkers [157,158] further extended their method to include the reaction of alkenes and benzene with lithium vapor. It was observed that both substitution of lithium for hydrogen and a minor side reaction addition of lithium to the double bonds occurred to produce polylithiated alkanes and alkenes.

A major limitation of the Lagow procedure for preparative purposes is the trapping of the lithiocarbon species in the lithium metal matrix, but techniques have been developed to minimize this problem [154]. The reaction products have been purified by grinding under argon and the resulting powder sieved resulting in removal of substantial amounts of lithium metal and approximately 85% pure lithiocarbon product. Subsequently the products were extracted with cold THF to remove LiCl and were then separated from lower density lithium rich particles by flotation on cold THF. Unfortunately, however, methods have not yet been developed to separate one lithium substituted hydrocarbon from a mixture of others.

3.8 Fragmentations

Decarboxylation of α-lithiated lithium salts of carboxylic acids theoretically should lead to geminal dilithium compounds. Warner and Le [159] attempted to prepare 1,1-dilithiocyclopropane *20* this way starting with lithium α-lithiocyclopropanecarb-

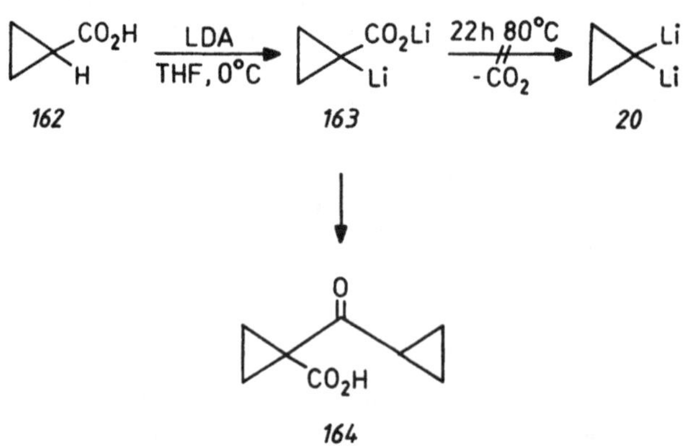

162 *163* *20* *164*

oxylate *163* readily accessible from cyclopropanecarboxylic acid *162*. The results, however, were negative. Jahngen and coworkers [160] have shown that self-condensation to the β-keto acid *164* was the only observed reaction at elevated temperatures. It is not known if this course of the reaction is general for all α-lithiated lithium carboxylates and further investigations seem promising.

Another fragmentation reaction especially useful for the synthesis of vinyllithium compounds is the well-known Shapiro reaction [161]. We were successful in preparing 2,3-dilithio-1,3-butadiene *166* by a double Shapiro reaction although in 12% yield only [117]. The main reaction product was 2-butyne.

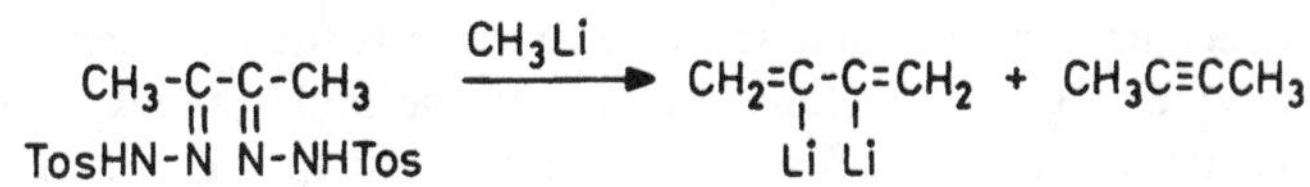

165 *166*

4 Properties

4.1 Reactions

Dilithiated hydrocarbons with two carbanionic centers isolated by at least two carbon atoms in general show identical reactivity as monolithium compounds and straightforward reactions with monofunctional electrophiles need not be discussed in this connection. The bifunctionality, however, offers the chance to prepare cyclic com-

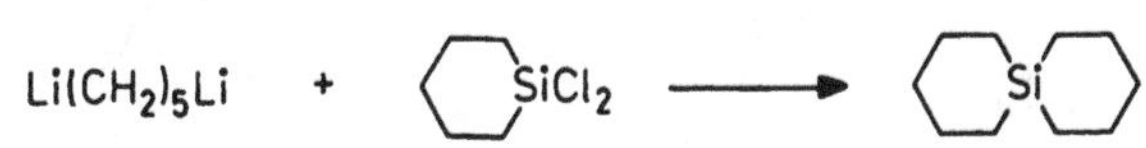

167 *168* (54%)

58 *169* (50%)

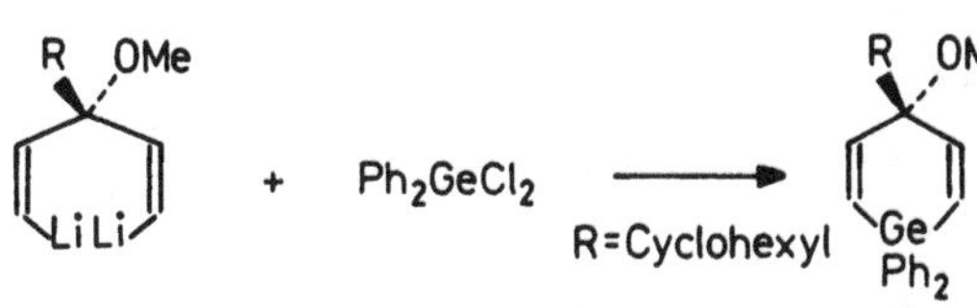

101 *170* (76%)

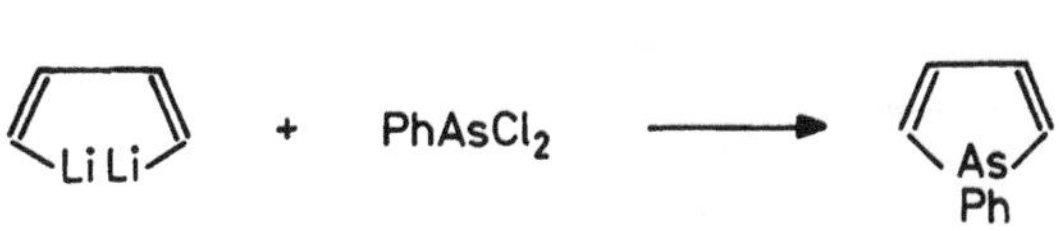

39 *171* (15%)

pounds. Examples are the sila-spirane *168* [79], the 1,1-dimethyl-1-sila-2,5-cyclohexadiene *169* [81], the 1-germa-2,5-cyclohexadiene derivative *170* [104], and the 1-phenylarsole *171* [80].

In special cases a stepwise introduction of two different electrophiles is possible [96], e.g.

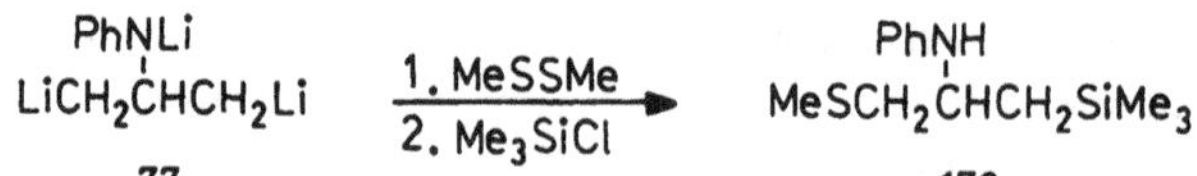

77 *172*

An intramolecular variant of this stepwise reaction is the treatment with gaseous carbon dioxide yielding cyclic ketones by an intramolecular attack of a still intact carbanionic center at an already formed carboxylate group, e.g. [79]:

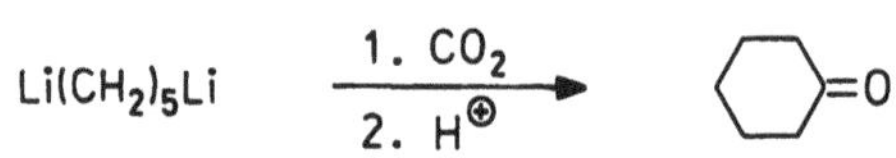

167 *173*

Cyclic compounds have also been prepared starting with the 2-methyleneallyl dianion *37* and 2,3-dimethylenebutadiene dianion *177* the latter yielding 2,3-disubstituted 1,3-butadienes [139, 142–144], e.g.

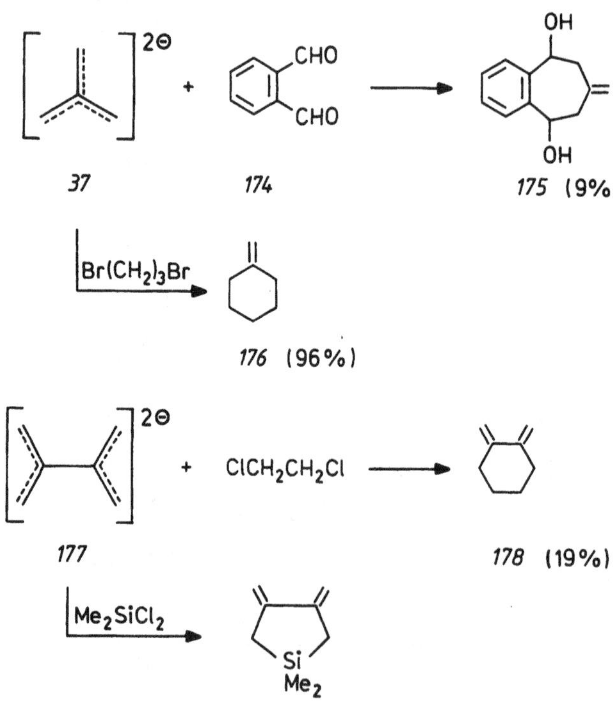

37 *174* *175* (9%)

176 (96%)

177 *178* (19%)

179 (30%)

We obtained similar butadiene derivatives by a different route starting with 2,3-dilithio-1,3-butadienes *123* [117]. The last reaction yielding an interesting thioketale of a 2,3-diisopropylidenecyclopropanone derivative *181* is related to the aforementioned synthesis of cyclic ketones with gaseous carbon dioxide.

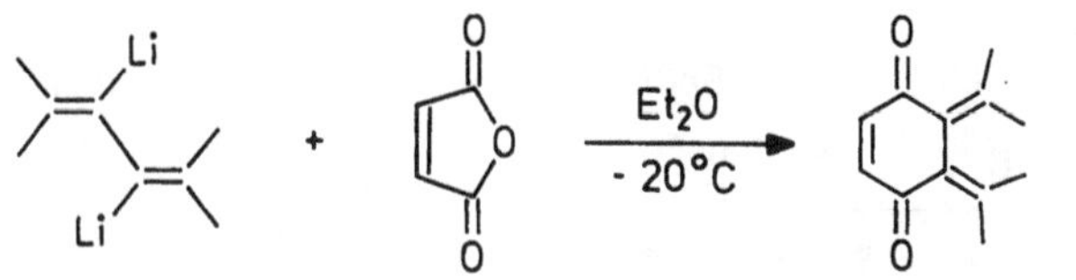

123a *180* (71%)

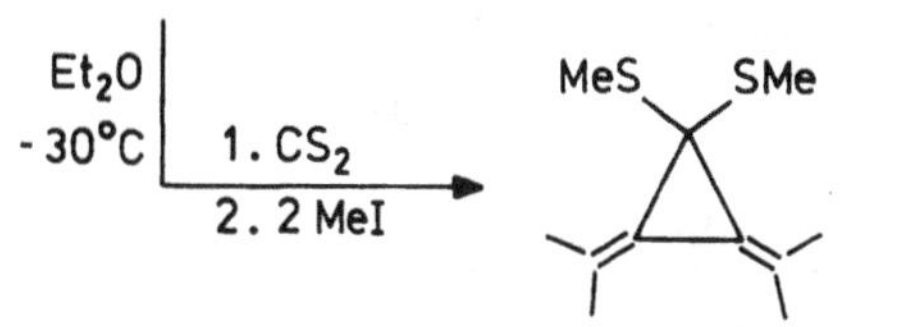

181 (85%)

With dichloromethylsilane, however, we obtained a dimer product, i.e. the 1,4-disilacyclohexane derivative *182* (*cis* + *trans*), a disila[6]radialene with four isopropylidene groups [162].

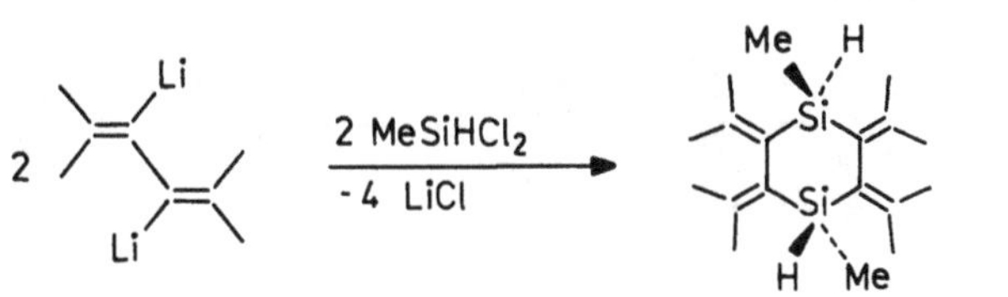

123a *182(trans)* (12%)

In contrast, isocentric polylithiated hydrocarbons often show reactivities quite atypical for normal lithiumorganic compounds. Already in 1971 Krohmer and Goubeau [163] reported that dilithiomethane *3* reacts with chlorotrimethylsilane only under forced conditions — 48 hours at 150 °C in a sealed tube — which is rather unusual for an alkyllithium compound. We found that the reaction can be performed much more conveniently by refluxing CH_2Li_2 in Me_3SiCl in the absence of a solvent, but anyhow it needs four days to be complete [25]!

$$\underset{3}{CH_2Li_2} \xrightarrow[-2\ LiCl]{2\ Me_3SiCl} \underset{51}{CH_2(SiMe_3)_2}$$

While dilithiomethane *3* finally delivers the expected product *51* in up to 88% yield, this is not the case with CLi_4 *5*; less than 5% of $C(SiMe_3)_4$ have been obtained besides numerous by-products [25].

With Me_3SnCl even dilithiomethane does not yield the expected product. Instead we obtained 72% hexamethyldistannane *183* besides 6% trimethylstannane *184* and 12% *186*, the product of chlorotrimethylstannane with metalated *183*. The latter of course could also have been formed by dimerization of trimethylstannyl radicals $Me_3Sn^{\cdot}$ [25].

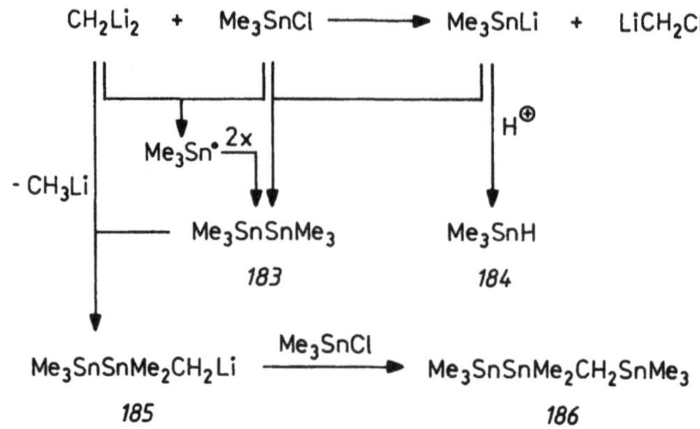

That dilithiomethane indeed prefers radical reactions was also evident from its reaction with cyclohexyl bromide in diethyl ether yielding 66% (net) bicyclohexyl besides 18% cyclohexane and 16% cyclohexene. The total yield was only 6% after two days at room temperature [25].

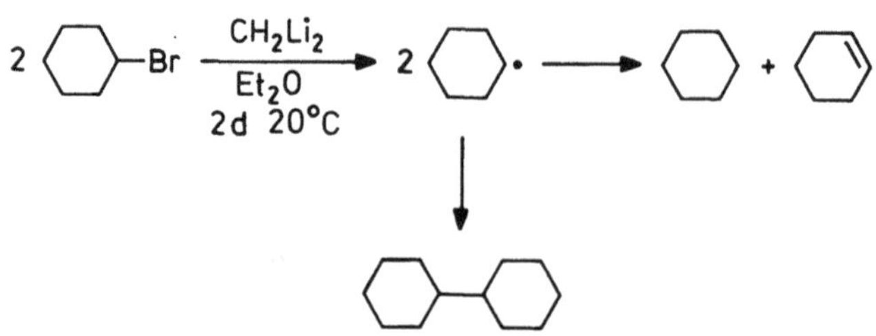

Another difference in reactivity was observed in connection with THF decomposition. While alkyllithium compounds add to the ethylene primarily formed [164] dilithiomethane adds to the enolate component [25].

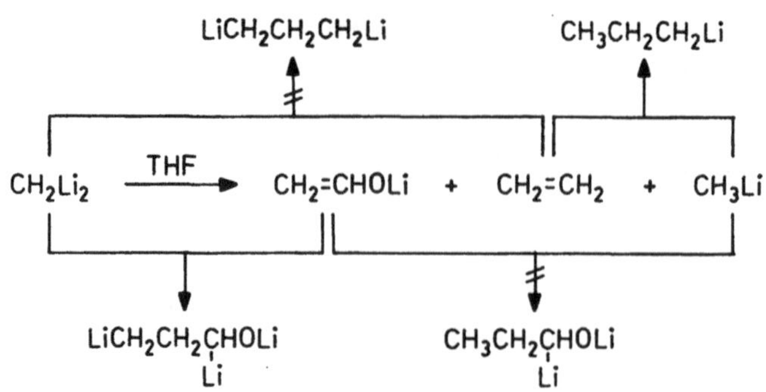

No reaction of dilithiomethane with benzaldehyde and benzophenone in diethyl ether could be detected even after two days at room temperature. With aliphatic ketones e.g. ethyl methyl ketone, on the other hand, only the products of aldol addition and aldol condensation have been found showing that CH_2Li_2 acts as a base and not as a nucleophile [25]. Carbonyl compounds therefore cannot be used in order to characterize isocentric polylithiated hydrocarbons.

The particular demands of polylithioalkanes on a derivatization reagent we finally found met by dimethyl disulfide delivering the corresponding methylthio derivatives in almost quantitative yield [25,97], e.g.

$$\underset{3}{CH_2Li_2} + 2\ MeSSMe \xrightarrow[-50^\circ C]{Et_2O} CH_2(SMe)_2 + 2\ LiSMe\downarrow$$

$$\underset{5}{CLi_4} + 4\ MeSSMe \xrightarrow[-50^\circ C]{Et_2O} C(SMe)_4 + 4\ LiSMe\downarrow$$

$$\underset{33}{CH_2{=}CHCH_2CHLi_2} \xrightarrow{2\ MeSSMe} CH_2{=}CHCH_2CH(SMe)_2$$

1,1-Dilithio-1-alkenes *81* on the other hand show a clean reaction with bromine even at —90 °C interestingly without addition to the double bond taking place [117], e.g.

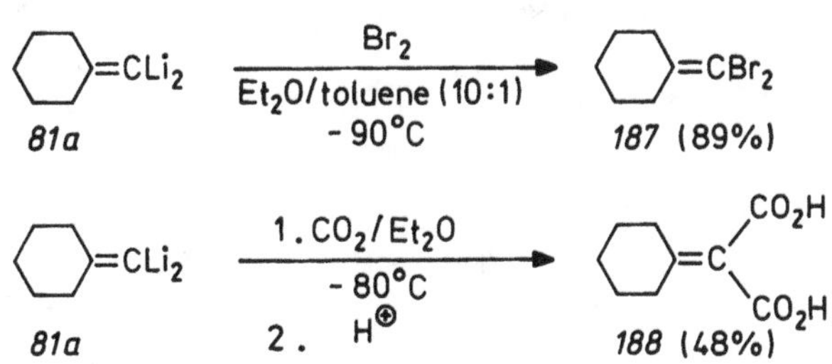

The same reagent upon carboxylation gave cyclohexylidenemalonic acid *188* in 48% yield [117].

Another remarkable reaction of a 1,1-dilithio-1-alkene *81c* is its smooth rearrangement into a five-membered ring-compound *189* with two isolated carbanionic centers [117]:

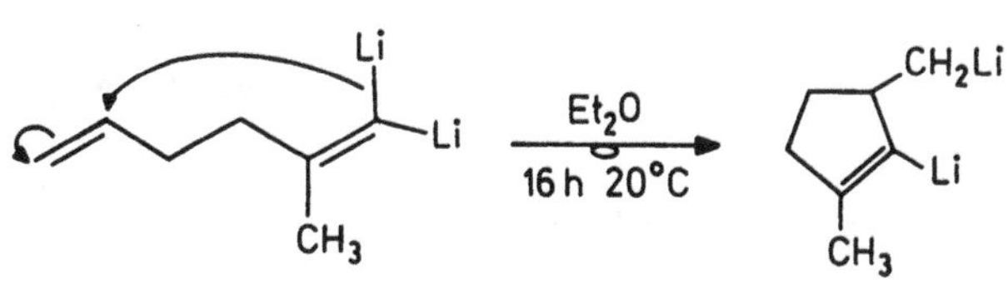

81c *189* (50%)

Especially interesting reactivity is shown by polylithiated alkynes. Not only can the lithiums usually be replaced stepwise by various electrophiles but also different reaction products often are obtained depending on the hardness or softness of the corresponding electrophiles in connection with steric demands [124, 125, 127, 129, 132, 133, 165–169] (Schemes 1–3).

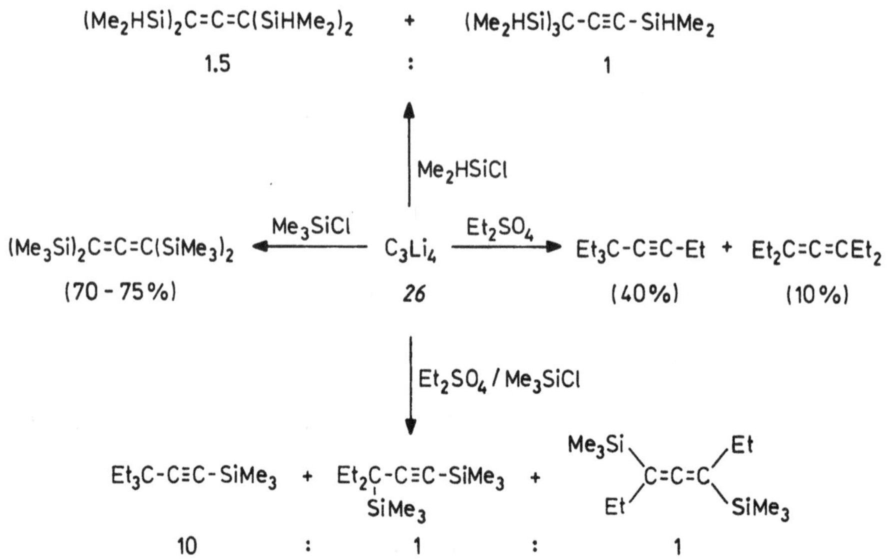

Scheme 1

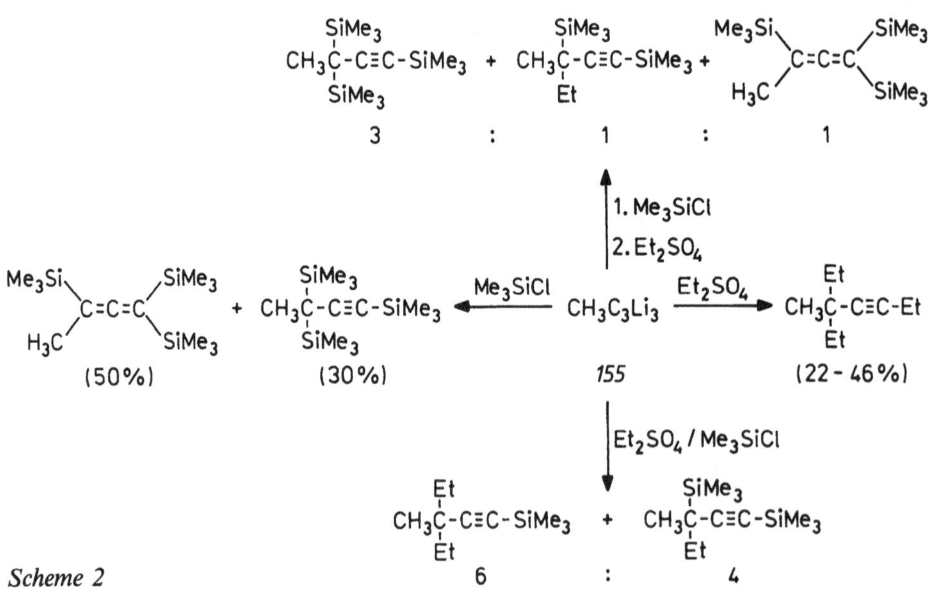

Scheme 2

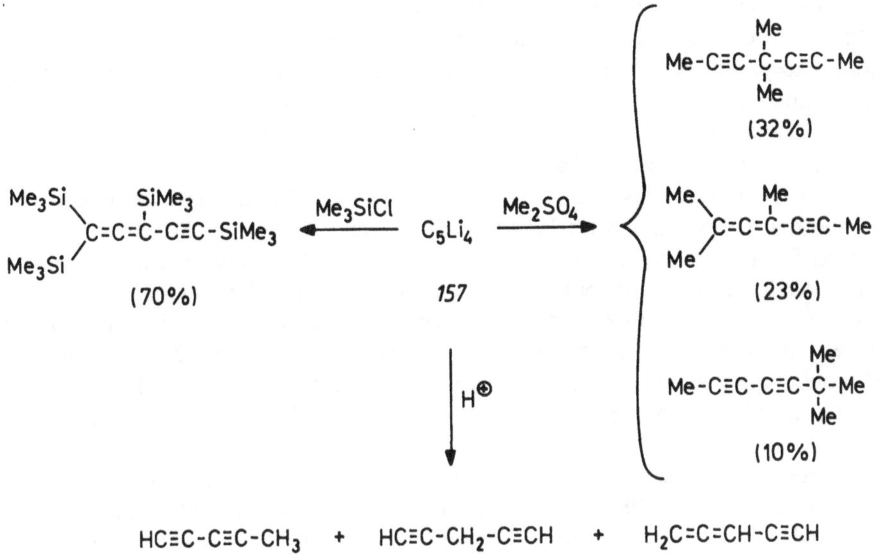

Scheme 3

4.2 Structures

Polylithium organic compounds appear to be electron-deficient polymeric material that are highly associated in the solid state. Conventional structure elucidation with isocentric polylithiated alkanes is impossible because neither crystals for X-ray structural analyses nor solvents for investigations in solution are available. We tried to dissolve dilithiomethane — prepared by the Ziegler procedure — in diethyl ether, THF, TMEDA, HMPT, and DMPU at temperatures up to 150 °C for several days, without success. A boiling 1 M solution of *tert*-butyllithium in cyclopentane even using ultrasound irradiation for one day was also not successful. On the other hand DMSO reacted with CH_2Li_2 extremely violently already under liquid nitrogen cooling. The main products were CH_3SSCH_3, CH_3SCH_3, and $CH_3SCH_2SCH_3$ [25].

Lagow in cooperation with Yannoni [170,171] achieved an excellent ^{13}C NMR spectrum of solid $CH_2{}^6Li_2$ using cross polarization magic angle spinning technique with ^6Li decoupling in addition to ^1H decoupling. A sharp singlet 26.5 ppm downfield from $CH_3{}^6Li$ was observed which implies that solid aggregate $(CH_2Li_2)_n$ may contain only one carbon environment and that the symmetry is very high. The downfield ^{13}C shift in CH_2Li_2 ($\delta = 10.5$ ppm) relative to that of methyllithium ($\delta = -16$ ppm) was attributed to an increase in charge density at the methylene carbon with increasing lithium substitution, although — in our view — an increase in charge density should lead to an upfield shift. The downfield ^{13}C shift perhaps might be due to rehybridization of carbon in this species from sp^3 to more or less sp^2 and the high charge density rather will be responsible for the downfield shift being so small.

The Debye-Scherrer powder X-ray diffraction data for dilithiomethane, reported by the same authors [170], unfortunately consist of 12 lines only, several of which overlap and except for one, all peaks are of low intensity. The low-symmetry crystal system and the limited number of lines made solving the crystal structure impossible. In an attempt to analyse the 12-line diffraction pattern a triclinic cell was obtained, which, however, cannot be considered as definitive.

X-ray photoelectron spectra of methyllithium and dilithiomethane were achieved by Lagow in cooperation with Hall [172]. They could show that the carbon charge increases from −1.02 in CH_3Li to −1.55 in CH_2Li_2 and that only one carbon and one lithium environment is present. This again suggests that the solid-state structure of CH_2Li_2 will be highly symmetric. The Li 1s binding energies for CH_3Li (54.0 eV) and CH_2Li_2 (53.9 eV) are closer to those of LiOH and Li_2O than to lithium salts containing noncoordinating anions like $ClO_4^{\ominus}$ and $PO_3^{2\ominus}$. Thus, the overall Li potential in these lithiocarbon compounds is similar to that for lithium in extended oxidic structures.

Of particular interest in connection with the structure elucidation of polylithiated alkanes are Lagow's excellent papers on flash vaporization mass spectroscopy of those compounds [90,154,173,174]. As already mentioned, polylithium organic compounds have no observable vapor pressure below 650 °C, even in high vacuum ($<10^{-6}$ torr). Thus, they decompose or rearrange completely long before they vaporize, giving lithium carbide as the principle lithiocarbon product. However, Lagow and coworkers [91] have observed that if flash heated — from room temperature to about 1500 °C in less than 3 seconds — CH_2Li_2 can be transported over a distance of 10 cm with less than 10% decomposition, and it was this experiment which finally led to the development of a new technique called flash vaporization mass spectroscopy. This way up to tetramer clusters $(CH_2Li_2)_4$ have been found in dilithiomethane. In addition to the monomer through tetramer clusters also those ions plus and minus a lithium atom were present. The same is true for other lithiomethanes, $CH_{4-n}Li_n$, although the mass spectra are often complicated by the fact that methods have not yet been developed to separate one lithium substituted methane from a mixture of others and assignment of certain fragmentation to specific precursors could be only speculative. On the other hand FAB mass spectra of pure samples of CH_2Li_2 and CLi_4 were unsuccessful [175].

Even lithiated pentacoordinate carbocations $CH_nLi_{5-n}^{\oplus}$ (n = 0–3) could be detected by flash vaporization mass spectroscopy [77,174] the stability of which had been predicted by Schleyer and coworkers [78].

IR-spectroscopy was also used as a tool for structure elucidation of polylithium organic compounds. Already in 1969 Krohmer and Goubeau [176] reported the IR-spectrum of dilithiomethane, which, however, cannot tell too much about its structure.

The metalation of propyne in hexane was followed IR-spectroscopically by West [125]. He observed characteristic changes in the IR-spectrum with the stepwise introduction of lithium:

$CH_3C{\equiv}CH$	$CH_3C{\equiv}CLi$	$C_3H_2Li_2$	C_3HLi_3	C_3Li_4
2130 cm^{-1}	2050 cm^{-1}	1870 cm^{-1}	1770 cm^{-1}	1675 cm^{-1}

Perlithiopropyne *26* (C_3Li_4) shows a strong absorption at 1675 cm^{-1} but no band in the C≡C stretching region between 2130 and 2190 cm^{-1}. The same is true for perlithiopentadiyne *157* (C_5Li_4) [132], and it was therefore believed that the polylithium derivatives of propyne and pentadiyne all have allenic structures in solution, i.e. *26b* and *157b* instead of acetylenic *26a* and *157a*.

$Li_3C-C{\equiv}CLi$	$Li_2C{=}C{=}CLi_2$
26a	*26b*
$Li_3C-C{\equiv}C-C{\equiv}CLi$	$Li_2C{=}C{=}C{=}C{=}CLi_2$
157a	*157b*

These interpretations, however, were not in agreement with later experimental results of the same research group [129, 133, 165] demanding the simultaneous presence of hard and soft carbanionic centers in those molecules. The distinction between propargylic (soft) and acetylenic (hard) lithium substituents was possible by reaction with the hard chlorotrimethylsilane and the softer dimethyl sulfate (cf. Schemes 1–3). Since that West and coworkers [129] favor a propargylide structure for these compounds, i.e. *23c* for dilithiopropyne ($C_3H_2Li_2$). It is reasonable to assume that the lithium ion and the C_3 chain due to attractive interactions will arrange in a way that the empty p orbitals of the metal ion will optimally overlap with the centers of negative charge at C1 and C3, the structure thus becoming identical with that precalculated by Schleyer and coworkers, *23* [40, 41].

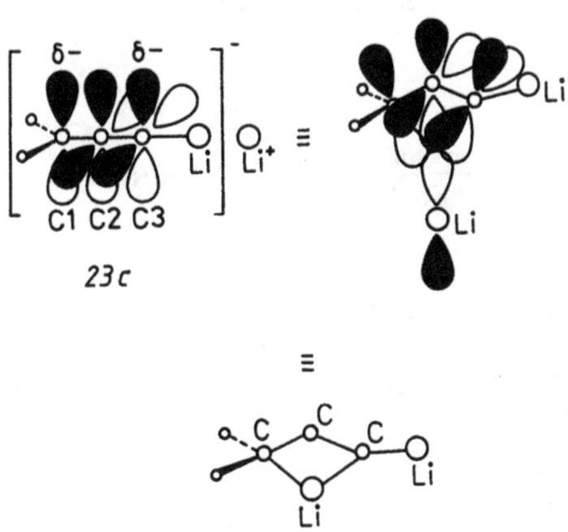

23c

23

We suggest another view which reduces the problem to the formulation of two resonance structures neglecting classical bond angles. This will lead qualitatively to the same results as pointed out above.

The previously observed bathochromic shift in the IR spectrum [125, 132] — according to West [129] due to a "lithium effect" — can now be explained very easily because the bridging lithium atom will abstract charge density from the C_3 fragment lowering its bond order thus shifting the IR frequency to lower wave numbers.

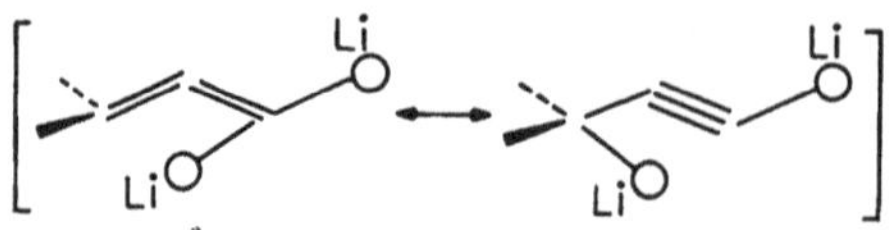

23b 23a

≡

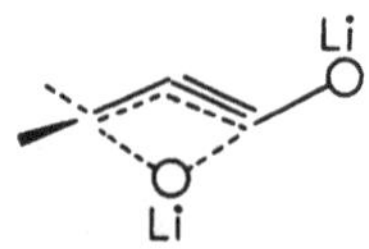

23

The resonance structures *23a* and *23b* show that C1 and C3 are centers of negative charge. The reaction with an electrophile therefore will occur sterically controlled at one of these two centers. This is why always mixtures of acetylenic and allenic products are obtained upon the reaction of polylithiated alkynes with reagents of low steric demand (see Scheme 1).

Further informations concerning the structure of polylithiated alkynes are found in a paper of Klein and Brenner [177]. By NMR spectroscopy they could show that dimetalation of compounds of type *190* yields a socalled "sesquiacetylenic" structure *191* with localized negative charges in the C_3 fragment. This obviously very favorable structure consists of two identical four-electron three-center bonds with symmetrical overlap.

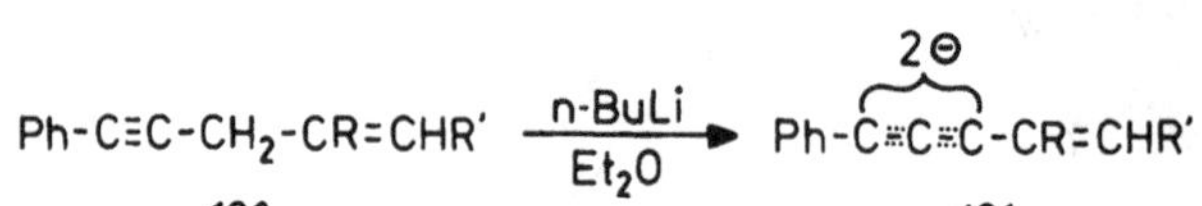

190 191

According to IR spectroscopic investigations [178] with compounds of type *192* a solvent dependent equilibrium between the sesquiacetylenic structure *192a* and the allenic structure *192b* was formulated. The result that sesquiacetylenes can be IR spectroscopically detected only in the presence of HMPT or TMEDA and partly in THF as the solvent, however, is not in agreement with the earlier report [177] of the same author where those species had been detected by NMR spectroscopy also in diethyl ether as the solvent.

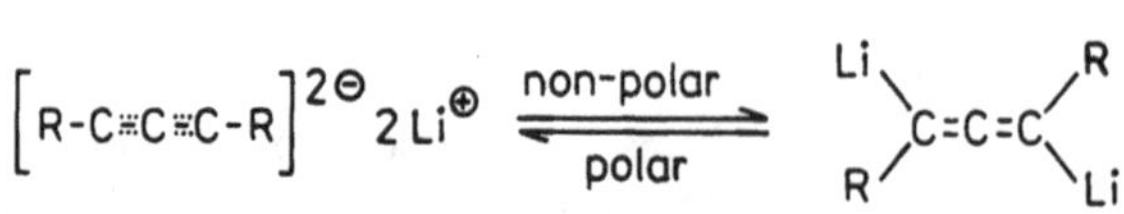

192 a 192 b

This contradiction can easily be eliminated by formulating the Klein compounds as bicyclic structures *192* with bridging lithium atoms in analogy to *23* (see above). The corresponding resonance structure can also be written down without any problems. According to calculations of Schleyer and coworkers [40] on the basic $C_3H_2Li_2$ system the twisted structure *24* (see Sect. 2.6) might be even more stable.

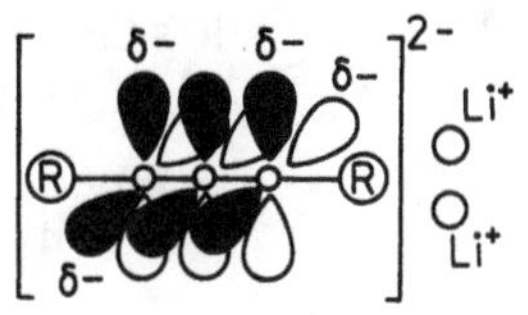

192a

≡

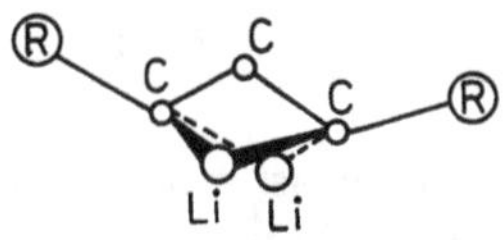

192

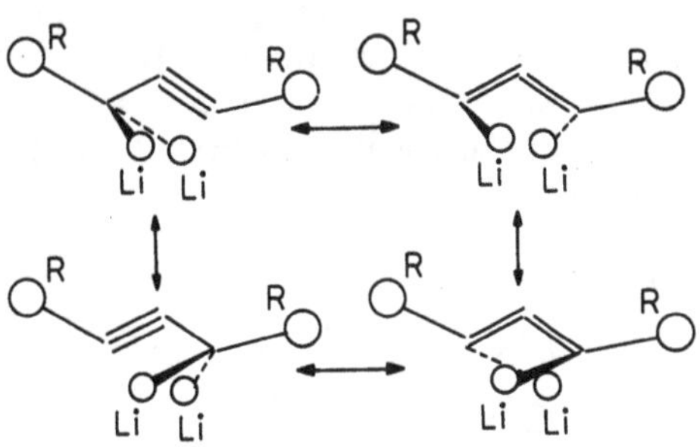

192

This doubly bridged structure in mind all experimental results easily can be explained. Thus the solvent dependence of the IR stretching frequencies will be due to the fact that in better solvating solvents the ionic character of the carbon lithium bond becomes higher, the decreasing covalent bond to the lithium atoms finally leading to a higher bond order within the C_3 fragment.

In summary the stepwise polylithiation of 1-alkynes can be formulated as shown in Scheme 4. The acetylenic proton, of course, is abstracted first. If $R^3 = H$, however, the second metalation takes place rapidly without the primarily formed acetylide being precipitated [128] (see Sect. 3.6). The thereby formed species bears the structure element corresponding to dilithiopropyne *23* [40, 129].

If it is also $R^2 = H$, a third metalation can take place yielding a "sesquiacetylide" containing the C_3 structure element of a sesquiacetylene *192* one of the ligands being replaced by a lithium atom [40, 125, 130, 177, 178] (see above).

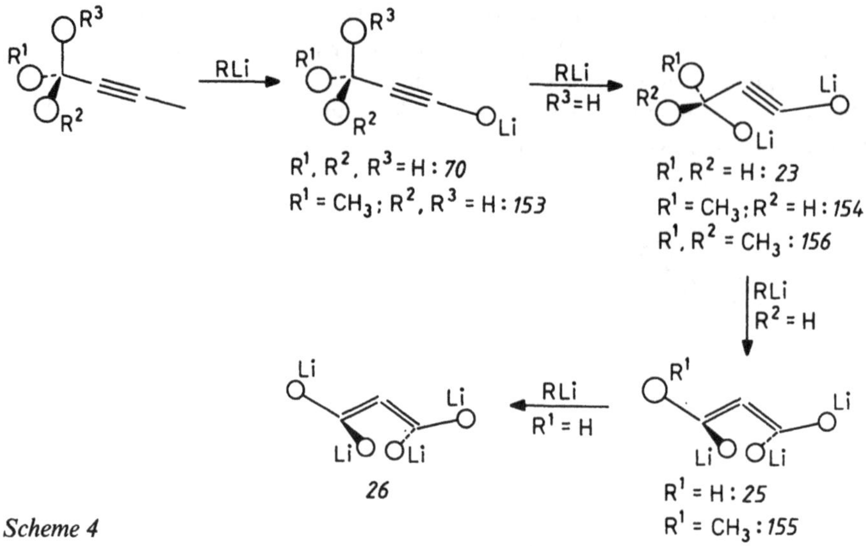

Scheme 4

Starting with propyne itself (R^1, R^2, R^3 = H) even the last proton can be removed and the product C_3Li_4 may be called a sesquiacetylide *26* with both ligands of the sesquiacetylene *192* being replaced by lithium atoms [40, 41, 124, 125, 129, 165].

Starting with 2-alkynes [125, 129] the second metalation yields a sesquiacetylene *192* one of the ligands being replaced by a hydrogen atom. A 1,3-hydrogen shift [130] therefore can take place giving a species identical with the corresponding dilithiated 1-alkyne, which for R = H (*23*) according to calculations [40] is 12.6 kcal/mol (52.7 kJ/mol) more stable than *24* (Scheme 5).

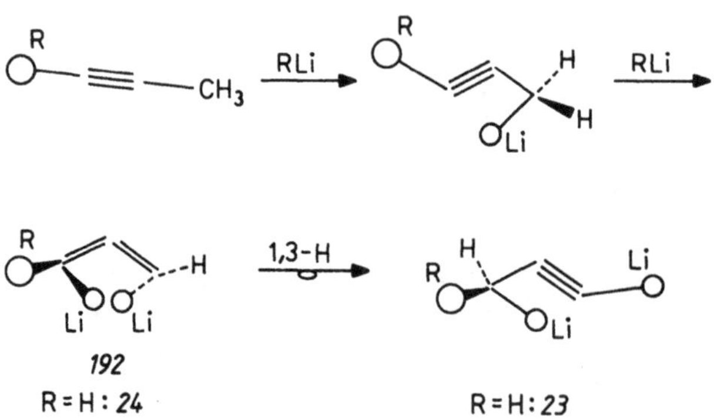

Scheme 5

This cannot happen with 3-alkynes and other internal alkynes which can be metalated only with much more difficulty and the reaction stops at the sesquiacetylene step i.e. after the introduction of two lithium atoms only [130, 177, 178].

Using perlithiopropyne (C_3Li_4) as an example its reactivity against hard and soft electrophiles can be easily explained assuming a sesquiacetylide structure *26*. In this structure the bridging lithium atoms will be soft while the other two are hard. Hard reagents like chlorosilanes therefore will preferrably react with the latter while the softer dialkyl sulfates attack the bridging positions first.

The reaction with chlorosilanes (Scheme 6) leads in three steps to a propargylide which finally — under steric control — yields allene or propyne derivatives. With

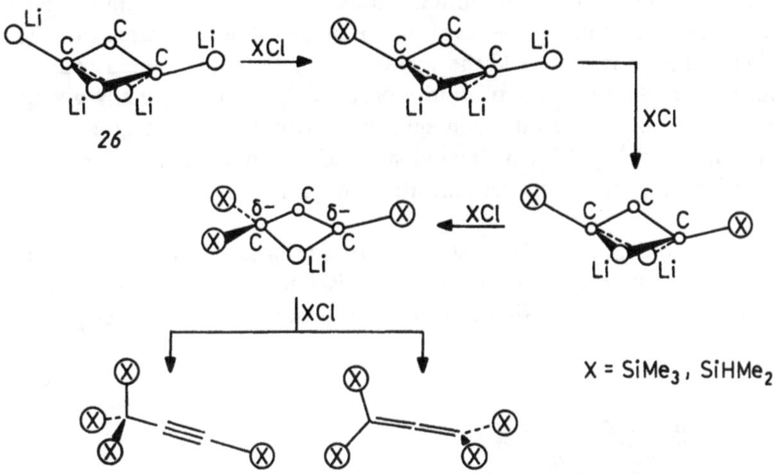

Scheme 6

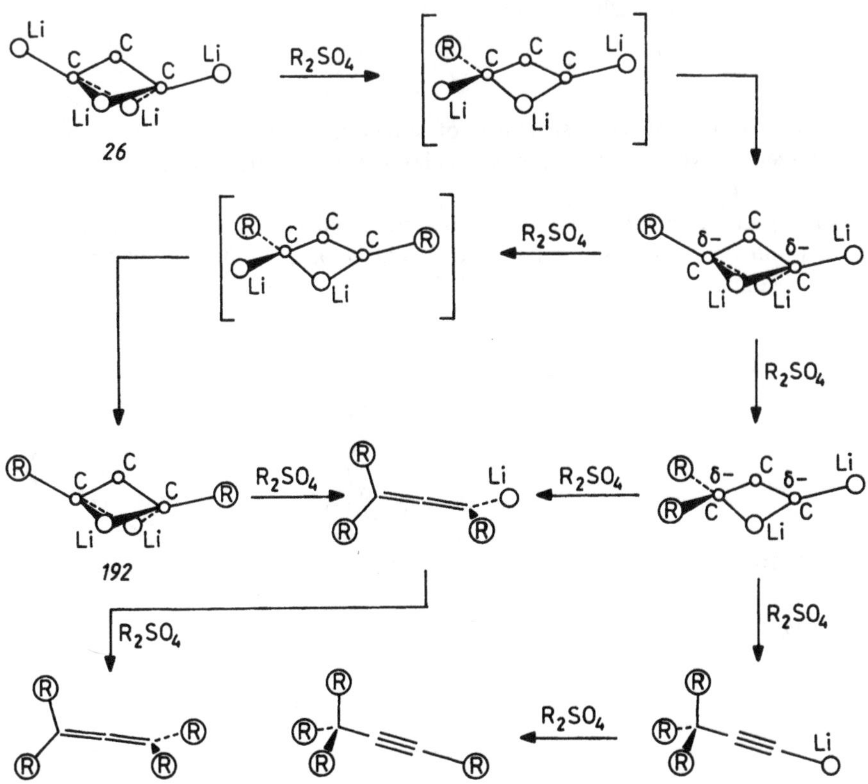

Scheme 7

chlorotrimethylsilane the corresponding allene is formed exclusively while with the smaller chlorodimethylsilane both compounds had been obtained (see Scheme 1) [124, 125, 129, 165].

Derivatization with dialkyl sulfates (Scheme 7), on the other hand, involves preferrable attack at one of the bridging lithium atoms followed by an intramolecular reorganization in order to built up the favorable sesquiacetylide structure again. The second step in a similar way may yield the corresponding sesquiacetylene *192* and/or a species with the structure element of dilithiopropyne *23* the latter finally giving a mixture of allene and alkyne derivatives in agreement with the experimental findings (see Schemes 1–3) [124, 125, 129, 165]. Kinetic and steric effects, of course, are important mainly when using more than one derivatization reagent — simultaneously or one after the other.

Unfortunately the only known X-ray structure of a lithiocarbon is that of lithium carbide *15c* [179]. Interestingly the elementary cell is built up by formally bicyclic C_2Li_2 units instead of linear ones similar to those calculated for dilithioacetylene in the gas phase *15b* (see Sect. 2.4) [6, 34].

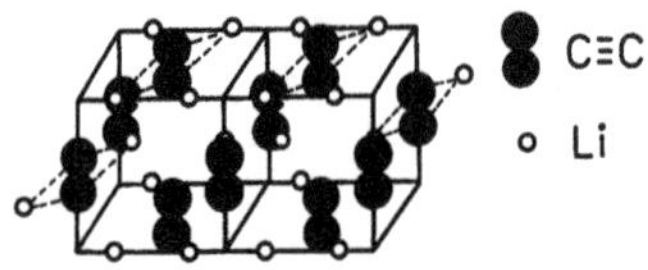

15c

The few other known X-ray structures of polylithiated aliphatic hydrocarbons are also in agreement with theoretical results as far as calculations on these systems have been performed at all.

This is especially true for the 1,4-dilithiobutatriene derivative *133* which turns out to have a dimeric *cis*-structure with two different kinds of lithium atoms and a true butatriene geometry *133a* (see Sect. 2.7) [7, 69].

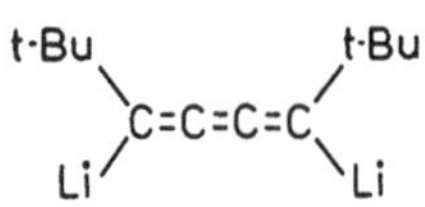

133

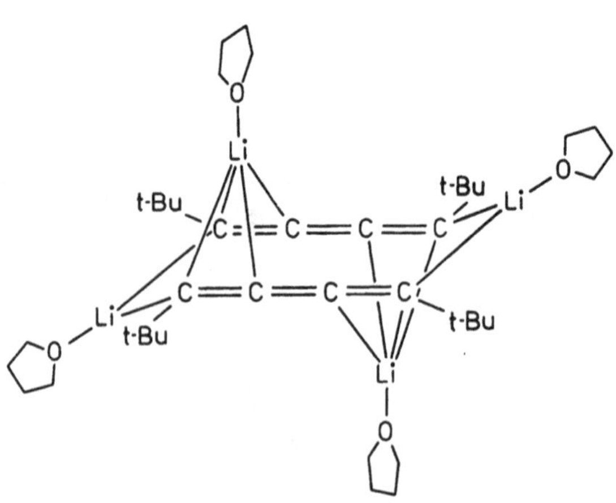

133a

Dilithiation of 2,2,8,8-tetramethyl-3,6-nonadiyne *193* gives a complicated tetrameric complex involving two crystallographically different diyne molecules in different environments with an octahedral array of six lithium atoms, and two lithium atoms without lithium-lithium contacts [7,180]. The complex appears to be held together by lithium-acetylenic π-interactions as well as lithium-lithium interactions, a geometrical arrangement anticipated by MNDO geometry optimization of C_5Li_4 *195*:

$$\text{t-Bu-C}\equiv\text{C-CH}_2\text{-C}\equiv\text{C-t-Bu} \xrightarrow{\text{n-BuLi}} \text{t-Bu-C}\equiv\text{C-CLi}_2\text{-C}\equiv\text{C-t-Bu}$$

193 *194*

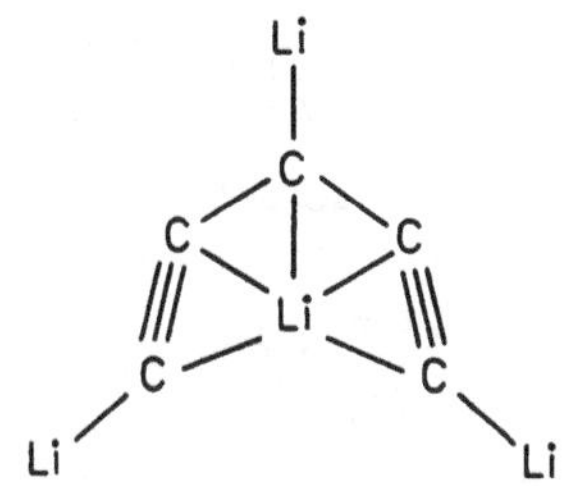

195

The 2,3-dilithio-1,3-butadiene derivative *123a* is also tetrameric involving two crystallographically different diene molecules in different environments [181]. Four lithium atoms form a tetralithiocyclobutane skeleton two bonds of which simultane-

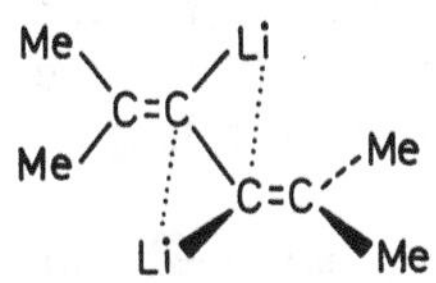

123a

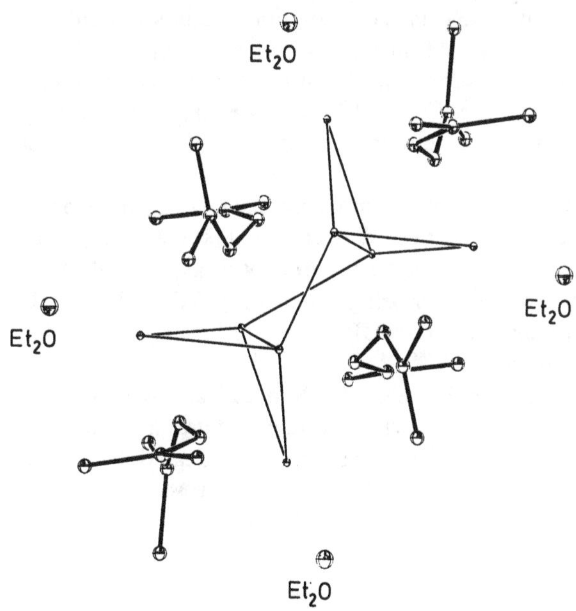

(123a)₄

ously being bridges of two tetralithiobicyclobutane systems. Only half of the lithium atoms are solvated by diethyl ether molecules.

According to the C—C bond lengths *123a* is a true butadiene derivative with C1–C2 = C3–C4 = 1.40 Å and C2–C3 = 1.50 Å on an average. The dihedral angle of the central bond is different for the terminal (91°) and inner (101°) units, respectively, i.e. there are two different gauche conformations present in the crystalline state.

Another molecule investigated already in 1975 by X-ray structure analysis should be mentioned: 1,6-dilithio-2,4-hexadiene *196* [138]. Again each lithium atom is bonded simultaneously to several carbon atoms in agreement with theoretical proposals that lithium is favoring multicenter bonds by accupying bridging positions.

$LiCH_2\text{-}CH{=}CH\text{-}CH{=}CH\text{-}CH_2Li$

196

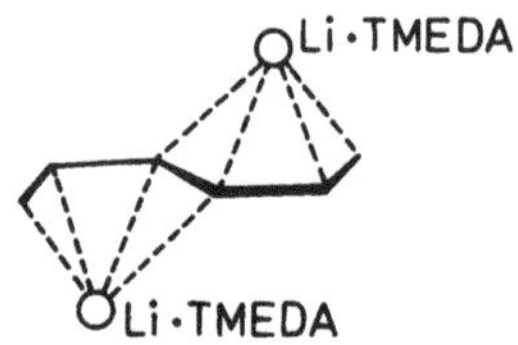

196a

In crystals showing polymorphy the positions of the lithium atoms can change as was recently found by Boche [182] who coined the name "lithiotropy" for this interesting phenomenon. For this and other reasons it is dangerous to deduce a structure in solution from an X-ray structure in the solid phase. Therefore NMR spectroscopic investigations in addition should be performed whenever possible. ^1H-, ^{13}C-, and ^6Li-NMR spectroscopy [183] for instance showed that the 2,3-dilithio-1,3-butadiene derivative *123a* in THF solution exhibits a similar behavior as in the solid state involving two lithium species and two differently bonded diene systems in equal amounts independent of temperature, concentration, and the addition of TMEDA. Moreover in a COSY 2D ^6Li NMR experiment crosspeaks between the two signals were obtained definitely proving that the two different lithium atoms belong to the same cluster [183].

^1H- and ^{13}C-NMR spectroscopic investigations have also been performed with the 1,4-dilithiobutatriene derivative *133* (see above) [69] and certain dilithiobicyclobutane derivatives [148] (see Sect. 3.6) as well as with "dilithium semibullvalenide" *131* (see Sect. 3.5) the latter being also investigated by ^6Li-NMR spectroscopy [118,119]. In addition numerous ^1H-NMR spectra of linear-, cyclic-, and Y-conjugated dianions have been reported [49,50,54–56,136,137,177].

Not always, however, NMR investigations deliver unequivocal results. For instance, the addition product of lithium to tetraphenylallene [184] not treated in the present review article was thoroughly investigated by two research groups [185,186] using ^1H, ^{13}C, ^6Li, and/or ^7Li NMR spectroscopy each group coming to quite different conclusions.

5 Acknowledgement

This work benefitted from stimulating discussion with Prof. P. v. R. Schleyer, Erlangen, whom is also thanked for information prior to publication. Financial support by the Fonds der Chemischen Industrie is gratefully acknowledged.

Received February 14, 1986

6 References

1. West, R.: Advan. Chem. Ser. *130*, 211 (1974)
2. Klein, J.: Propargylic metalation, in: The Chemistry of the Carbon—Carbon Triple Bond, (Patai, S., Ed.), p. 343, Wiley, New York 1978
3. Klein, J.: Tetrahedron *39*, 2733 (1983)
4. Lagow, R. J., Gurak, J. A.: Chem. Future, Proc. 29th IUPAC Congr. 1983 (Grünewald, H., Ed.), p. 107, Pergamon, Oxford 1984
5. Chinn, Jr., J. W., Gurak, J. A., Lagow, R. J.: Lithium: Curr. Appl. Sci., Med., Technol. (Bach, R. O., Ed.), p. 291, Wiley, New York 1985
6. Schleyer, P. v. R.: Pure Appl. Chem. *55*, 355 (1983)
7. Schleyer, P. v. R.: ibid. *56*, 151 (1984)
8. Streitwieser, Jr., A.: Acc. Chem. Res. *17*, 353 (1984)
9. Collins, J. B., Dill, J. D., Jemmis, E. D., Apeloig, Y., Schleyer, P. v. R., Seeger, R., Pople, J. A.: J. Am. Chem. Soc. *98*, 5419 (1976)
10. Dilithiomethane has been calculated independently by Nilssen, E. W., Skancke, A.: J. Organomet. Chem. *116*, 251 (1976)
11. Laidig, W. D., Schaefer III, H. F.: J. Am. Chem. Soc. *100*, 5972 (1978)
12. Bachrach, S. M., Streitwieser, Jr., A.: ibid. *106*, 5818 (1984)
13. Alvarado-Swaisgood, A. E., Harrison, J. F.: J. Phys. Chem. *89*, 62 (1985)
14. Jemmis, E. D., Schleyer, P. v. R., Pople, J. A.: J. Organomet. Chem. *154*, 327 (1978)
15. Kos, A. J., Jemmis, E. D., Schleyer, P. v. R., Gleiter, R., Fischbach, U.,Pople, J. A.: J. Am. Chem. Soc. *103*, 4996 (1981)
16. Radom, L., Stiles, P. J., Vincent, M. A.: J. Mol. Struct. *48*, 431 (1978)
17. Rautenstrauch, V.: Angew. Chem. *87*, 254 (1975); Angew. Chem. Int. Ed. Engl. *14*, 259 (1975)
18. Bogdanović, B., Wermeckes, B.: Angew. Chem. *93*, 691 (1981); Angew. Chem. Int. Ed. Engl. *20*, 684 (1981)
19. Schleyer, P. v. R., Kos, A. J., Kaufmann, E.: J. Am. Chem. Soc. *105*, 7617 (1983)
20. Kuus, H.: Tartu Riikliku Ulikooli Toim. *219*, 245 (1968); Chem. Abstr. *71*, 49155 (1969)
21. Kuus, H.: Uch. Zap. Tartu. Gos. Univ. *193*, 130 (1966); Chem. Abstr. *69*, 67443 (1968)
22. Dewar, M. J. S., Thiel, W.: J. Am. Chem. Soc. *99*, 4899; 4907 (1977); Dewar, M. J. S., Rzepa, H. S.: ibid. *100*, 58; 777 (1978); Dewar, M. J. S., McKee, M. L.: ibid. *99*, 5231 (1977); Dewar, M. J. S., McKee, M. L., Rzepa, H. S.: ibid. *100*, 3607 (1978). Lithium parametrization by Thiel, W. and Clark, T., unpublished
23. Maercker, A., Theis, M., Kos, A. J., Schleyer, P. v. R.: Angew. Chem. *95*, 755 (1983); Angew. Chem. Int. Ed. Engl. *22*, 733 (1983)
24. Houk, K. N., Rondan, N. G.: unpublished results; cf. Houk, K. N., Rondan, N. G., Schleyer, P. v. R., Kaufmann, E., Clark,, T.: J. Am. Chem. Soc. *107*, 2821 (1985)
25. Theis, M.: Ph.D. Thesis, University of Siegen 1985
26. Kos, A., Poppinger, D., Schleyer, P. v. R., Thiel, W.: Tetrahedron Lett. *21*, 2151 (1980)
27. Apeloig, Y., Schleyer, P. v. R., Binkley, J. S., Pople, J. A.: J. Am. Chem. Soc. *98*, 4332 (1976)
28. Nagase, S., Morokuma, K.: ibid. *100*, 1661 (1978)
29. Laidig, W. D., Schaefer III, H. F.: ibid. *101*, 7184 (1979)
30. Apeloig, Y., Clark, T., Kos, A. J., Jemmis, E. D., Schleyer, P. v. R.: Israel J. Chem. *20*, 43 (1980)

31. Maercker, A., Graule, T., Demuth, W.: unpublished results
32. Schleyer, P. v. R., Kaufmann, E., Kos, A. J., Clark, T., Pople, J. A.: Angew. Chem. *98*, 164 (1986); Angew. Chem. Int. Ed. Engl. *25*, 169 (1986)
33. Seyferth, D., Vick, S. C.: J. Organomet. Chem. *144*, 1 (1978)
34. Apeloig, Y., Schleyer, P. v. R., Binkley, J. S., Pople, J. A., Jorgensen, W. L.: Tetrahedron Lett. *1976*, 3923
35. Ritchie, J. P.: ibid. *23*, 4999 (1982)
36. Seetz, J. W. F. L., Schat, G., Akkerman, O. S., Bickelhaupt, F.: J. Am. Chem. Soc. *104*, 6848 (1982)
37. Klein, J., Medlik-Balan, A.: J. Chem. Soc., Chem. Commun. *1975*, 877
38. Schleyer, P. v. R., Kos, A. J.: ibid. *1982*, 448
39. Kost, D., Klein, J., Streitwieser, Jr., A., Schriver, G. W.: Proc. Natl. Acad. Sci. USA *79*, 3922 (1982)
40. Jemmis, E. D., Chandrasekhar, J., Schleyer, P. v. R.: J. Am. Chem. Soc. *101*, 2848 (1979)
41. Jemmis, E. D., Poppinger, D., Schleyer, P. v. R., Pople, J. A.: ibid. *99*, 5796 (1977)
42. Frenking, G.: Chem. Phys. Lett. *111*, 529 (1984)
43. Lansbury, P. T., Pattison, V. A., Clement, W. A., Sidler, J. D.: J. Am. Chem. Soc. *86*, 2247 (1964)
44. Maercker, A., Bsata, M., Buchmeier, W., Engelen, B.: Chem. Ber. *117*, 2547 (1984)
45. Dill, J. D., Greenberg, A., Liebman, J. F.: J. Am. Chem. Soc. *101*, 6814 (1979)
46. Douglas, J. E., Rabinovitch, B. S., Looney, F. S.: J. Chem. Phys. *23*, 315 (1955)
47. Kos, A. J., Stein, P., Schleyer, P. v. R.: J. Organomet. Chem. 280, Cl (1985)
48. Klein, J., Medlik, A.: J. Chem. Soc., Chem. Commun. *1973*, 275
49. Bates, R. B., Beavers, W. A., Greene, M. G., Klein, J. H.: J. Am. Chem. Soc. *96*, 5640 (1974)
50. Klein, J., Medlik-Balan, A., Meyer, A. Y., Chorev, M.: Tetrahedron *32*, 1839 (1976)
51. Gund, P.: J. Chem. Educ. *49*, 100 (1972)
52. Bates, R. B., Hess, Jr., B. A., Ogle, C. A., Schaad, L. J.: J. Am. Chem. Soc. *103*, 5052 (1981)
53. Inagaki, S., Hirabayashi, Y.: Chem. Lett. *1982*, 709
54. Mills, N. S., Shapiro, J., Hollingsworth, M.: J. Am. Chem. Soc. *103*, 1263 (1981)
55. Rusinko III, A., Mills, N. S., Morse, P.: J. Org. Chem. *47*, 5198 (1982)
56. Mills, N. S.: J. Am. Chem. Soc. *104*, 5689 (1982)
57. Clark, T., Wilhelm, D., Schleyer, P. v. R.: Tetrahedron Lett. *23*, 3547 (1982)
58. Wilhelm, D., Clark, T., Schleyer, P. v. R., Buckl, K., Boche, G.: Chem. Ber. *116* 1669 (1983)
59. Wilhelm, D., Clark, T., Schleyer, P. v. R.: Tetrahedron Lett. *24*, 3985 (1983)
60. Schleyer, P. v. R., Kos, A. J., Wilhelm, D., Clark, T., Boche, G., Decher, G., Etzrodt, H., Dietrich, H., Mahdi, W.: J. Chem. Soc., Chem. Commun. *1984*, 1495
61. Wilhelm, D., Clark, T., Schleyer, P. v. R.: J. Chem. Soc., Perkin Trans. II *1984*, 915
62. Agranat, I., Skancke, A.: J. Am. Chem. Soc. *107*, 867 (1985)
63. Boche, G., Etzrodt, H., Marsch, M., Thiel, W.: Angew. Chem. *94*, 141 (1982); Angew. Chem. Int. Ed. Engl. *21*, 132 (1982); Angew. Chem. Suppl. *1982*, 345
64. Boche, G., Etzrodt, H., Marsch, M., Thiel, W.: Angew. Chem. *94*, 141 (1982); Angew. Chem. Int. Ed. Engl. *21*, 133 (1982); Angew. Chem. Suppl. *1982*, 355
65. Kos, A. J., Schleyer, P. v. R.: J. Am. Chem. Soc. *102*, 7928 (1980)
66. Seeger, R., Dujardin, R., Maercker, A.: unpublished results
67. Schleyer, P. v. R., Wilhelm, D.: unpublished results
68. Radom, L., Stiles, P. J., Vincent, M. A.: J. Mol. Struct. *48*, 259 (1978)
69. Neugebauer, W., Geiger, G. A. P., Kos, A. J., Stezowski, J. J., Schleyer, P. v. R.: Chem. Ber. *118*, 1504 (1985)
70. Ritchie, J. P.: J. Am. Chem. Soc. *105*, 2083 (1983)
71. Disch, R. L., Schulman, J. M., Ritchie, J. P.: ibid. *106*, 6246 (1984)
72. Rauscher, G., Clark, T., Poppinger, D., Schleyer, P. v. R.: Angew. Chem. *90*, 306 (1978); Angew. Chem. Int. Ed. Engl. *17*, 276 (1978)
73. Zefirov, N. S., Kirin, V. N., Yur'eva, N. M., Koz'min, A. S., Kulikov, N. S., Luzikov, Yu. N.: Tetrahedron Lett. *1979*, 1925
74. Schleyer, P. v. R., Würthwein, E.-U., Kaufmann, E., Clark, T., Pople, J. A.: J. Am. Chem. Soc. *105*, 5930 (1983)

75. Waite, J., Papadopoulos, M. G.: J. Mol. Struct. *125*, 155 (1984)
76. Reed, A. E., Weinhold, F.: J. Am. Chem. Soc. *107*, 1919 (1985)
77. Jemmis, E. D., Chandrasekhar, J., Würthwein, E.-U., Schleyer, P. v. R., Chinn, Jr., J. W., Landro, F. J., Lagow, R. J., Luke, B., Pople, J. A.: ibid. *104*, 4275 (1982)
78. Schleyer, P. v. R., Tidor, B., Jemmis, E. D., Chandrasekhar, J., Würthwein, E.-U., Kos, A. J., Luke, B. T., Pople, J. A.: ibid. *105*, 484 (1983)
79. West, R., Rochow, E. G.: Naturwissenschaften *40*, 142 (1953); J. Org. Chem. *18*, 1739 (1953)
80. Ashe III, A. J., Drone, F. J.: Organometallics *4*, 1478 (1985)
81. Jutzi, P., Baumgärtner, J., Schraut, W.: J. Organomet. Chem. *132*, 333 (1977)
82. Siegel, H., Hiltbrunner, K., Seebach, D.: Angew. Chem. *91*, 845 (1979); Angew. Chem. Int. Ed. Engl. *18*, 785 (1979)
83. Seebach, D., Siegel, H., Gabriel, J., Hässig, R.: Helv. Chim. Acta *63*, 2046 (1980)
84. Seebach, D., Siegel, H., Müllen, K., Hiltbrunner, K.: Angew. Chem. *91*, 844 (1979); Angew. Chem. Int. Ed. Engl. *18*, 784 (1979)
85. Warner, P. M., Herold, R. D.: J. Org. Chem. *48*, 5411 (1983)
86. Warner, P. M., Chang, S.-C., Koszewski, N. J.: ibid. *50*, 2605 (1985)
87. Hiyama, T., Takehara, S., Nozaki, H.: Tetrahedron Lett. *1974*, 3295
88. Warner, P. M., Le, D.: unpublished results
89. Ziegler, K., Nagel, K., Patheiger, M.: Z. Anorg. Allg. Chem. *282*, 345 (1955)
90. Gurak, J. A., Chinn, Jr., J. W., Lagow, R. J.: J. Am. Chem. Soc. *104*, 2637 (1982)
91. Shimp, L. A., Morrison, J. A., Gurak, J. A., Chinn, Jr., J. W., Lagow, R. J.: ibid. *103*, 5951 (1981)
92. Kawa, H., Manley, B. C., Lagow, R.: ibid. *107*, 5313 (1985)
93. Schlenk, W., Holtz, J.: Ber. Dtsch. Chem. Ges. *50*, 271 (1917)
94. Wittig, G., Bickelhaupt, F.: Chem. Ber. *91*, 883 (1958)
95. Barluenga, J., Villamaña, J., Fañanás, F. J., Yus, M.: J. Chem. Soc., Chem. Commun. *1982*, 355
96. Barluenga, J., Fañanás, F. J., Villamaña, J., Yus, M.: J. Chem. Soc., Perkin Trans. I *1984*, 2685
97. Maercker, A., Theis, M.: Angew. Chem. *96*, 990 (1984); Angew. Chem. Int. Ed. Engl. *23*, 995 (1984)
98. Maercker, A., Dujardin, R.: Angew. Chem. *96*, 222 (1984); Angew. Chem. Int. Ed. Engl. *23*, 224 (1984)
99. Maercker, A., Dujardin, R.: Organometallic Syntheses *3*, 356 (1986)
100. Maercker, A., Dujardin, R.: Angew. Chem. *97*, 612 (1985); Angew. Chem. Int. Ed. Engl. *24*, 571 (1985)
101. The synthesis claimed in the patent literature appears to be highly unlikely: Eberly, K. C.: U.S. Patent *1960*, 2,947,793; Chem. Abstr. *55*, 382 (1961)
102. Maercker, A., Theis, M.: Organometallic Syntheses *3*, 378 (1986)
103. Jutzi, P., Baumgärtner, J.: J. Organomet. Chem. *148*, 257 (1978)
104. Märkl, G., Hofmeister, P.: Tetrahedron Lett. *1976*, 3419
105. Reich, H. J., Yelm, K. E., Reich, I. L.: J. Org. Chem. *49*, 3438 (1984)
106. Mitchel, T. N., Amamria, A.: J. Organomet. Chem. *252*, 47 (1983)
107. Farah, D., Karol, T. J., Kuivila, H. G.: Organometallics *4*, 662 (1985)
108. Kauffmann, T.: Top. Curr. Chem. *92*, 109 (1980)
109. Kauffmann, T.: Angew. Chem. *94*, 401 (1982); Angew. Chem. Int. Ed. Engl. *21*, 410 (1982)
110. Cainelli, G., Dal Bello, G., Zubiani, G.: Tetrahedron Lett. *1965*, 3429
111. Cainelli, G., Dal Bello, G., Zubiani, G.: ibid. *1966*, 4315
112. Zweifel, G., Arzoumanian, H.: ibid. *1966*, 2535
113. Kulik, W., Verkruijsse, H. D., deJong, R. L. P., Hommes, H., Brandsma, L.: ibid: *24*, 2203 (1983)
114. Schlenk, W., Bergmann, E.: Liebigs Ann. Chem. *463*, 1 (1928)
115. Skinner, D. L., Peterson, D. J., Logan, T. J.: J. Org. Chem. *32*, 105 (1967)
116. Bogdanović, B.: Angew. Chem. *97*, 253 (1985); Angew. Chem. Int. Ed. Engl. *24*, 262 (1985)
117. Maercker, A., Dujardin, R.: unpublished results; Dujardin, R.: Ph. D. Thesis, University of Siegen 1986
118. Goldstein, M. J., Wenzel, T. T., Whittaker, G., Yates, S. F.: J. Am. Chem. Soc. *104*, 2669 (1982)

119. Goldstein, M. J., Wenzel, T. T.: J. Chem. Soc., Chem. Commun. *1984*, 1654; 1655
120. Maercker, A., Graule, T., Girreser, U.: Angew. Chem. *98*, 174 (1986); Angew. Chem. Int. Ed. Engl. *25*, 167 (1986)
121. Manceron, L., Andrews, L.: J. Am. Chem. Soc. *107*, 563 (1985)
122. Cf.: Blaukat, U., Neumann, W. P.: J. Organomet. Chem. *49*, 323 (1973)
123. Duboudin, J. G., Jousseaume, B.: Synth. Commun. *9*, 53 (1979); Duboudin, J. G., Jousseaume, B., Alexakis, A., Cahiez, G., Villieras, J., Normant, J. F.: C.R. Acad. Sci. Paris *285*, 29 (1977)
124. West, R., Carney, P. A., Mineo, I. C.: J. Am. Chem. Soc. *87*, 3788 (1965)
125. West, R., Jones, P. C.: ibid. *91*, 6156 (1969)
126. Jaffé, F.: J. Organomet. Chem. *23*, 53 (1970)
127. Bhanu, S., Scheinmann, F.: J. Chem. Soc., Chem. Commun. *1975*, 817
128. Eberly, K. C., Adams, H. E.: J. Organomet. Chem. *3*, 165 (1965)
129. Priester, W., West, R., Chwang, T. L.: J. Am. Chem. Soc. *98*, 8413 (1976)
130. Klein, J., Becker, J. Y.: Tetrahedron *28*, 5385 (1972)
131. Becker, J. Y.: Tetrahedron Lett. *1976*, 2159
132. Chwang, T. L., West, R.: J. Am. Chem. Soc. *95*, 3324 (1973)
133. Priester, W., West, R.: ibid. *98*, 8426 (1976)
134. Klein, J., Becker, J. Y.: J. Chem. Soc., Perkin II *1973*, 599
135. Brandsma, L., Verkruÿsee, H. D., Hommes, H.: J. Chem. Soc., Chem. Commun. *1982*, 1214
136. Klein, J., Medlik-Balan, A.: Tetrahedron Lett. *1978*, 279
137. Barfield, M., Bates, R. B., Beavers, W. A., Blacksberg, I. R., Brenner, S., Mayall, B. I., McCulloch, C. S.: J. Am. Chem. Soc. *97*, 900 (1975)
138. Arora, S. K., Bates, R. B., Beavers, W. A., Cutler, R. S.: ibid. *97*, 6271 (1975)
139. Bahl, J. J., Bates, R. B., Beavers, W. A., Mills, N. S.: J. Org. Chem. *41*, 1620 (1976)
140. Bahl, J. J., Bates, R. B., Beavers, W. A., Launer, C. R.: J. Am. Chem. Soc. *99*, 6126 (1977)
141. Bahl, J. J., Bates, R. B., Gordon III, B.: J. Org. Chem. *44*, 2290 (1979)
142. Bates, R. B., Beavers, W. A., Gordon III, B., Mills, N. S.: ibid. *44*, 3800 (1979)
143. Bates, R. B., Gordon III, B., Keller, P. C., Rund, J. V., Mills, N. S.: ibid. *45*, 168 (1980)
144. Bates, R. B., Gordon III, B., Highsmith, T. K., White, J. J.: ibid. *49*, 2981 (1984)
144a. Bates, R. B.: Dianions and Polyanions, in: Comprensive Carbanion Chemistry, Part A, (Buncel, E., Durst, T., Ed.), p. 1, Elsevier, Amsterdam 1980
145. Cf.: Kaiser, E. M., Petty, J. D., Knutson, P. L. A.: Synthesis *1977*, 509
146. Applequist, D. E., Saurborn, E. G.: J. Org. Chem. *37*, 1676 (1972)
147. Murata, I., Nakazawa, T., Kato, M., Tatsuoka, T., Sugihara, Y.: Tetrahedron Lett. *1975*, 1647
148. Schlüter, A.-D., Huber, H., Szeimies, G.: Angew. Chem. *97*, 406 (1985); Angew. Chem. Int. Ed. Engl. *24*, 404 (1985)
149. Hérold, A.: Bull. Soc. Chim. Fr. *1955*, 999
150. Chung, C., Lagow, R. J.: J. Chem. Soc., Chem. Commun. *1972*, 1078
151. Sneddon, L. G., Lagow, R. J.: ibid. *1975*, 302
152. Shimp, L. A., Lagow, R. J.: J. Org. Chem. *44*, 2311 (1979)
153. Landro, F. J., Gurak, J. A., Chinn, Jr., J. W., Newman, R. M., Lagow, R. J.: J. Am. Chem. Soc. *104*, 7345 (1982)
154. Landro, F. J., Gurak, J. A., Chinn, Jr., J. W., Lagow, R. J.: J. Organomet. Chem. *249*, 1 (1983)
155. Shimp, L. A., Lagow, R. J.: J. Am. Chem. Soc. *101*, 2214 (1979)
156. Shimp, L. A., Lagow, R. J.: ibid. *95*, 1343 (1973)
157. Morrison, J. A., Chung, C., Lagow, R. J.: ibid. *97*, 5015 (1975)
158. Shimp, L. A., Chung, C., Lagow, R. J.: Inorg. Chim. Acta *29*, 77 (1978)
159. Warner, P. M., Le, D.: J. Org. Chem. *47*, 893 (1982)
160. Jahngen, E. G. E., Phillips, D., Kobelski, R. J., Demko, D. M.: ibid. *48*, 2472 (1983)
161. Adlington, R. M., Barrett, A. G. M.: Acc. Chem. Res. *16*, 55 (1983)
162. Maercker, A., Brauers, F.: unpublished results
163. Krohmer, P., Goubeau, J.: Chem. Ber. *104*, 1347 (1971)
164. Maercker, A., Theysohn, W.: Liebigs Ann. Chem. *746*, 70 (1971)
165. Priester, W., West, R.: J. Am. Chem. Soc. *98*, 8421 (1976)
166. Sagar, A. J. G., Scheinmann, F.: Synthesis *1976*, 321

167. Bhanu, S., Khan, E. A., Scheinmann, F.: J. Chem. Soc., Perkin I *1976*, 1609
168. Khan, G. R., Pover, K. A., Scheinmann, F.: J. Chem. Soc., Chem. Commun. *1979*, 215
169. Pover, K. A., Scheinmann, F.: J. Chem. Soc., Perkin I *1980*, 2338
170. Gurak, J. A., Chinn, Jr., J. W., Lagow, R. J., Steinfink, H., Yannoni, C. S.: Inorg. Chem. *23*, 3717 (1984)
171. Gurak, J. A., Chinn, Jr., J. W., Lagow, R. J., Kendrick, R. D., Yannoni, C. S.: Inorg. Chim. Acta *96*, L75 (1985)
172. Meyers, G. F., Hall, M. B., Chinn, Jr., J. W., Lagow, R. J.: J. Am. Chem. Soc. *107*, 1413 (1985)
173. Chinn, Jr., J. W., Lagow, R. J.: Organometallics *3*, 75 (1984)
174. Chinn, Jr., J. W., Lagow, R. J.: J. Am. Chem. Soc. *106*, 3694 (1984)
175. Guilhaus, M., Brenton, A. G., Beynon, J. H., Theis, M., Maercker, A.: Org. Mass. Spectrom. *20*, 592 (1985)
176. Krohmer, P., Goubeau, J.: Z. Anorg. Chem. *39*, 238 (1969)
177. Klein, J., Brenner, S.: J. Am. Chem. Soc. *91*, 3094 (1969)
178. Klein, J., Becker, J. Y.: J. Chem. Soc., Chem. Commun. *1973*, 576
179. Juza, R., Wehle, V., Schuster, H.-U.: Z. Anorg. Allg. Chem. *352*, 252, (1967)
180. Setzer, W. N., Schleyer, P. v. R.: Adv. Organomet. Chem. *24*, 353 (1985)
181. Maercker, A., Dujardin, R., Engelen, B., Buchmeier, W., Jung, M.: unpublished results
182. Boche, G., Etzrodt, H., Massa, W., Baum, G.: Angew. Chem. *97*, 858 (1985); Angew. Chem. Int. Ed. Engl. *24*, 863 (1985)
183. Günther, H., Moskau, D., Dujardin, R., Maercker, A.: Tetrahedron Lett. *27*, 2251 (1986)
184. Dowd, P.: J. Chem. Soc., Chem. Commun. *1965*, 568
185. Bernard, J., Schnieders, C., Müllen, K.: ibid. *1985*, 12
186. Rajca, A., Tolbert, L. M.: J. Am. Chem. Soc. *107*, 2969 (1985)

Heteroatom Directed Aromatic Lithiation Reactions for the Synthesis of Condensed Heterocyclic Compounds

Nurani Sivaramakrishna Narasimhan and Raghao Shivaram Mali

Garware Research Centre, Department of Chemistry, University of Poona, Pune-411007, India.

1 Introduction . . . 65

2 Phthalides . . . 73

3 Phthalans . . . 90

4 Phthalic Anhydrides . . . 92

5 2,3-Dihydro-isoindol-1-ones . . . 93

6 Isocoumarins and Dihydroisocoumarins . . . 95

7 Isochroman-3-ones . . . 105

8 Homophthalic Anhydrides . . . 109

9 Coumarins . . . 113

10 Condensed Furans . . . 118

11 Oxaphenalenes . . . 122

12 Isoquinolines . . . 123

13 Isoquinolones . . . 126

14 Phenanthridines . . . 129

15 Dibenzothio-, ox- and di-azepines . . . 130

16 Naphthastyrils and Related Compounds . . . 132

17 Acridones . . . 132

18 Furoquinolines 134

19 Pyranoquinolines 135

20 Condensed Pyrimidines 136

21 Indoles 138

22 Azacarbostyrils and Anthramycin 138

23 Miscellaneous Heterocycles 139

24 References 141

Heteroatom directed aromatic lithiation reaction[1-4)] is now widely used[1] for the synthesis of condensed heterocyclic compounds. New synthesis of condensed heterocyclic compounds are now known which can provide compounds not readily available by the usual acid catalysed methods. In this chapter these newer methods and their applications to several natural products are presented. Also presented, in some cases, are other non-lithiation methods which can provide comparison with the lithiation method and especially to bring about the latter's superiority in specific cases.

Condensed carbocyclic compounds can also be synthesised by aromatic lithiation reaction. This topic, however, is excluded from the present chapter.

1 Four articles have appeared recently relating to the topic of the present review. The article by Beak and Snieckus, *Accounts Chem. Res.*, *15*, 306 (1982) is essentially a review of their work. Snieckus's article (*Heterocycles*, 1980, *14*, 1649) reviews the synthesis of phthalides and use of these for the synthesis of isoquinoline alkaloids, anthraquinones and ellipticine. Watanabe's article (Yuki Gosei Kagaku Kyokaishi, *41*, 728 (1983) is more comprehensive and describes the synthesis of phthalides, homophthalic anhydrides, anthraquinones, anthracyclines, isocoumarins, coumarins, isoquinolines, berbines, lignans and ellipticine. Narasimhan and Mali's article (*Synthesis*, *957*, 1983) is also more exhaustive and gives a list of synthetic methods without discussing the detailed synthetic strategies.

1 Introduction

Preparation of an appropriately disubstituted aromatic compound is the first step in the synthesis of fused aromatic heterocyclic compounds. Thus a method for the synthesis of benzofurans, coumarins etc. involves the preparation of the salicyl aldehydes as the first step. Standard methods are then used to add one or more carbon atoms and to effect cyclisation to complete the synthesis.

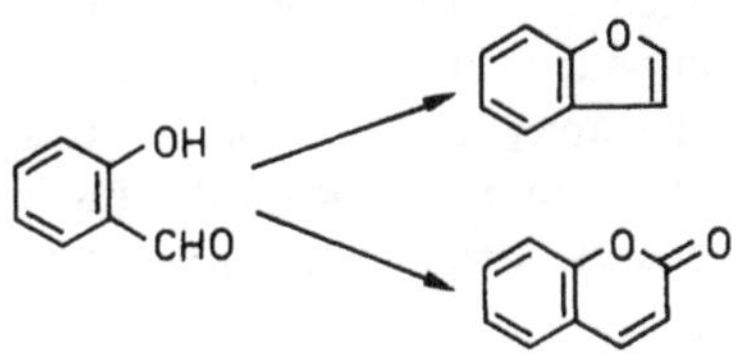

An alternate approach to the fused heterocyclic compounds is the intramolecular cyclisation at an aromatic position. The Bischler-Napieralski reaction [5] for the synthesis of the isoquinoline ring system illustrates this approach.

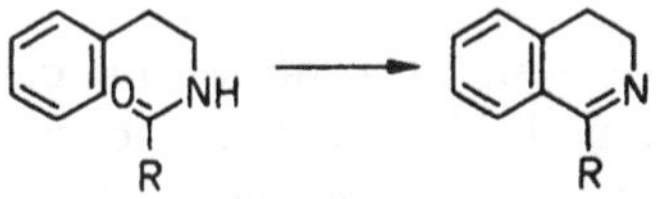

Both the approaches mentioned above have intrinsic problems. This may be illustrated with the synthesis of certain specific compounds like 5-methoxy coumarin, naphtho(2,3-b)furan and 8-methoxy isoquinoline.

The starting compound for the synthesis of 5-methoxy coumarin would be the resorcinol derivative (1), which after demethylation followed by Perkin condensation can afford 5-methoxy coumarin.

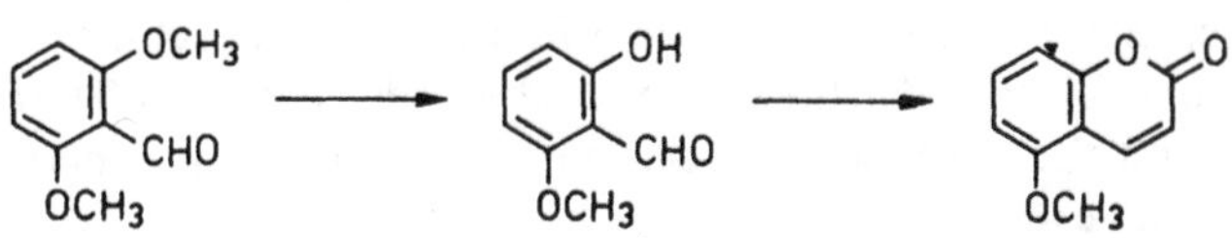

1

The usual method of synthesis of methoxy substituted benzaldehyde derivatives is by formylation of the methoxy substituted benzene derivatives. The formylation is generally effected by acid catalysed electrophilic substitution reactions like the Vilsmeyer-Haack reaction. When resorcinol dimethylether is formylated under these conditions, the product obtained is the derivative (*2*) [6] and not (*1*).

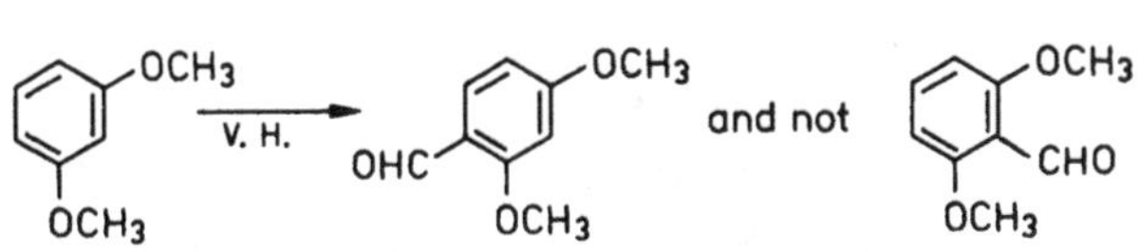

2 *1*

The synthesis of naphtho(2,3-b)-furan, similarly, is not achieved by the usual methods, since the required starting compound (*3*) is not readily available. Thus formylation of 2-naphtholmethyl ether, by Vilsmeyer-Haack reaction, furnishes [7] (*4*) and not (*3*) which is needed for the synthesis.

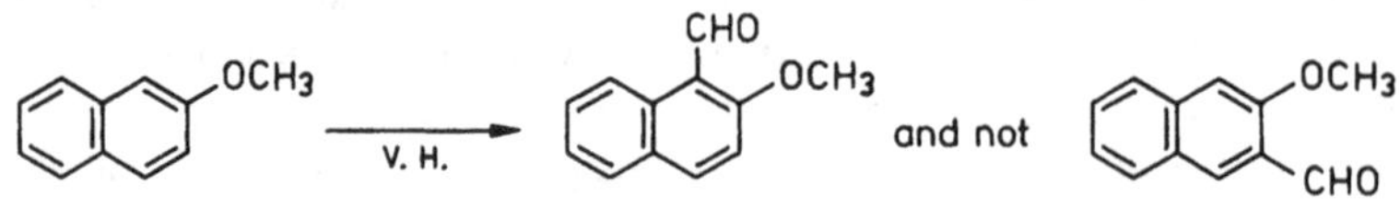

4 *3*

The failure of the usual acid-catalysed cyclisation methods to furnish specific methoxy substituted heteroaromatic compounds may be illustrated with the synthesis of 8-methoxy isoquinoline by Bischler-Napieralski reaction. In this, the starting compound is the acyl derivative of *meta* methoxy β-phenylethylamine. This, in the acid catalysed ring closure reaction, furnishes [5] the 6-methoxy isoquinoline (*5*) rather than the 8-methoxy isomer (*6*).

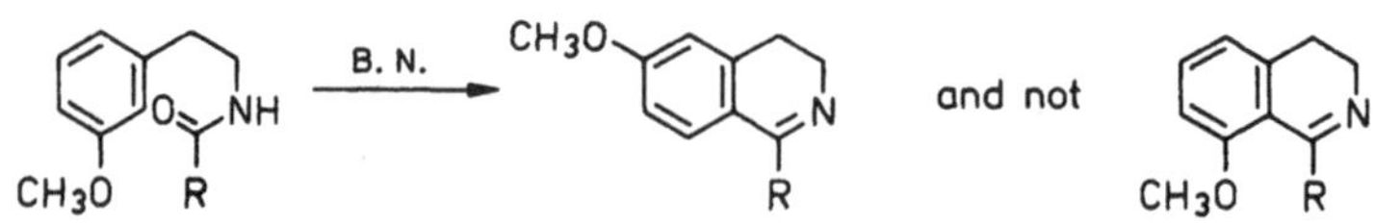

5 *6*

The reason for the failure of the above methods to furnish the desired specific methoxy substituted derivatives lies in the mechanism of the above reactions. These, being typical acid catalysed aromatic electrophilic substitution reactions, are well known to occur more at the *para* position than at the *ortho*. In the cyclisation reaction, the preference to cyclisation at the *para* position is even greater, quite often almost exclusive.

A general strategy to circumvent the above orientation problem in the above approaches is to block the more reactive position by a labile group and carry out the substitution reaction in the usual way, which would now occur at the less reactive position. In later stages the blocking group is removed. In such strategies either the blocking group is introduced at the stage prior to the crucial substitution or cyclisation reaction, or it is included in the starting compounds themselves. The first approach is illustrated with the synthesis of a berbine alkaloid (*7*), while the second with a coumarin derivative (*8*).

Tetrahydropalmatine (*7*) [8]

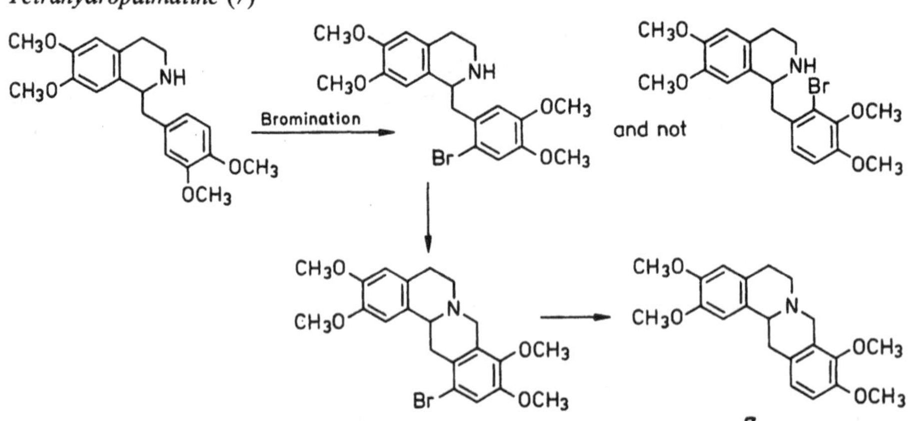

7

5-Methoxycoumarin (8) [9)]

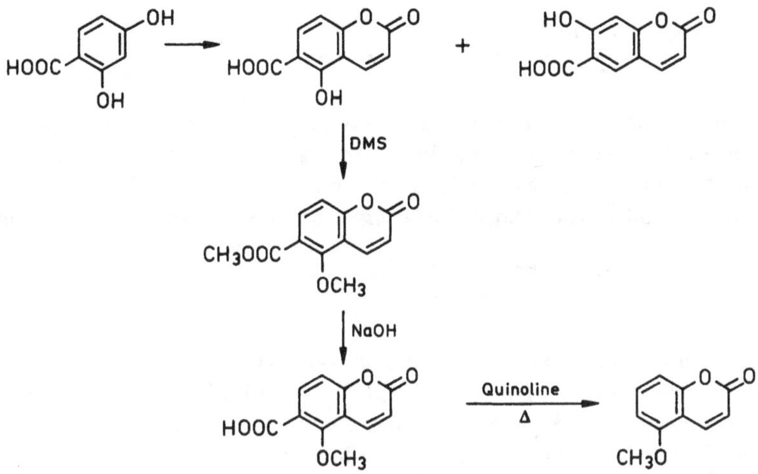

8

The limitations of the first method is that quite often the blocking group, being labile, is knocked-off during the electrophilic substitution reaction. With the second, the problem could be the ready availability of the starting compounds and removal of the blocking group.

While prevention of substitution or cyclisation reaction at an electrophilically more reactive site is one of the techniques to promote reaction at the less reactive site, strategies to specifically *direct* substitution at the less reactive position are also known. Phenol, for example, is specifically formylated at *ortho* position in alkaline medium [10)]. Cinnamic acid is specifically nitrated at *ortho* position by N_2O_5 [11)]. It is possible that in such reactions, the first step is the complexation of the reagent with the substituent already present, which then favours the attack of the reagent at the sterically close (to the complexing group) aromatic position which is *ortho* to the substituent already present.

An alternate approach to effect substitution at the less active (*ortho*) position would be to promote carbanion character at that position. Aromatic lithiation reactions indeed can serve this purpose by introduction of lithium atoms at *ortho* positions [4)].

Aromatic Lithiation Reactions

The direct replacement of an aromatic hydrogen by lithium, on treatment with an alkyl or aryl lithium compound, is known as aromatic lithiation reaction [4)].

[An aromatic lithium compound can also be synthesised by a halogen-metal exchange reaction [1)]. In the present chapter we are not concerned with this, although the organolithium compounds obtained by this way can be put to the same synthetic use as the ones obtained by lithiation reaction.

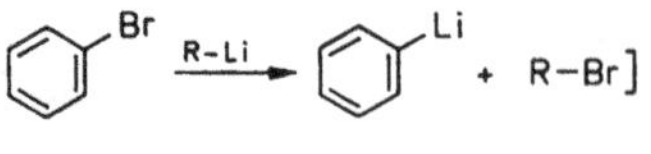

The aromatic lithiation reaction was originally discovered by Wittig [12] and by Gilman [6]. A typical aromatic lithiation reaction is that of anisole. In this reaction anisole, on treatment with BuLi is converted exclusively to the *ortho*-lithio derivative. No *para* product is obtained. The *ortho*-lithiation has been extensively investigated in

recent times, and several groups, besides $-OCH_3$, are now known to direct lithiation at the *ortho* position. These are OH, NH_2, CONHMe, CH_2NMe_2 etc. [6-9].

A mechanism to explain exclusive *ortho*-lithiation was first put forward by Roberts and Curtin [13]. In this mechanism, the lithiating R—Li reagent complexes first with groups like OCH_3 and then abstracts the *ortho* hydrogen, through its basic end, to generate the aromatic carbanion. A lithium atom is then introduced at this place.

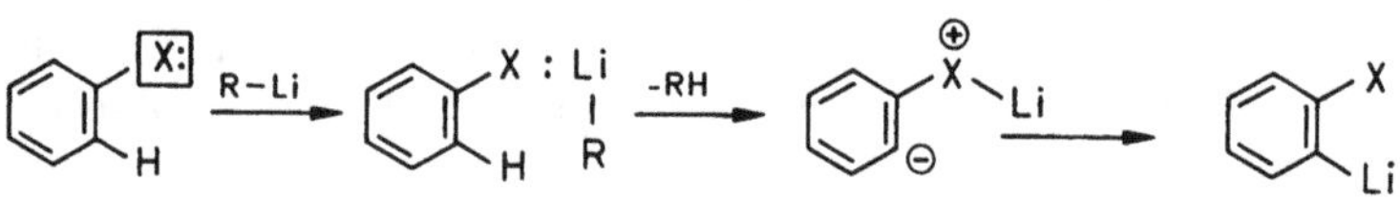

The attack being by the complexed RLi reagent, the aromatic carbanion is formed only at the sterically close position, which is *ortho* in the above case.

An examination of the Roberts-Curtin mechanism would indicate that the primary requirement for *ortho* lithiation to occur would be the presence of a substituent on the aromatic ring, which can complex with the lithiating agent. The primary requirement for a substituent to complex with the lithiating agent would be the presence in it of a heteroatom with an unshared electron pair. This, indeed is found to be the case and the variety of groups, which are known to bring about *ortho* lithiation reaction, contain such a heteroatom. Perhaps, the most interesting feature to be noted here is that the lithiation directing substituents include both those which are *ortho*/*para*-directing and *meta*-directing in acid catalysed electrophilic substitution reaction. In lithiation reactions, all these groups are only *ortho*-directing.

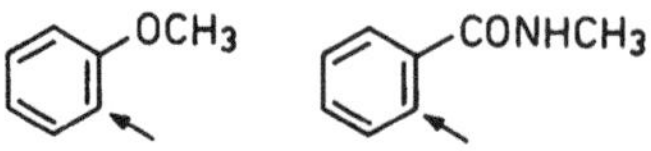

Other distinguishing features of an aromatic lithiation reaction, which have emerged during the last 20 years, are as follows:

1. Lithiation occurs not only at *ortho* but also at other *sterically close* positions [14-16].

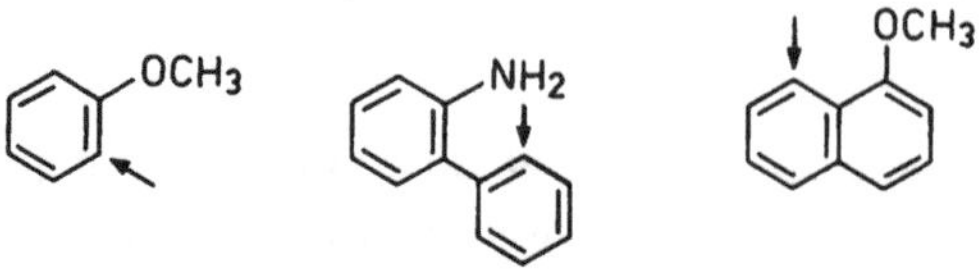

2. Lithiation occurs at *ortho* (or sterically close) position even when the heteroatom is one or two atoms away from the aromatic ring [17].

3. When two groups capable of complexation with the lithiating agent are present, lithiation occurs *ortho* (or sterically close) to that group which complexes better with the lithiating agent [18].

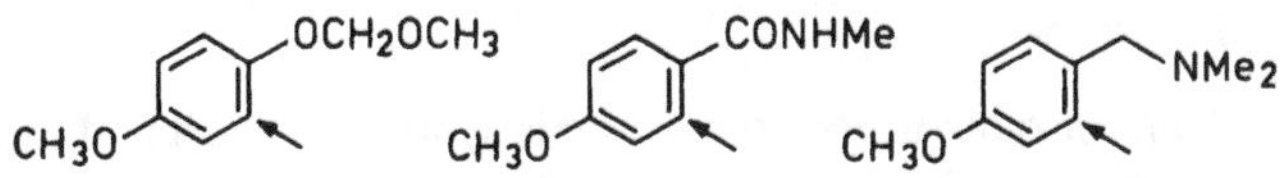

4. When two hydrogens capable of being displaced by lithium are present, lithiation occurs predominantly at that position which carries the most acidic hydrogen [15, 16].

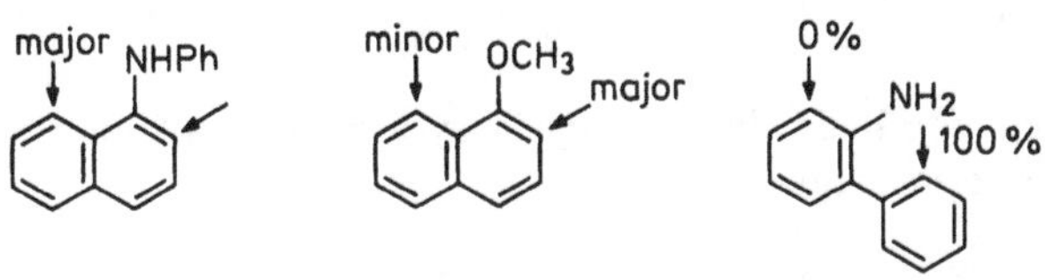

The acidity, in these cases, is determined by the inductive effect of the groups present. Resonance effects are not involved in determination of the acidity, since the electrons of the carbanion in the intermediate are present in the σ orbital which is orthogonal to the π orbital of the aromatic system.

In summary, there are two striking trends in the lithiation reactions mentioned above. One is that lithiation occurs *ortho* (or sterically close) to all groups, irrespective of whether they are electron withdrawing or electron donating, provided they have atoms with unshared electron pair not more than two atoms away from the aromatic ring. The other is that lithiation occurs at the position which, among those which are *ortho* or sterically close *to the better complexing group*, carries the more acidic hydrogen. The interesting fact is that the latter, quite often, corresponds to the less active position in the usual acid catalyzed electrophilic aromatic substitution reactions. Lithiation reactions then provide a method to introduce lithium atom[2] and promote electrophilic reactivity at these positions which are normally not sufficiently active to give substitution in any meaningful yield. Indeed, quite often, through lithiation reaction exclusive substitution at these positions can be achieved. The orientation in the lithiation reaction can, then, be used for the synthesis of several heterocyclic compounds and natural products not available by the usual methods. This aspect of heterocyclic synthesis is then discussed in this chapter. The chapter describes the synthesis of the basic ring system through aromatic lithiation reaction, and synthesis of specific methoxy substituted compounds and natural products, which cannot be synthesised or can be synthesised only through lengthy routes by the usual methods.

2 A few cases of dilithiation are also reported[19]. Experimental conditions, however, can be defined to lead to mono lithiation.

Heteroatom Directed Aromatic Lithiation Reaction for the Synthesis of Fused Heterocyclic Compounds

In general, three approaches are possible for the synthesis of fused heterocyclic compounds.

In the first a disubstituted aromatic compound is synthesised first. The second ring is then built on this by standard methods.

In the second a monosubstituted compound is synthesised first and by a cyclisation reaction the second ring is built.

A third approach is to build the fused cyclic system by an addition reaction on a reactive aromatic intermediate.

Synthesis of Disubstituted Aromatic Compounds by Aromatic Lithiation Reactions

A fused aromatic heterocycle, for synthetic purpose, may be considered to belong to three types. In one the heteroatom is common to both the rings (*9*), in the second it is directly attached to the bridge position (*10*) and in the third it is one or more atoms away (*11*).

9 10 11 12

A synthesis for the benzene fused heterocycles of the type *10* and *11*, which can be also applied to other fused heterocycle *12* would involve disubstituted starting compounds of the type *13* and *14* respectively, which by standard methods can be converted

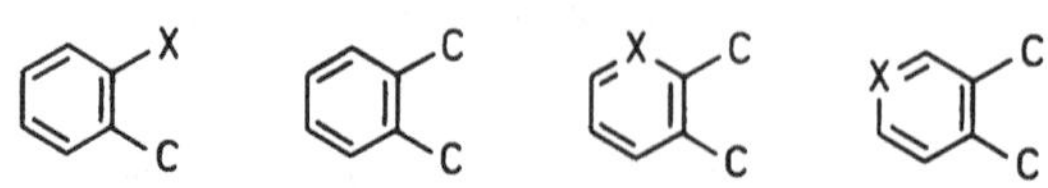

13 14 15 16

to *10* and *11*. Other disubstituted starting compound which can provide *10* and *11* are *15* and *16*. In general however the disubstituted benzene compounds *13* and *14* are used rather than the disubstituted heterocyclic compounds *15* and *16*.

For the synthesis of the disubstituted aromatic compounds of type (*13*), the starting compound could be the ones carrying a carbon substituent in the first instance or a heteroatom substituent.

The usual carbon substituents which function as lithiation directing groups [4)] are COOH, CHO, CH_2OH, CH_2NH_2, $CH_2CH_2NH_2$. These substituents are suitably modified to prevent the reaction of the lithiating agents with the directing groups themselves. The electrophilic reagents used used for the introduction of the heteroatom substituents are O_2, peroxide, CH_3ONH_2 etc. [4)]. The approach (type A) is illustrated by the following examples.

Type A

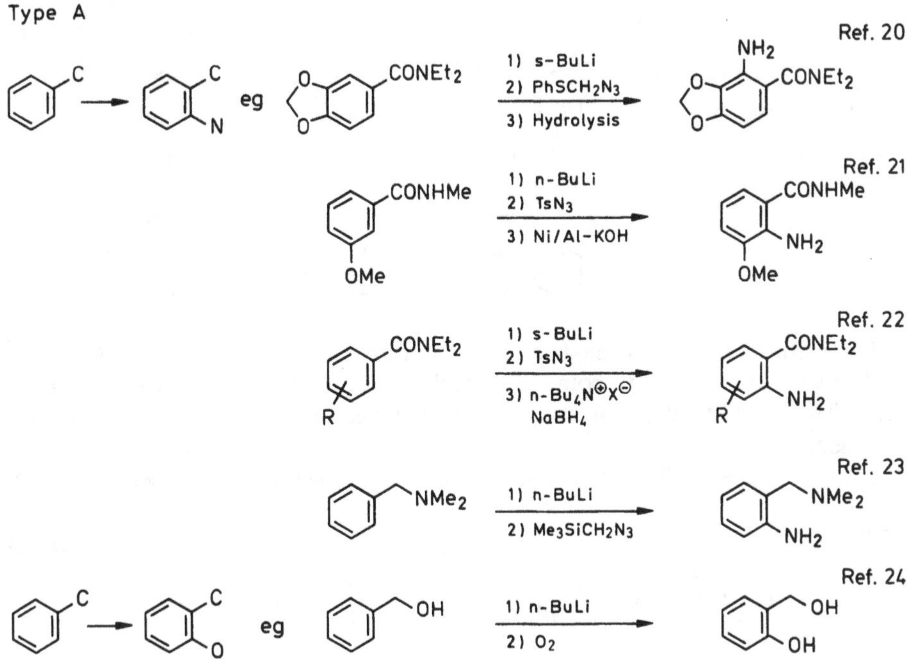

The heteroatom substituents which function as lithiation directing groups[4] are OH, NH_2, SH etc. These groups are again modified[3] to prevent reaction of the lithiating agent with them. The electrophilic reagents used, to introduce an *ortho* carbon substituent [4)], are R–I, 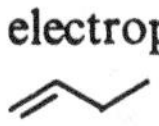I, R–CH–CH–R (O), R–CHO, =O, CO_2, ClCOOEt

The approach (type B) is illustrated by the following examples.

---

3 NH_2 group is not converted to its acetyl a benzoyl derivatives, since alkyl lithium reagents would preferentially lithiate at the CH_3 of $COCH_3$ or aromatic ring of COPh. Conversion to the —CO-t-Bu derivative can, however, lead to lithiation at the desired position [28)].

Type B

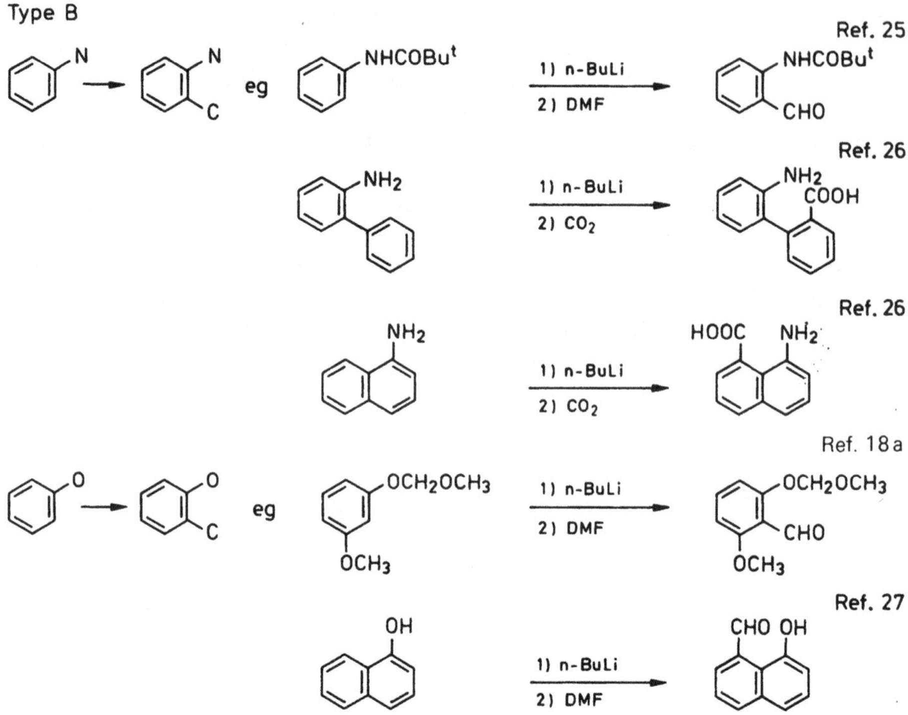

For synthesis of disubstituted compounds of type *14*, the starting compounds are carbon substituted aromatic compounds and the electrophiles are also carbon derivatives. The same lithiation directing groups and electrophiles, as mentioned earlier, are used.

Some examples to illustrate the above approach (type C) are given below.

Type C

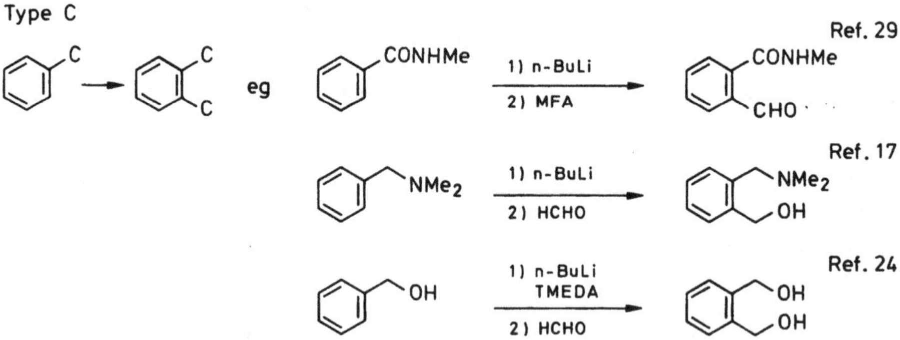

Synthesis of Heterocyclic Compounds

Having obtained the appropriate disubstituted compounds, further steps in the synthesis of the heterocyclic compounds involve standard reactions like condensation, cyclisation etc. In the discussion that follows synthesis of individual heterocyclic ring system, through aromatic lithiation reaction, is first described. This is then illustrated by synthesis of aromatic ring-substituted compounds and natural products which are either not readily synthesised or synthesised only through lengthy routes[4]. In some cases the non-lithiation route for the synthesis of heterocyclic compounds is also given either to provide comparison with the lithiation route or to bring out the novelty of the reagents used.

2 Phthalides

The phthalides occur in several natural products[31]. They are also intermediates for the synthesis of a variety of natural products like naphthols, anthraquinone natural products, anthracyclic antibiotics, the phthalide isoquinolines, protoberberine alkaloids etc.

The synthesis of phthalides by lithiation route is according to type C and involves, in the first step, the synthesis of the disubstituted compound of the type *14*. This is achieved either by incorporation of a —COOH group ortho to $-\overset{|}{\underset{|}{C}}-OH$ group or a $-\overset{|}{\underset{|}{C}}-OH$ group ortho to COOH group[24,32],

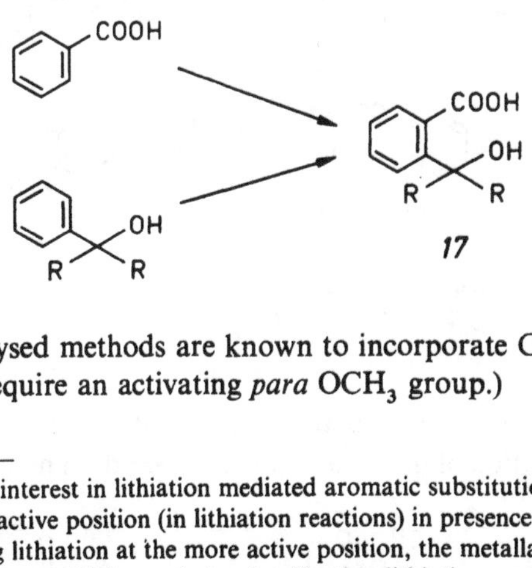

17

(Acid catalysed methods are known to incorporate CH_2OH ortho to COOH. They, however, require an activating *para* OCH_3 group.)

4 A point of interest in lithiation mediated aromatic substitution reaction is, how to effect substitution at a less active position (in lithiation reactions) in presence of the more active. In one approach, after effecting lithiation at the more active position, the metallation mixture is treated with $ClSiMe_3$ which introduces $-SiMe_3$ at that point. Further lithiation now occurs at the second position. After reacting with a suitable electrophile, the $SiMe_3$ group is replaced by H by acid cleavage of the C—Si bond. The following example illustrates the approach[30].

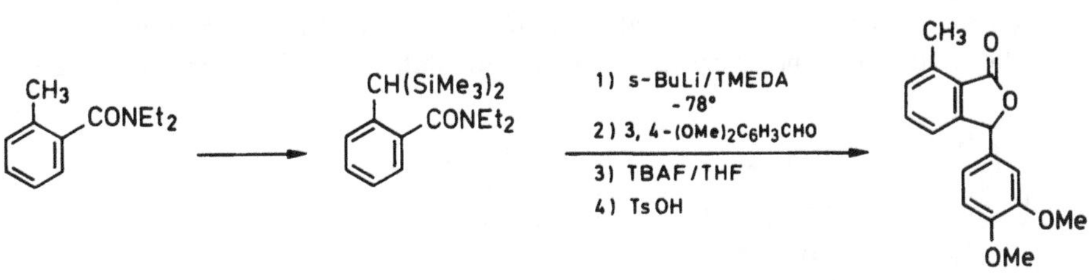

Lithiation of Aromatic Carboxylic Acids and their Derivatives

The carboxylic acid group, itself, reacts with lithiating agents such as BuLi. The chief reaction here is the formation of ketones. The aromatic carboxylic acid is then converted to suitable derivative before lithiation. The esters of aromatic acids also react with BuLi. However, the secondary and tertiary amides lead to aromatic lithiation reactions. The secondary amides form the N-lithio salts in the first instance on treatment with BuLi, but using excess BuLi (2.5 eq) and THF as the solvent the *ortho* lithio derivatives can be obtained. The tertiary amides, under similar conditions, give only compounds corresponding to reaction with the amide group. However, on treatment with a hindered lithiating agent such as *sec*-BuLi the *ortho*lithio derivatives are obtained.

The *ortho* lithio compounds on treatment with an aldehyde or ketone (CH_2O, RCHO, RCOR′) furnish the intermediates for the phthalides.

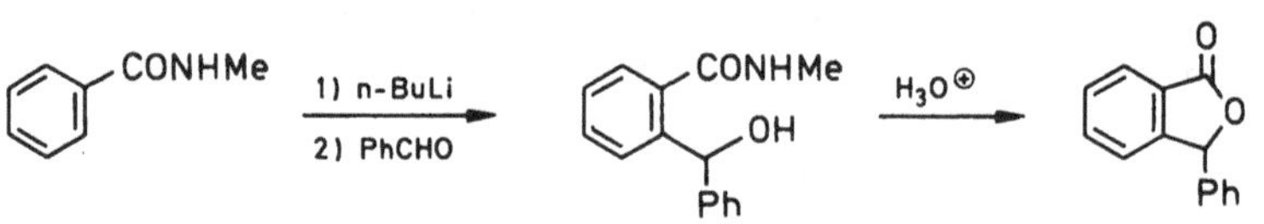

Ref. 32

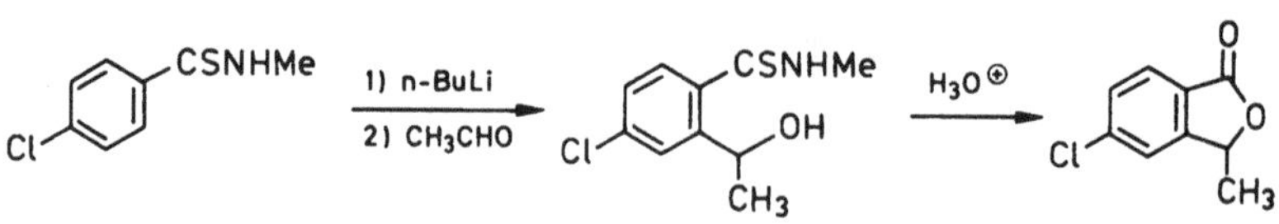

Ref. 33

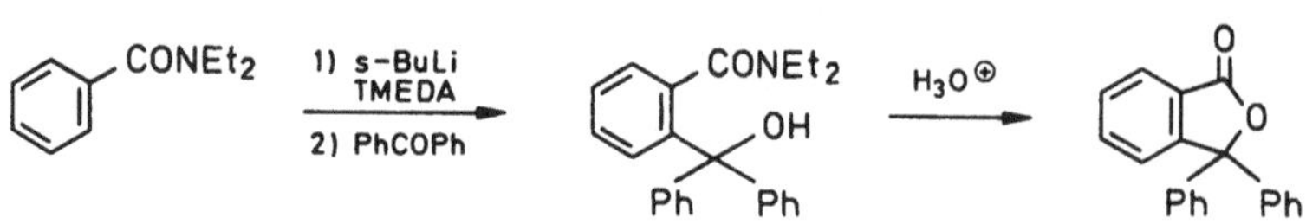

Ref. 34

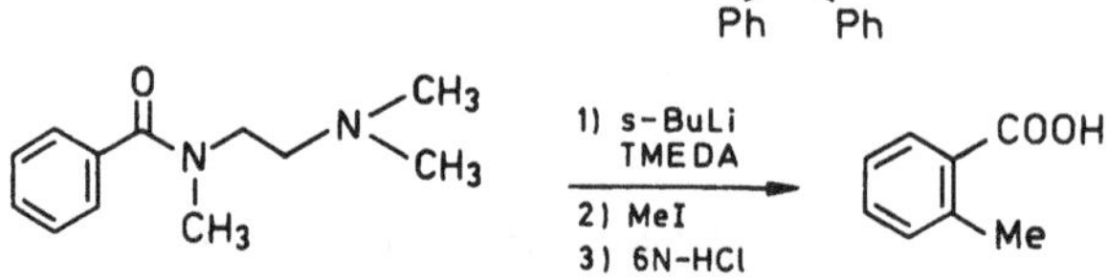

Ref. 35

Another carboxylic acid derivative which is also used as a lithiation directing group is the oxazole [4]. The *ortho* lithiation of these compounds proceeds in high yield [36].

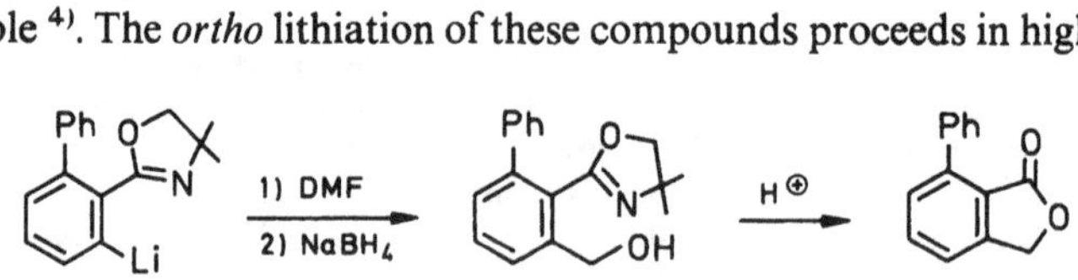

The general methods, indicated above, can be extended to the synthesis of benzophthalides and other heterocyclic analogues of phthalides. These are given below.

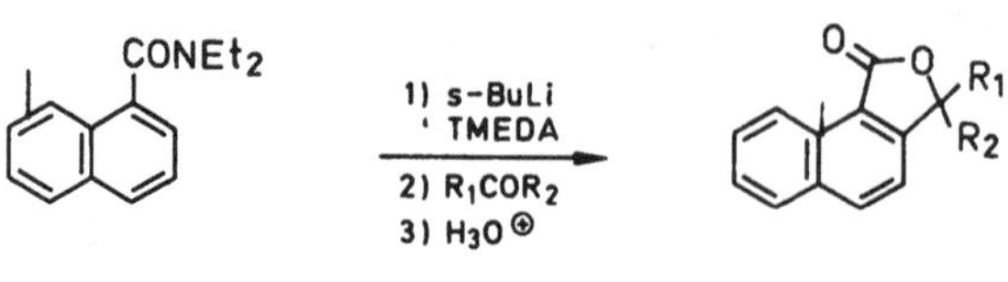

Ref. 37

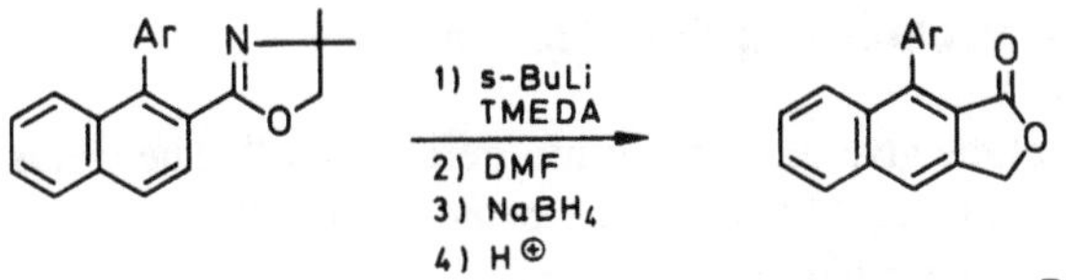

Ref. 36

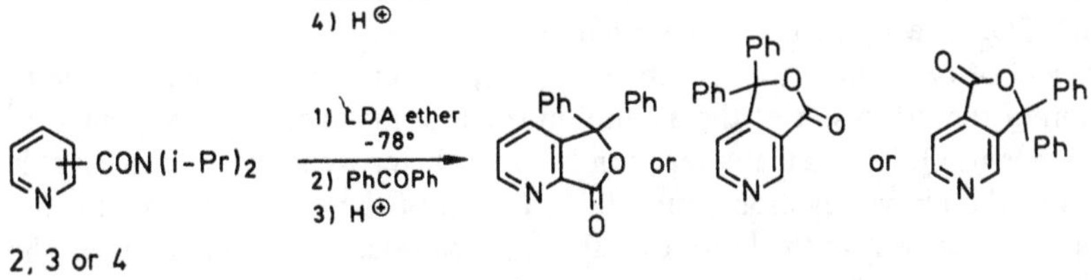

Ref. 38

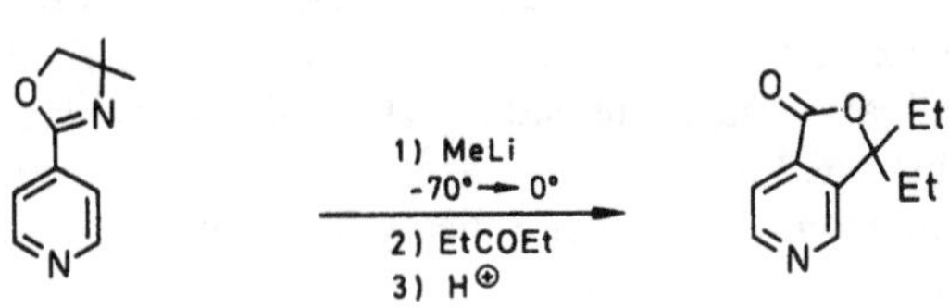

Ref. 39

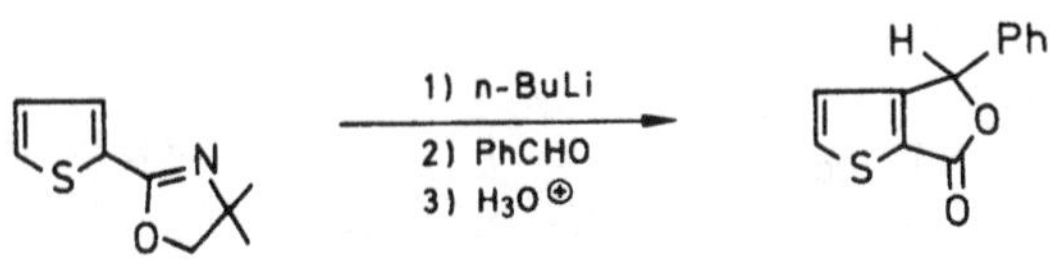

Ref. 40

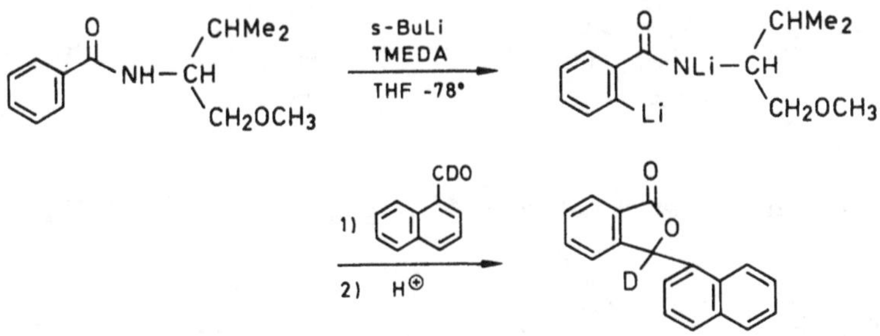

Ref. 41

When two amide groups are present one secondary and the other tertiary, lithiation occurs to a larger extent ortho to the secondary [41].

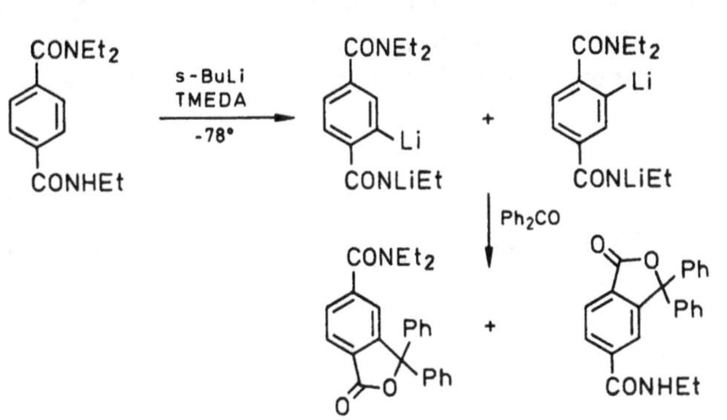

Lithiation of Aromatic Methylalcohols and their Derivatives

Lithiation of aromatic methyl alcohols give the *ortho* lithio derivatives [4]. The yields are poor, but can be improved by adding TMEDA. The *ortho* lithio derivatives on reaction with CO_2 directly furnish the phthalides.

The low yield of *ortho* lithiation, in the above experiments, is presumably due to the formation of the lithio salt of the alcohol in the first instance on treatment with the lithiating agent. Quite often the reaction becomes heterogeneous. A conceivable modification of the above reaction would be lithiation of the corresponding methyl ether derivative. However, in the latter case the reaction takes a different course. The chief reaction now is the formation of the benzylic carbanion which undergoes the Wittig rearrangement [4]. When the benzylic position is fully substituted by a carbon residue (such as alkyl or aryl), *ortho* lithiation occurs in better yield, and further reaction with CO_2 furnishes the 3,3-disubstituted phthalides.

The synthesis of phthalides, starting from the aromatic methyl alcohols, may be illustrated by the following examples.

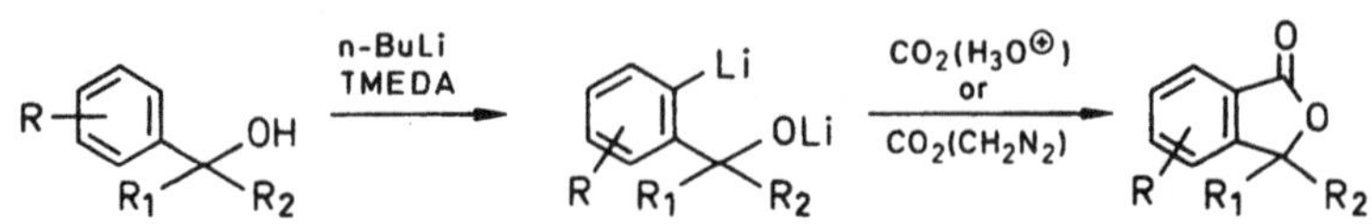

Ref. 24, 42

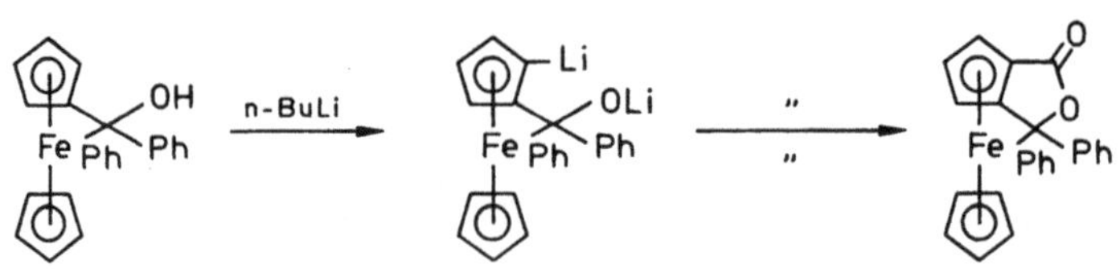

Ref. 43

The reactivity in the above cases can be improved by complexation with $Cr(CO)_6$ [44].

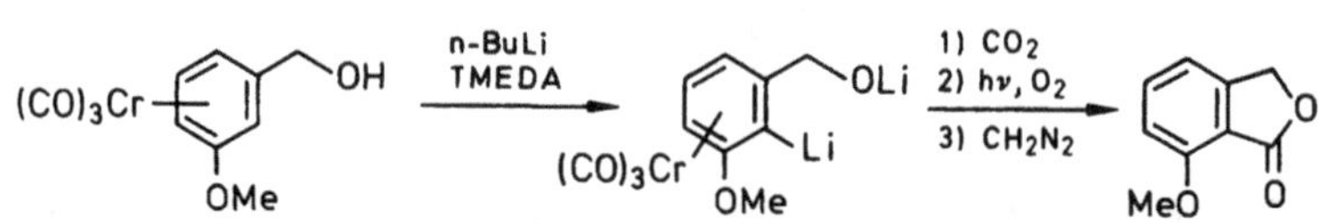

A recent method to introduce a carboxyl group *ortho* to hydroxy methyl group may be mentioned here. This involves thallation of a benzyl alcohol [45]. The reaction is governed by orientation rules observed for acid catalysed electrophilic substitution reaction. Thus *m*-methoxy benzyl alcohol gives 5-methoxy phthalide [45].

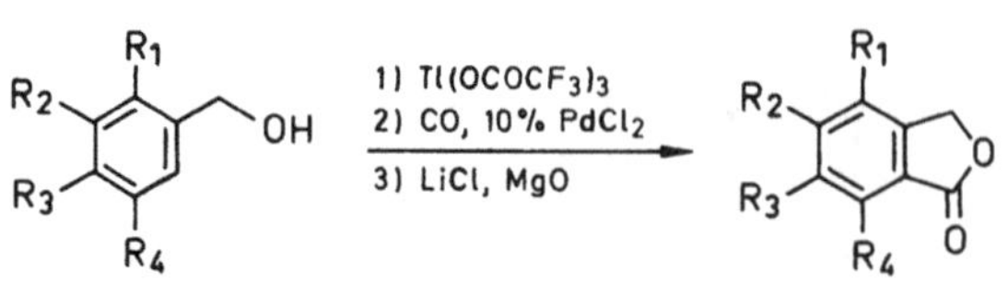

R_1	R_2	R_3	R_4
H	H	H	H
H	OMe	H	H
H	OH	H	H
H	Cl	H	H
H	H	CH_3	H
OMe	OMe	H	H
OMe	H	H	OMe

Lithiation of Aromatic Methylamines and their Derivatives

Aromatic methyl amines, as the N,N-dimethyl derivatives, are readily lithiated at *ortho* positions [17]. Indeed the $-CH_2NMe_2$ group is a very good lithiation directing group. On treatment of the lithio derivative with ClCOOEt, introduction of COOEt group takes place *ortho* to $-CH_2NMe_2$. The dimethyl amino group is also displaced by $-Cl$ to give the chloro ester which on thermal cyclisation furnishes the phthalide.

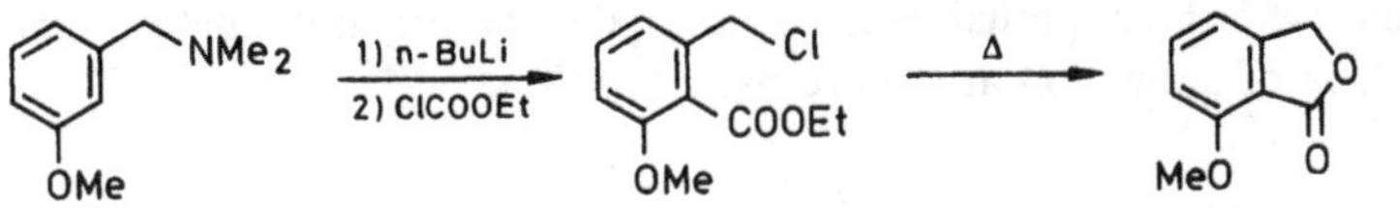

Ref. 46

Lithiation of Aromatic Aldehydes

The aldehydes are converted to an appropriate derivative, which are then used as the starting compounds. At a later stage they are regenerated and reduced to the alcohol or oxidised to the acid for obtaining the phthalides.

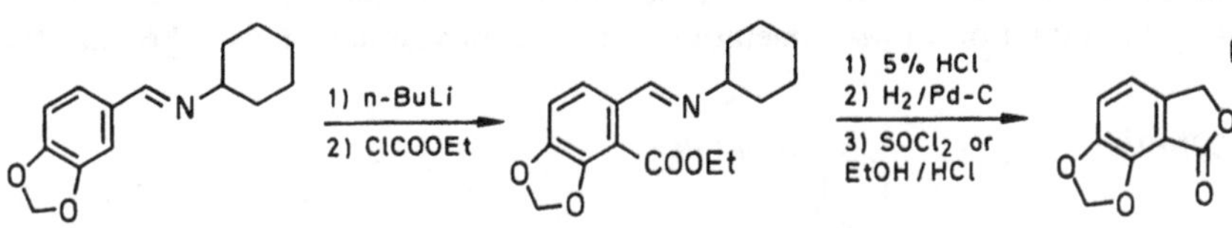

Ref. 47

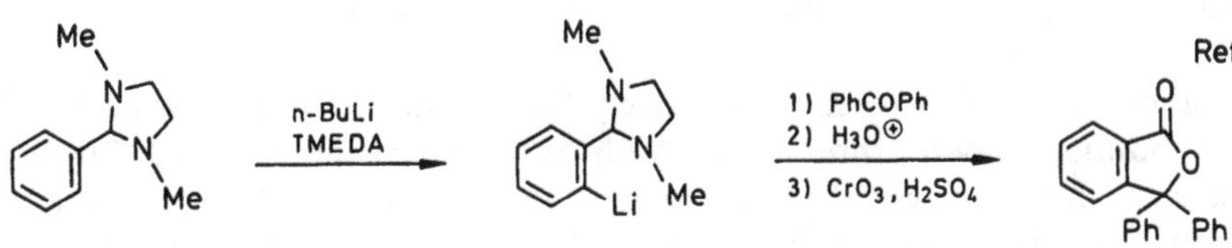

Ref. 48

An interesting protection of an aldehyde group in a lithiation reaction is to convert it to α-amino alkoxide. The alkoxide, here, is the *ortho* directing group [42,49].

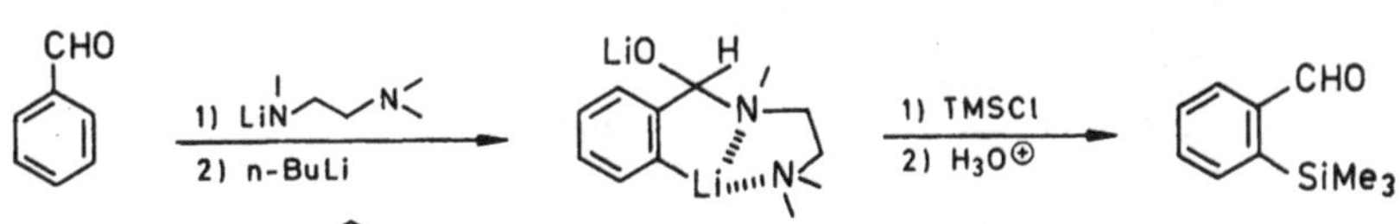

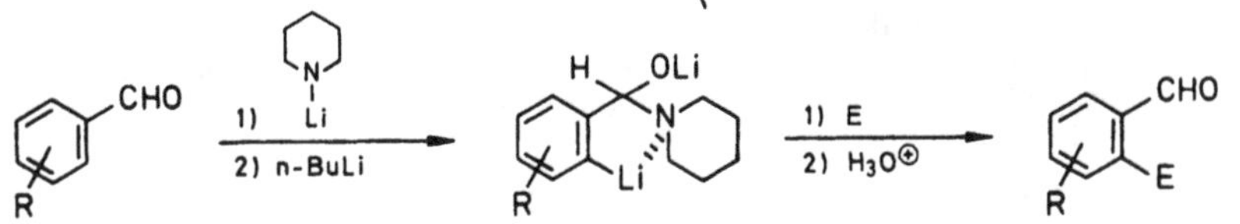

Aromatic Ring Substituted Phthalides

6-Methoxy phthalide could be synthesised by the usual method of phthalide synthesis, involving hydroxymethylation of *meta*-methoxy benzoic acid. The orientation of the methoxyl group in presence of acid catalysts favours this reaction. The 5- and 7-methoxy phthalides are not available by this method.

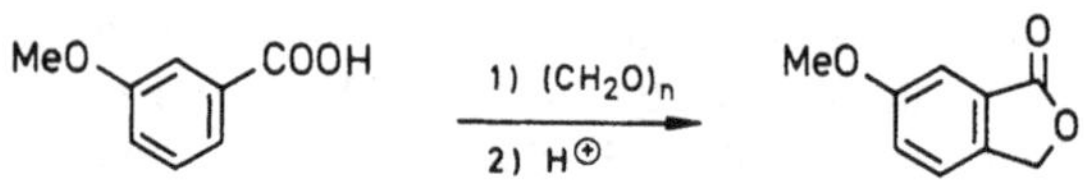

Ref. 50

5-Methoxy phthalide is available from *m*-methoxy benzyl alcohol via organo thallation reaction. 6-Methoxy phthalide is also available via organothallation reaction, despite the lack of activation at the aromatic position.

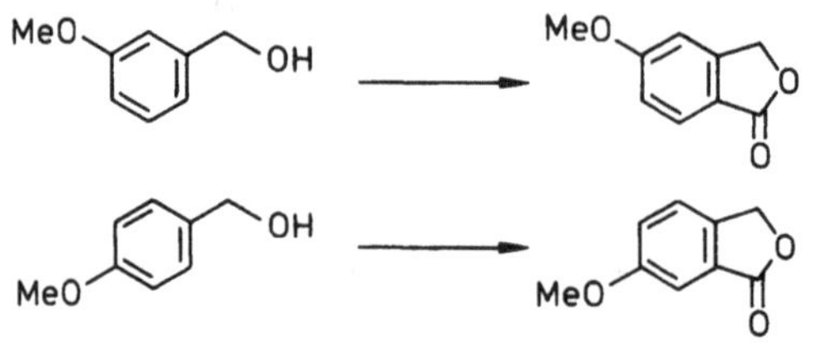

Ref. 45

Ref. 45

A simple synthesis of 4-methoxy phthalide is only *via* an organolithium intermediate derived by the lithiation of *meta*-methoxy N-methyl benzamide. This is because the

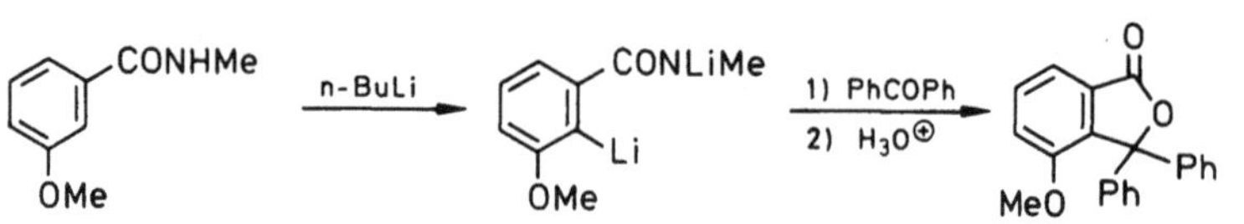

Ref. 51, 52

orientation in lithiation reaction exclusively favours introduction of lithium at the middle position. Indeed, three of the methoxy phthalides i.e. 4,5- and 7-methoxy phthalides can be obtained through aromatic lithiation reaction from appropriately substituted aromatic carboxylic acid derivatives or aromatic methyl alcohols or

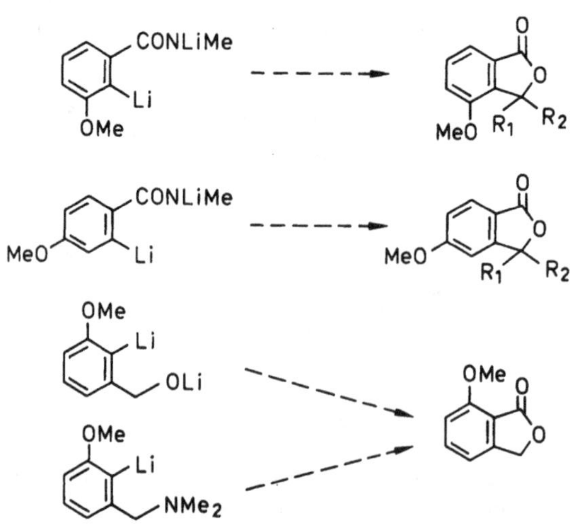

Ref. 51, 52

Ref. 51, 52

Ref. 42, 44, 46

methyl amines. The fourth phthalide i.e. 6-methoxy phthalide, of course, can be synthesised readily from *m*-methoxy benzoic acid by acid catalysed methods. Its synthesis by lithiation method has not been attempted. It can be readily prepared by lithiation of *p*-methoxy N,N-dimethyl benzylamine as follows. Here the lithiation reaction occurs preferentially *ortho* to the dimethylamino methyl group because the group complexes better with the alkyl lithium reagent.

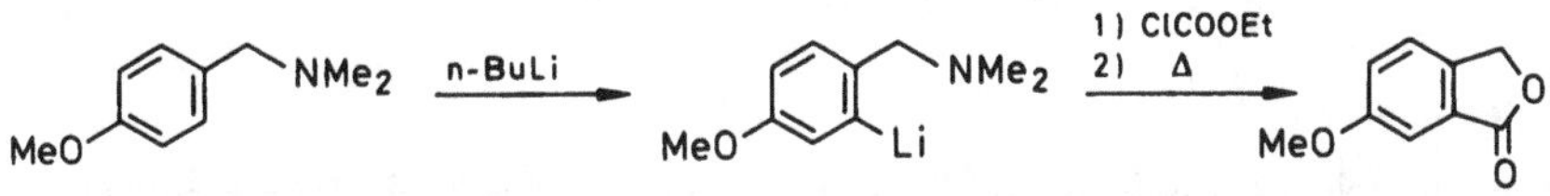

The principle enunciated above may be extended to obtain other polysubstituted phthalides. The substituents are chosen with due consideration of the trends in orientation in aromatic lithiation reactions. A list of substituted phthalides, synthesised by lithiation reaction, has recently been published [4,53].

Naturally Occurring Phthalides

The naturally occurring phthalides have been recently reviewed by Dean [31a] and others [31b]. In this chapter only those obtained recently, and which have been synthesised through lithiation reactions are given.

5-Methoxy-7-hydroxyphthalide (18) and 5,7-dimethoxyphthalide (19)

5-Methoxy-7-hydroxyphthalide (*18*) and 5,7-dimethoxyphthalide (*19*) have been isolated from the over ground parts of *Helichrysum italicum* [54].

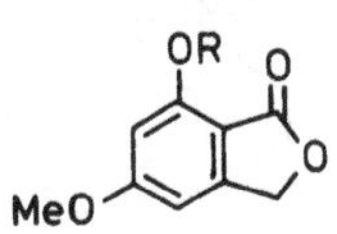

18 R = H

19 R = Me

The latter has been synthesised from 3,5-dimethoxy benzyl alcohol by lithiation route [55]. Thus, lithiation of 3,5-dimethoxy benzyl alcohol (using n-BuLi in presence of TMEDA) followed by reaction with excess of CO_2 and later treatment with acid gives *19*.

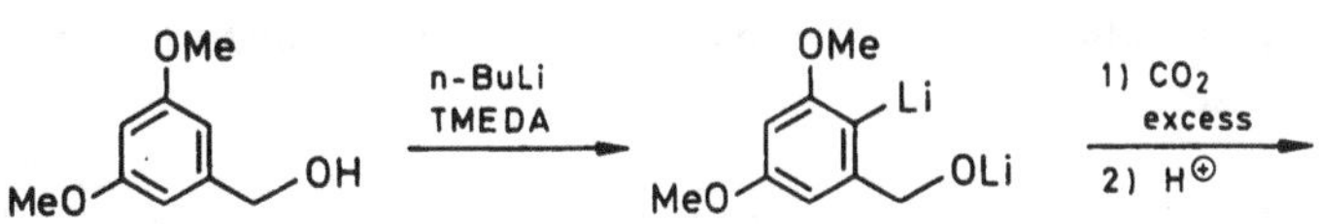

19

Meconine (20)

Meconine (*20*) is isolated from *Papaver somniferum* L. [56]. Three lithiation routes are known for its syntheses which are depicted in the following chart.

Ref. 57

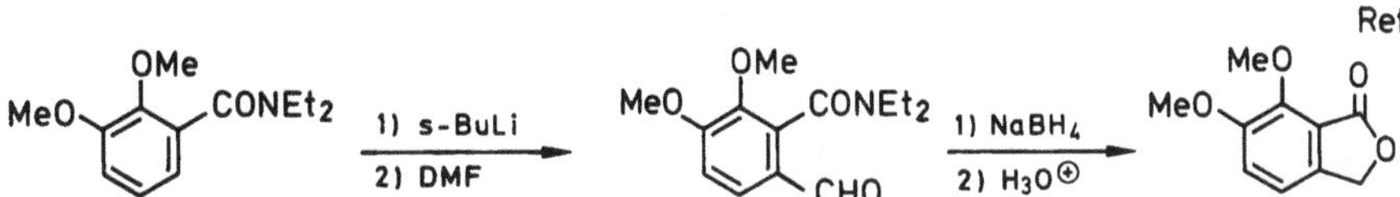

20

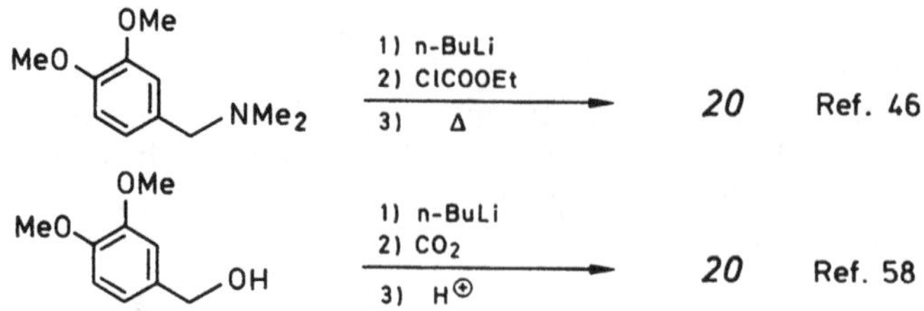

Isoochracinic acid (21)

Isoochracinic acid (*21*) is isolated [59] from *Alternaria kikuchiana*. Three routes are available for its synthesis which involve directed metallation reaction.

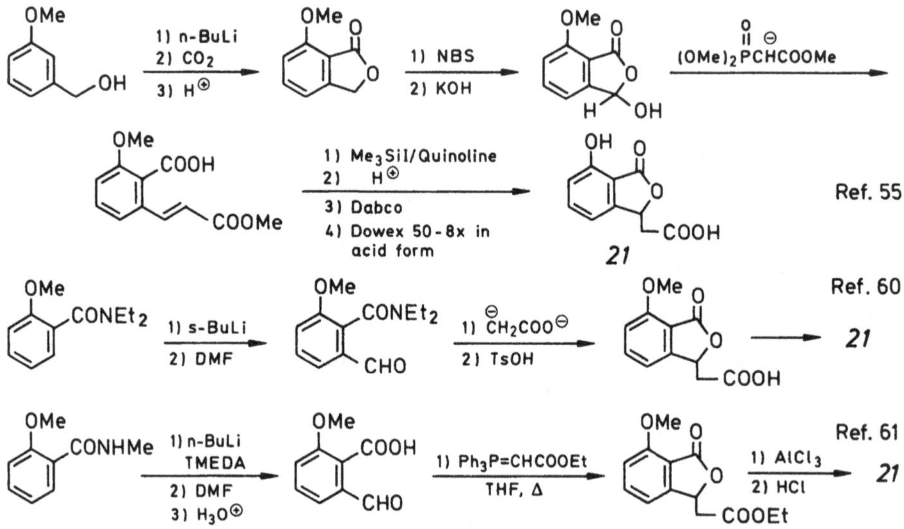

Another synthesis [62] of isoochracinic acid (*21*), utilises an interesting application of Wittig reaction. 3-Methoxyphthalic anhydride on reaction with carbobenzyloxymethylenetriphenyl phosphorane, followed by hydrogenation and demethylation furnishes 21. It may be noted that 3-methoxyphthalic anhydride can be readily obtained by aromatic lithiation of N,N-diethyl *m*-methoxy benzamide followed by treatment with CO_2 (vide supra).

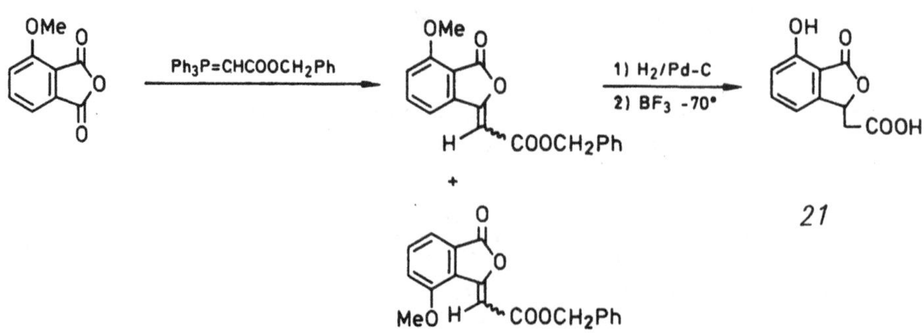

The naturally occurring Z-butylidene phthalide also can be synthesised [63] from the phthalaldehydic acid using Wittig reaction.

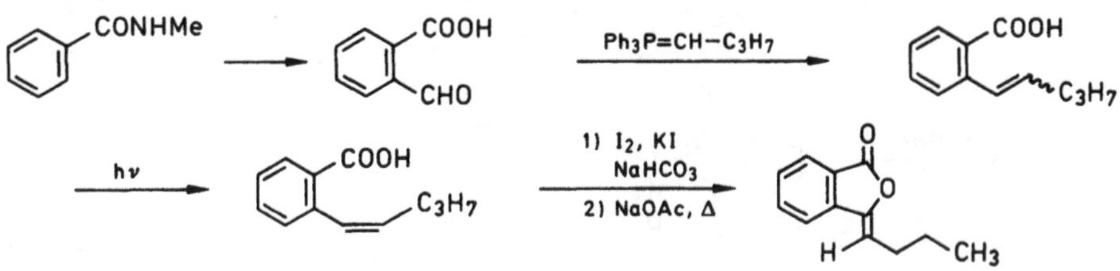

Phthalides as Synthons

Phthalides are valuable synthons for obtaining several biologically active compounds and natural products. Naphthols, anthraquinones and anthraquinone antibiotics, isoquinolones, phthalide isoquinolines, indolizidine and quinolizidine alkaloids, and berbine alkaloids have been synthesised with phthalides as the starting compounds. Some of the syntheses, which have interesting chemistry, are described below. The required phthalides, in some cases, have been synthesised through aromatic lithiation reactions, while in others by more conventional methods.

1. Naphthols

Michael addition of phthalide anions with α,β-unsaturated carbonyl compounds, followed by cyclisation provide substituted naphthols.

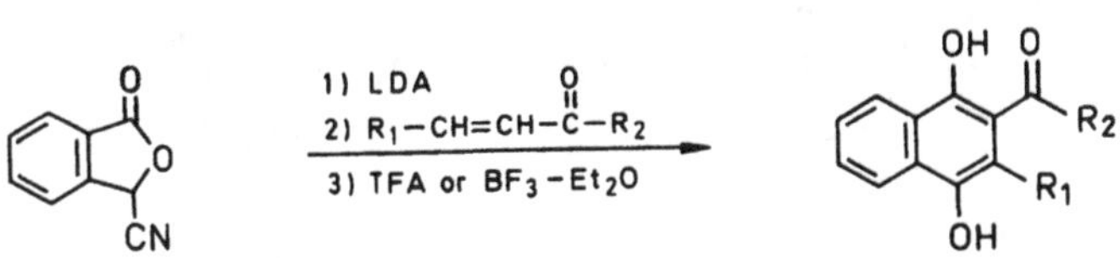

Ref. 64

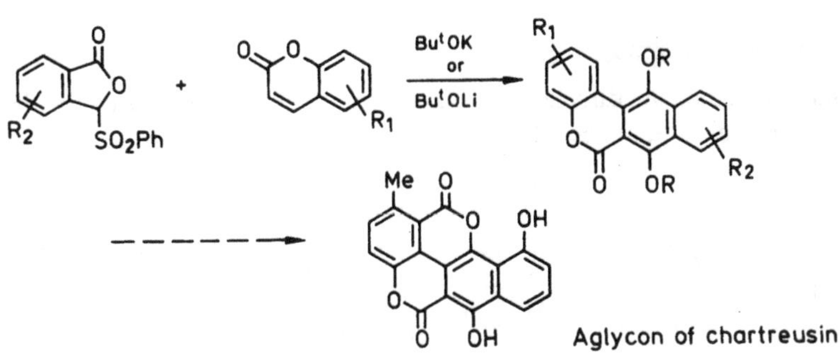

Ref. 65

2. Lignans

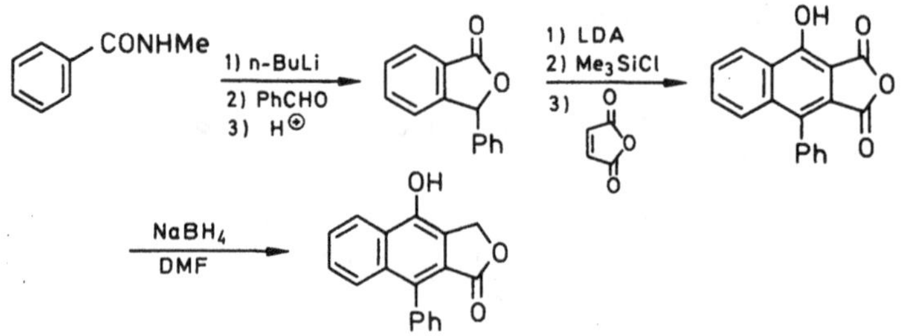

Ref. 66a

Ref. 66b

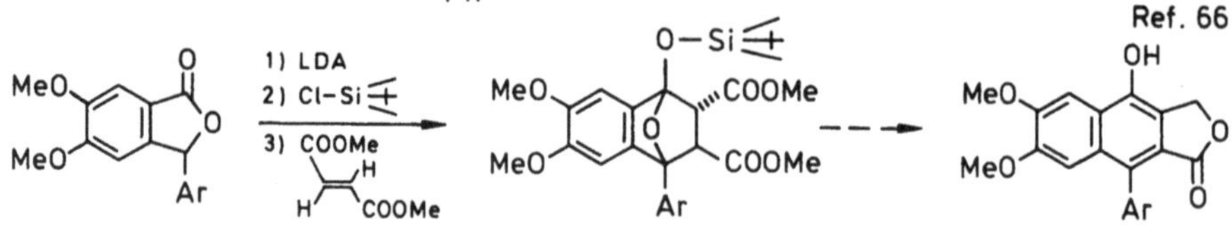

Ref. 66c

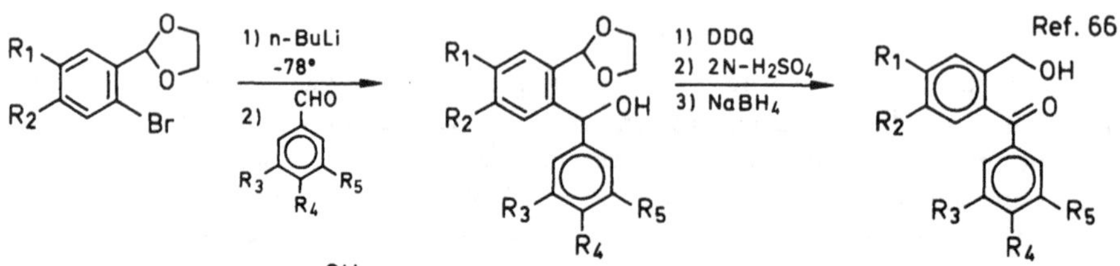

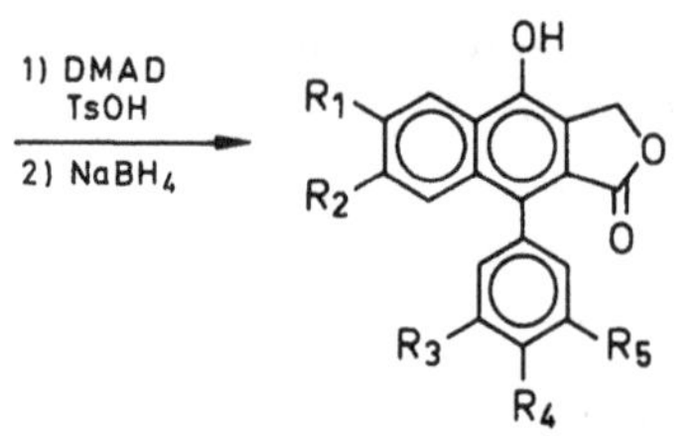

	R_1	R_2	R_3	R_4	R_5
a	OCH_3	OCH_3	H	OCH_3	H
b	$O-CH_2-O$		H	$O-CH_2-O$	
					Taiwanin-E
c	$O-CH_2-O$		OCH_3	OCH_3	OCH_3
					Dehydropodophyllotoxin

Ref. 66d

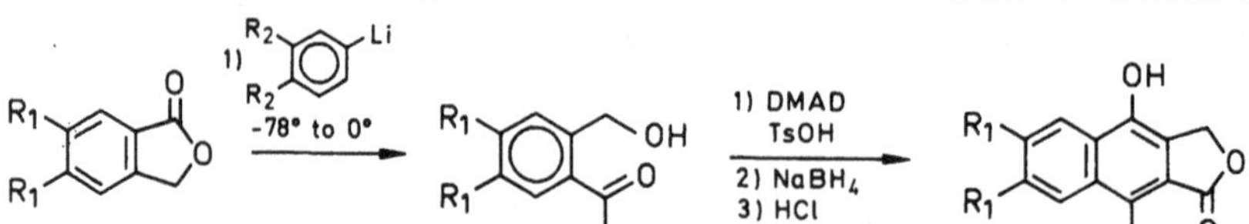

	R_1	R_2
a	H	H
b	OMe	OMe

3. Anthraquinones and Anthracyclines

Several phthalides obtained via directed metallation reactions have been used for the synthesis of various anthraquinones [67].

Ref. 68

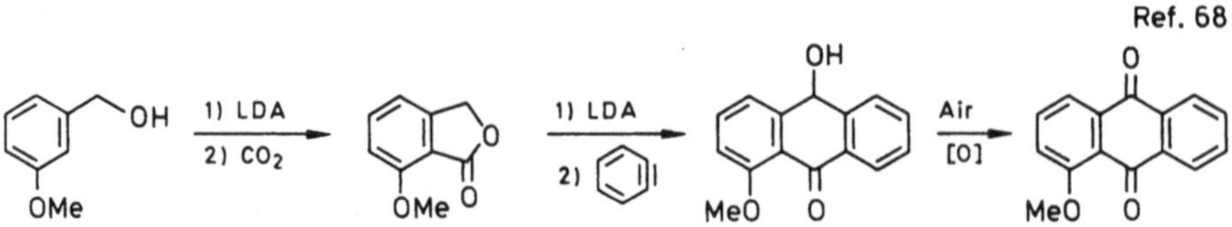

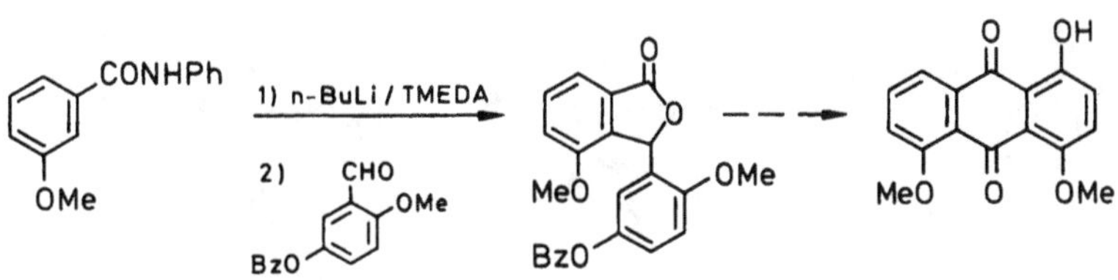

Ref. 69

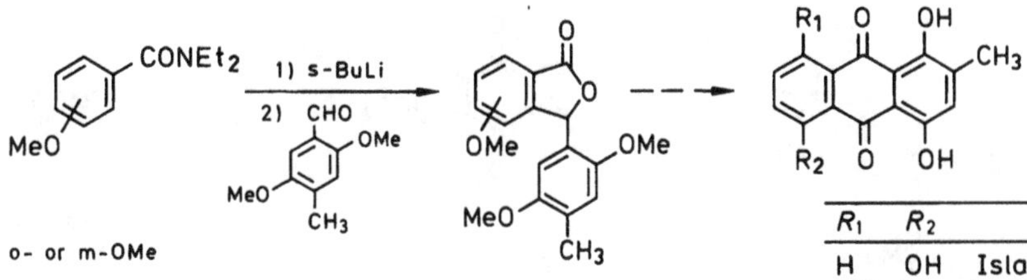

Ref. 70

o- or m-OMe

R_1	R_2	
H	OH	Islandicine
OH	H	Digitopurpone

Ref. 71

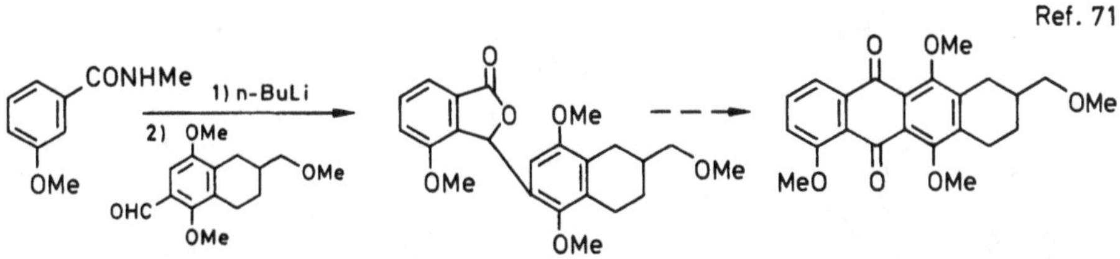

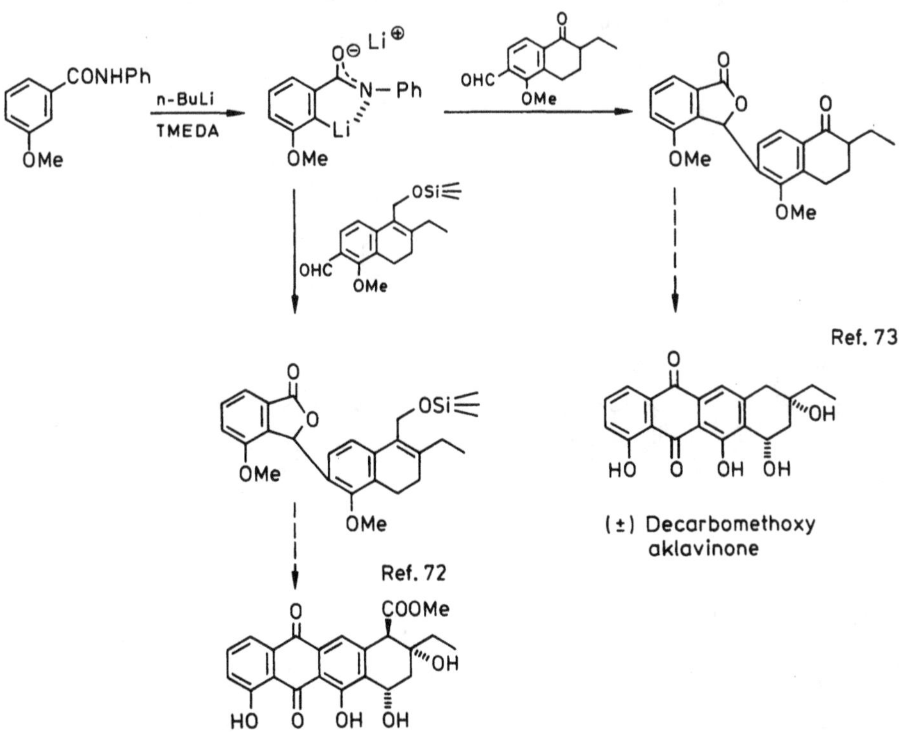

Ref. 73

(±) Decarbomethoxy aklavinone

Ref. 72

(±) Aklavinone

Ref. 74

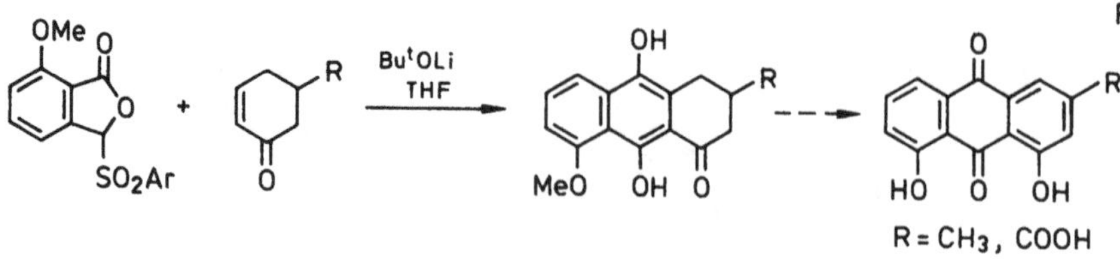

Ref. 75

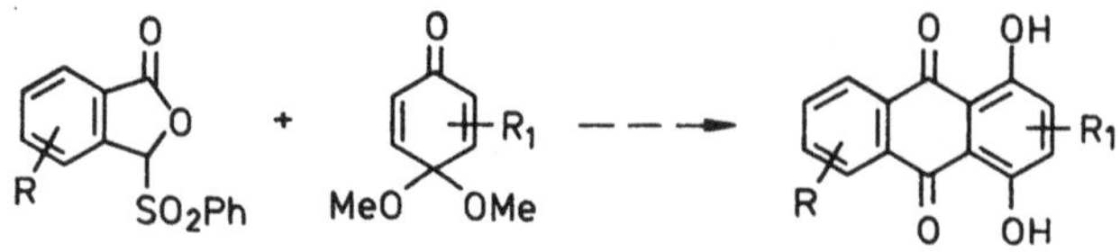

Ref. 76

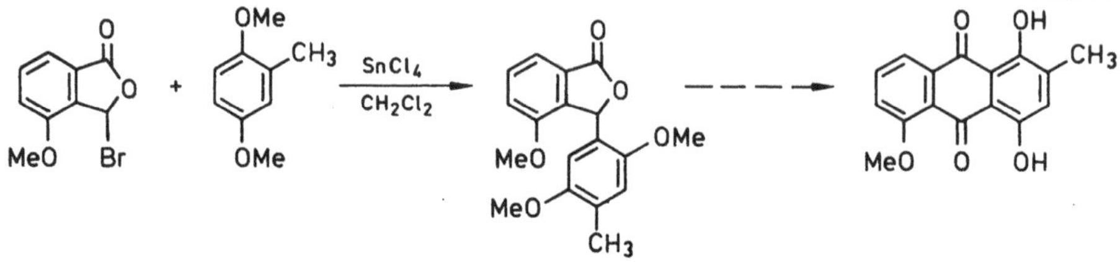

Ref. 77

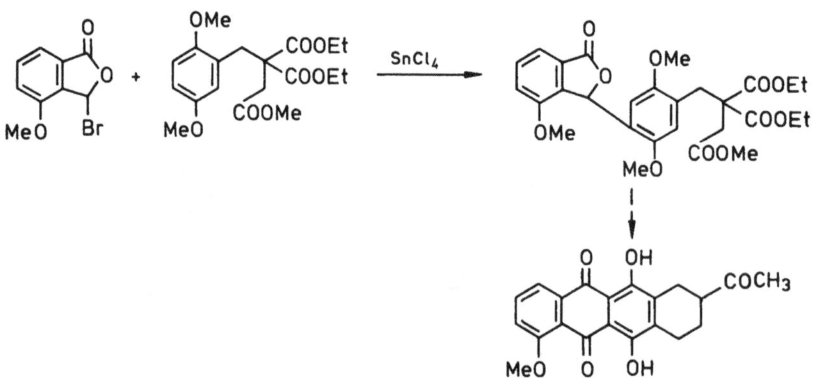

Ref. 78

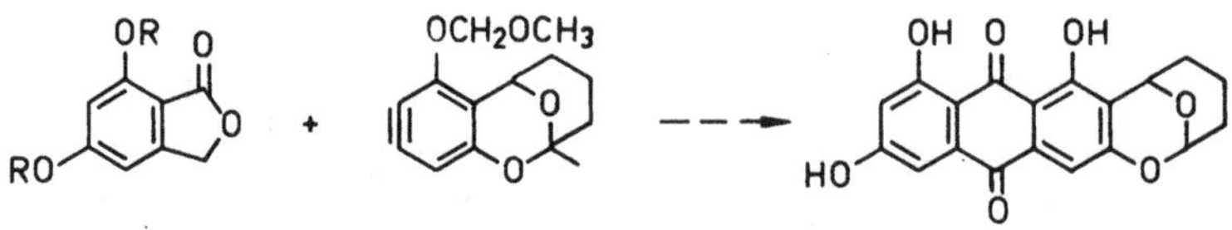

Ref. 79

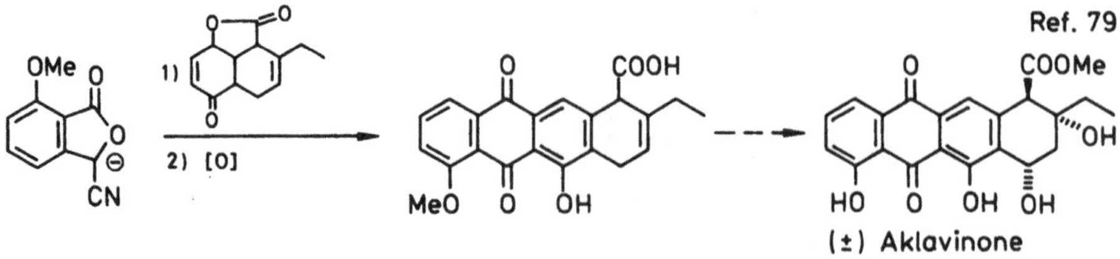

Ref. 80

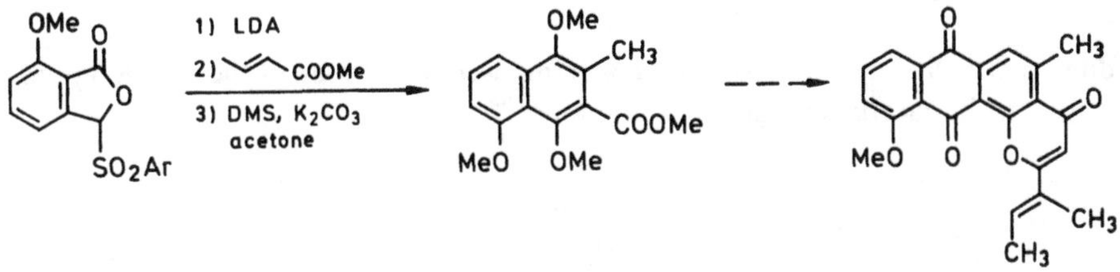

o-methylkidamycinone

Ref. 81

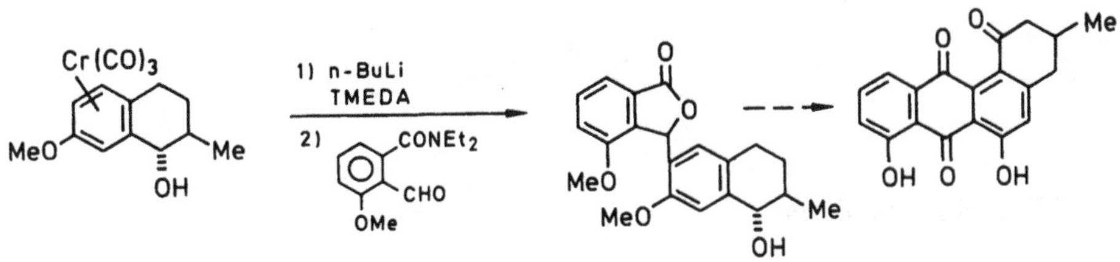

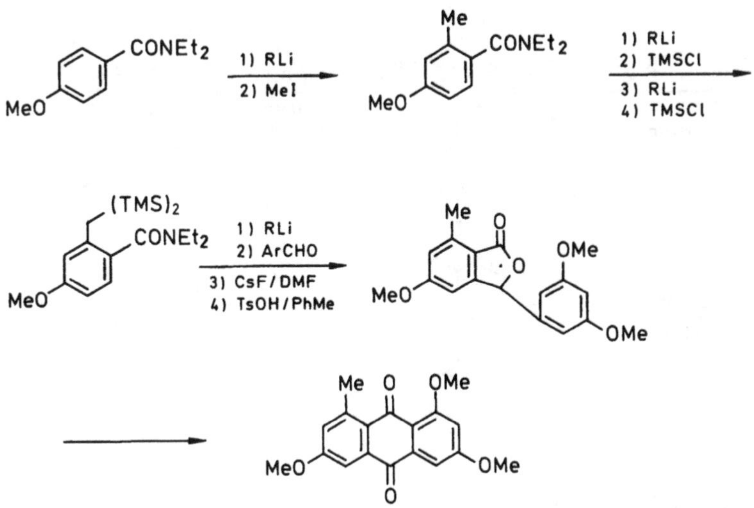

Ref. 82

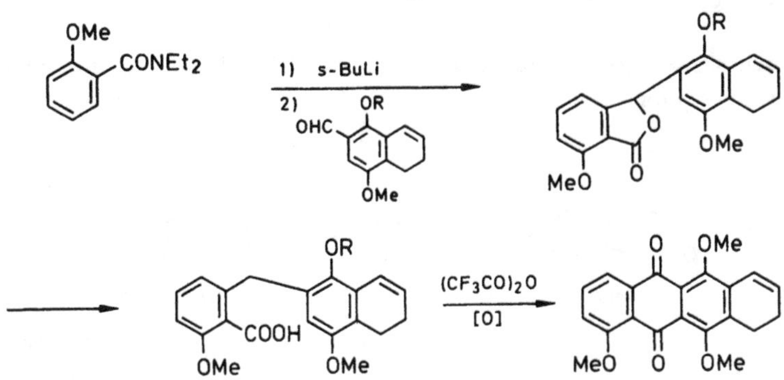

Ref. 83

4. Isoquinolones

Addition of phthalide anion to Schiffs base followed by reaction of the addition product with trifluroacetic acid give isoquinolones [84].

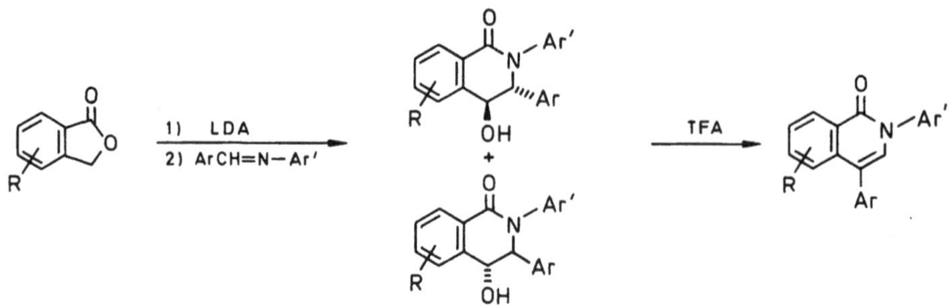

5. 4-Phenylisoquinolines

The alkaloid cherilline has been synthesised by the following method [85].

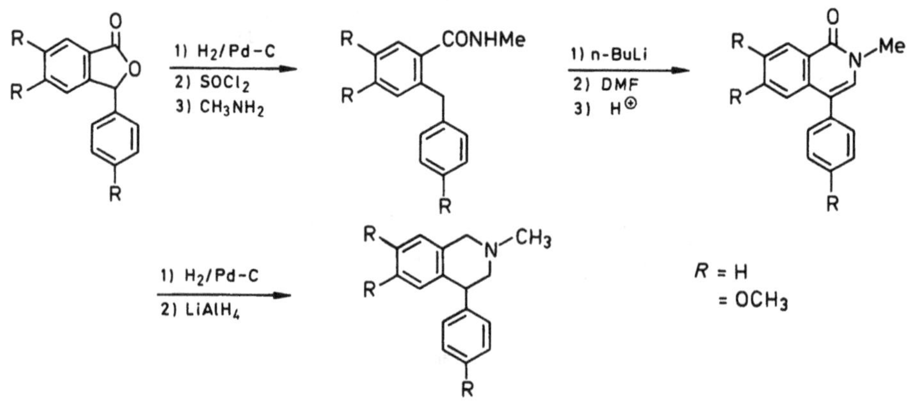

6. Phthalide Isoquinoline Alkaloids

These have been synthesised from phthalides by reaction with 3,4-dihydroisoquinolium salts [86]. The yields are poor.

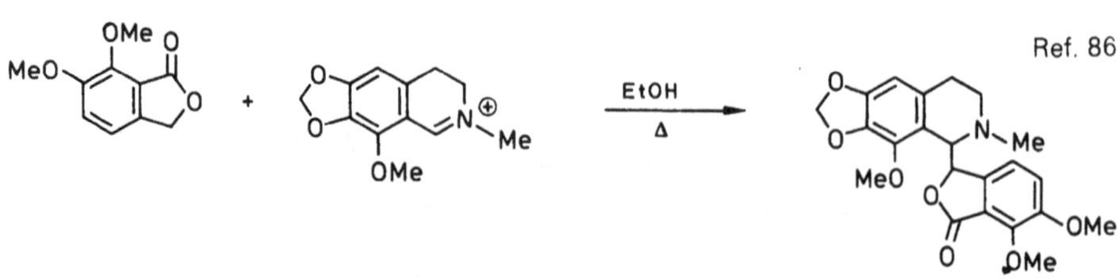

Ref. 87

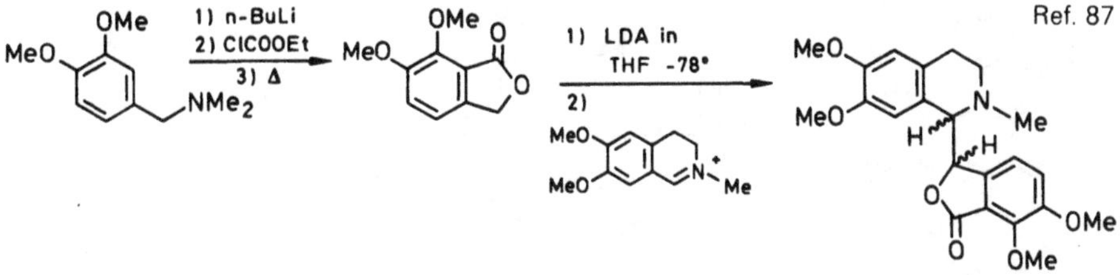

Cordrastine I and II

A new electroreductive method for the synthesis of tetrahydro-phthalide-isoquinolines is also known [88]. If isoquinolinium methiodide is used in the reaction the 1,2-dihydrophthalide isoquinoline is obtained [89].

Ref. 88

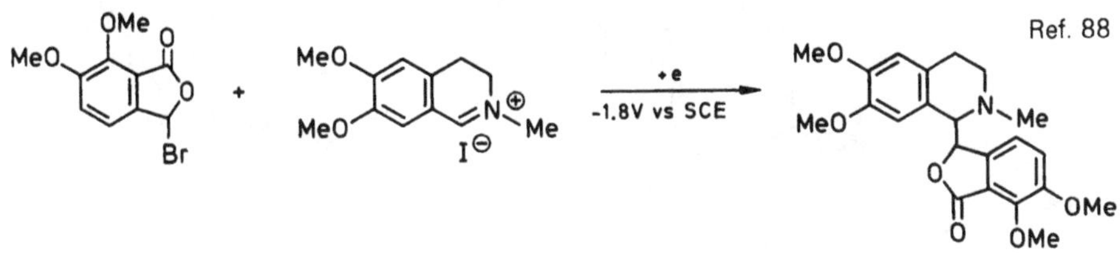

Ref. 89

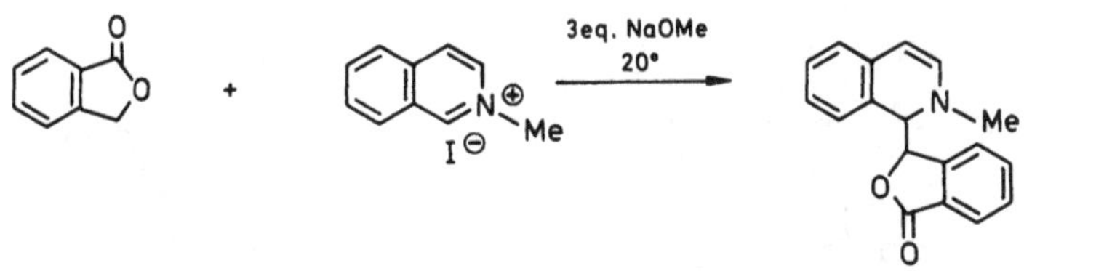

7. (±) Sesbanine [90]

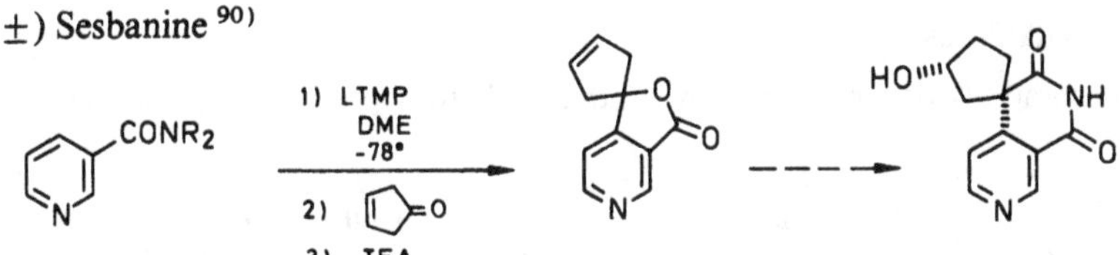

R = i-Pr
= Et

(±) Sesbanine

8. Indolizidine and Quinolizidine Alkaloids

A novel synthesis of antofine *22*, an indolizidine alkaloid, and cryptopleurine *23*, a quinolizidine alkaloid, involves lithiation of trimethoxy-phenanthrene-carbox mide. The key step in both these cases is the generation of substituted phthalides [91,92].

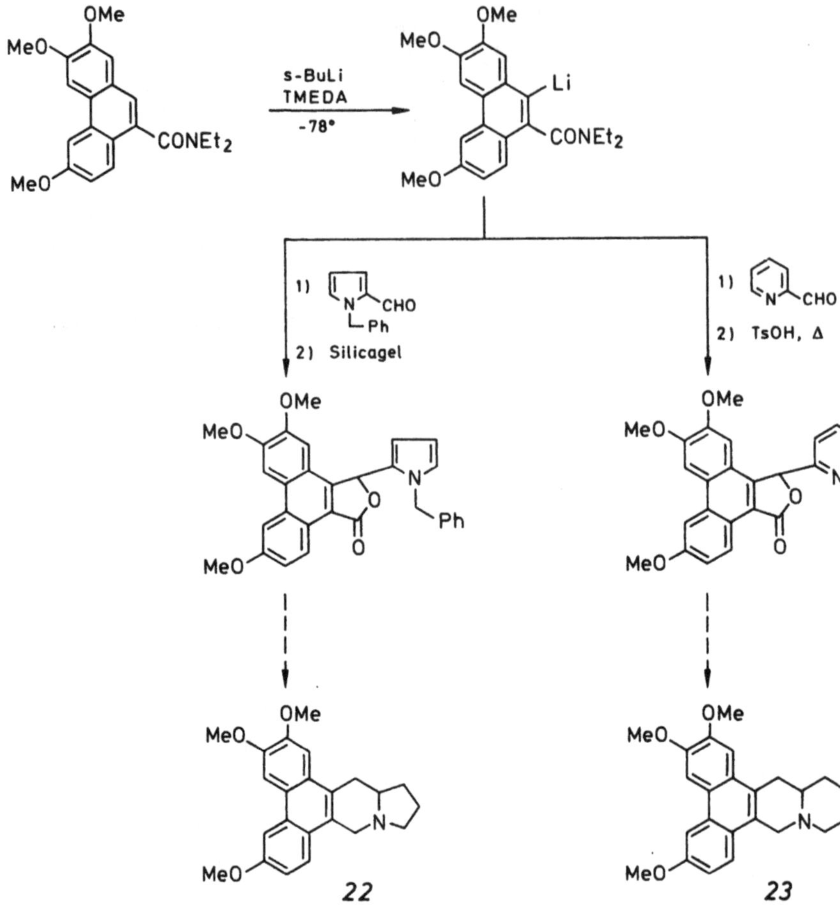

9. Berbine Alkaloids

Several berbine alkaloids have been synthesised from phthalides as shown below:

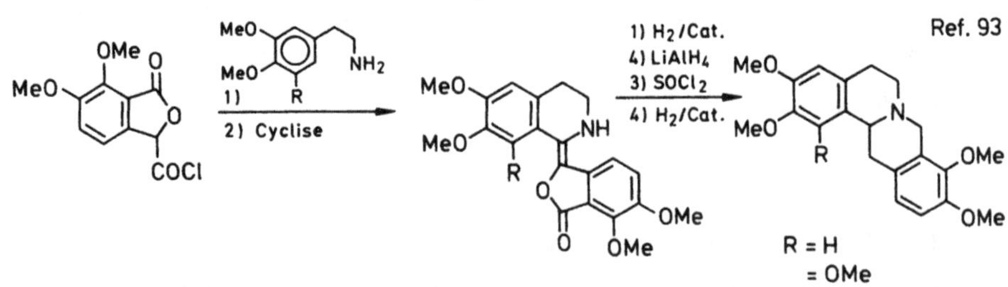

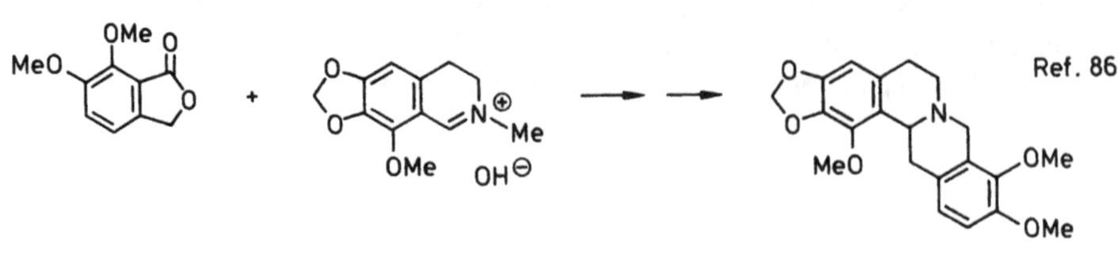

Ref. 94

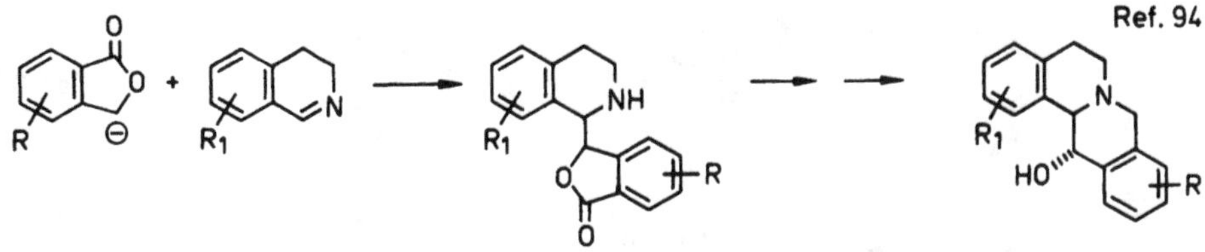

10. Macrolide Antibiotics

Milbemycin B_3 (24), a novel macrolide antibiotic has been synthesised as shown in the following scheme [95]. The route involves both, lithiation and Wittig reactions. It may be noted that the lithiation reaction occurs at the less hindered site.

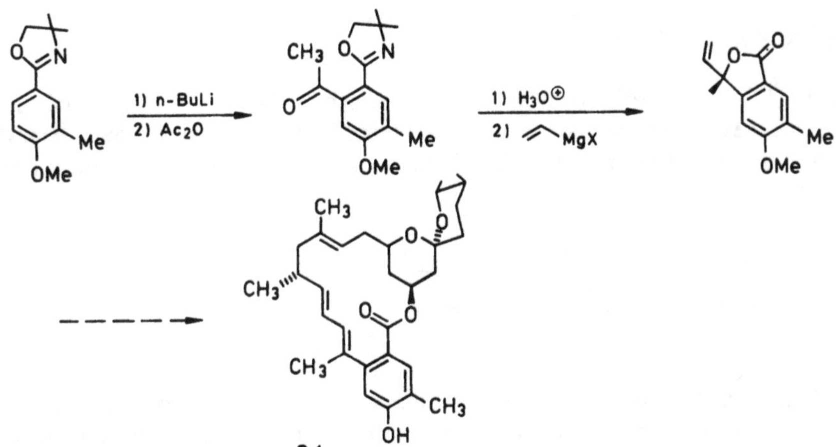

24

3-Hydroxyphthalides

3-Hydroxyphthalide is the cyclic lactol of phthalaldehydic acid. This compound can be readily obtained by lithiation of a benzoic acid derivative (N-methyl amide, N,N-diethylamide or oxazole derivative) and treatment either with N-methylformanilide (MFA) or DMF. When an N-methylamide is the starting compound quite often, the intermediate 25 can be isolated which can be hydrolysed to 3-hydroxyphthalide [29,61,96].

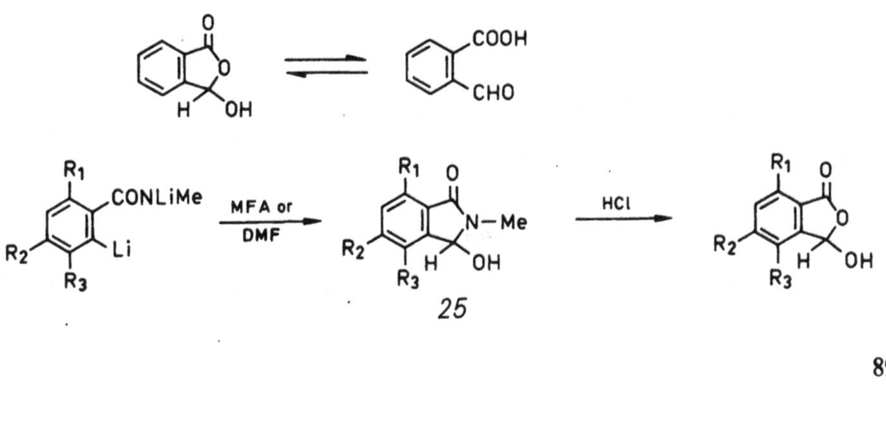

25

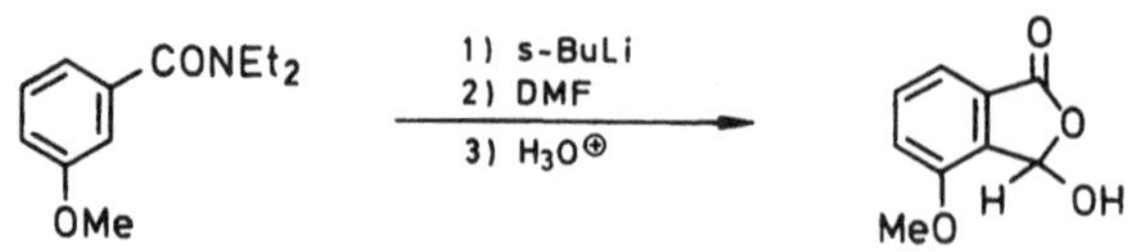

Ref. 97

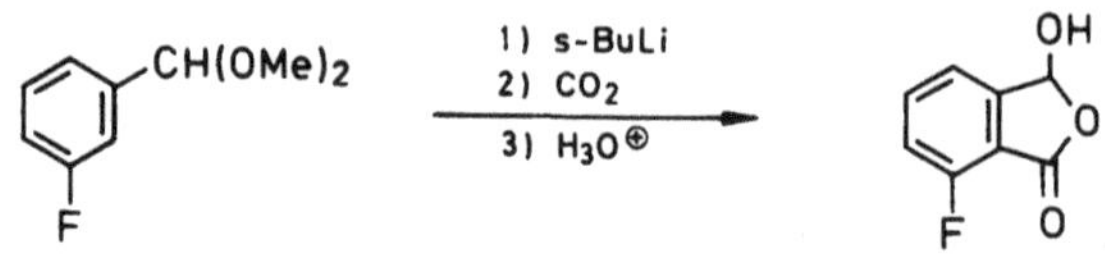

Ref. 97

The 3-hydroxyphthalide is a valuable synthon for isocoumarins (vide infra).

3 Phthalans

Phthalans (1,3-dihydroisobenzofuran) can be obtained by the reduction of phthalides. A direct synthesis of the phthalans can be achieved by lithiation of aromatic benzyl alcohols followed by treatment with an aldehyde or a ketone. This is illustrated below [24].

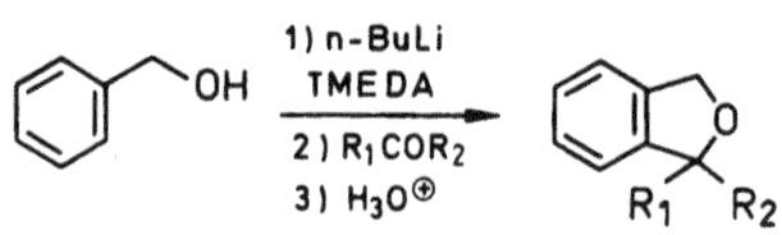

An alternate method is by lithiation of aromatic N,N-dimethylbenzylamine followed by treatment with an aldehyde or ketone and cyclisation. This method is better since the yields of lithiation of N,N-dimethylbenzylamine, unlike the yield in the lithiation of the benzylalcohol, is better. The cyclisation, where the dimethyl group is displaced by oxygen, also proceeds in satisfactory yield. Several fused phthalans synthesised by this method, are presented below.

1)

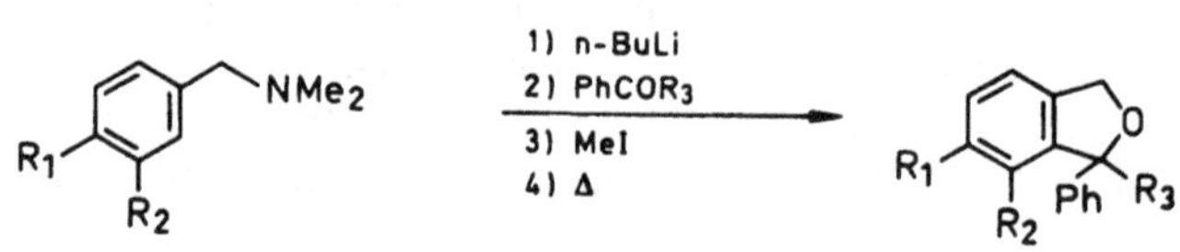

Ref. 98, 99

2)

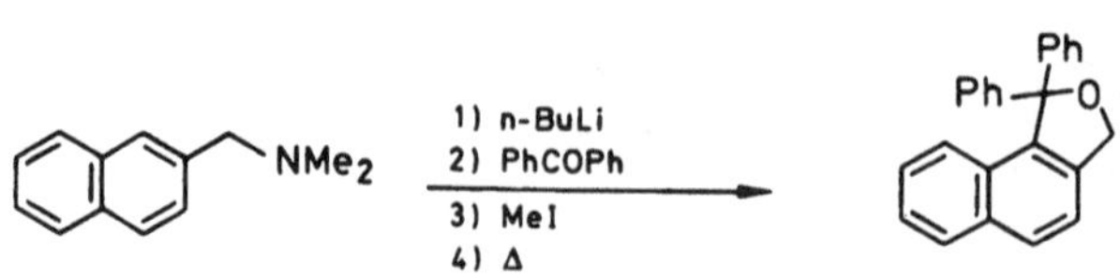

Ref. 100

3)

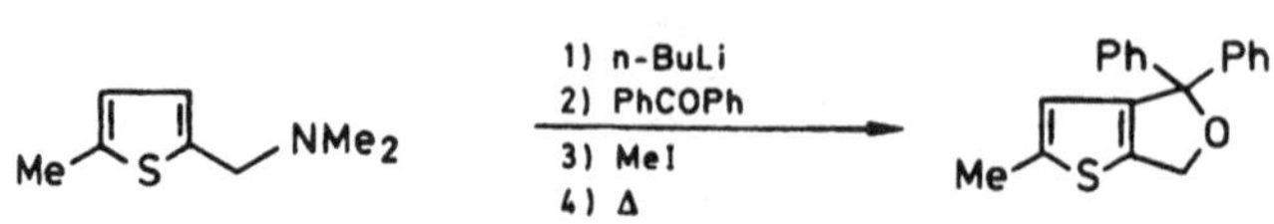

Ref. 101

4) 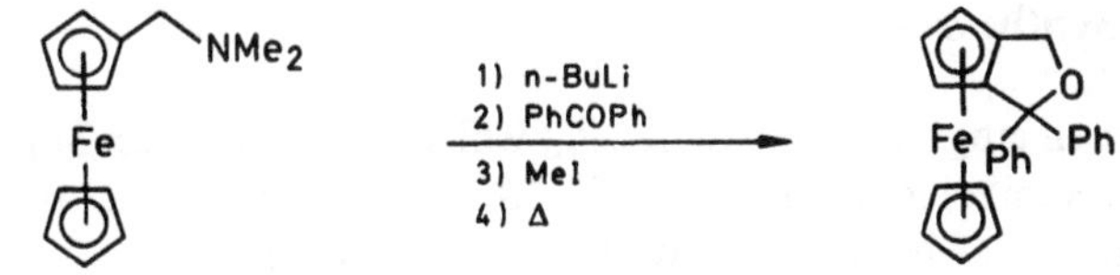 Ref. 102

5) 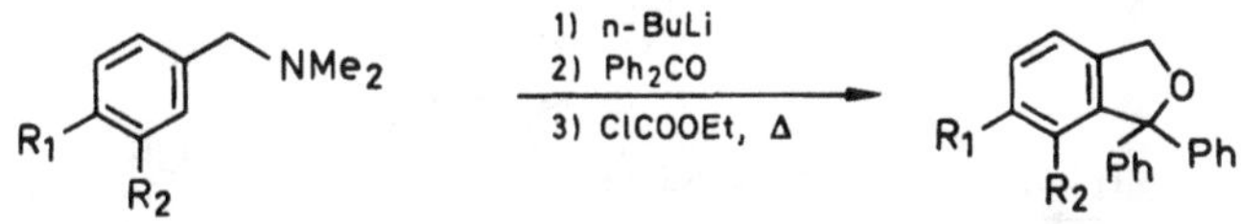 Ref. 103

R_1	R_2
OMe	H
OMe	OMe
$-OCH_2O-$	

1-Hydroxy- and 1-amino-phtalans

The synthesis of 1-hydroxy and 1-aminophthalans would involve lithiation of an aromatic ketone or an imine. Although the ketones and the imines can react with the normal lithiating agents such as BuLi, choice of a hindered base (which would not combine with a keto group or imino group or combine only reversibly) can lead to aromatic lithiation. This is illustrated by the following examples.

1) 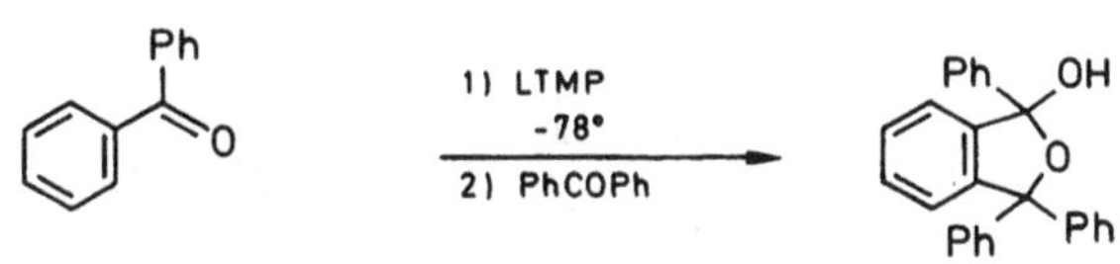 Ref. 104

2) 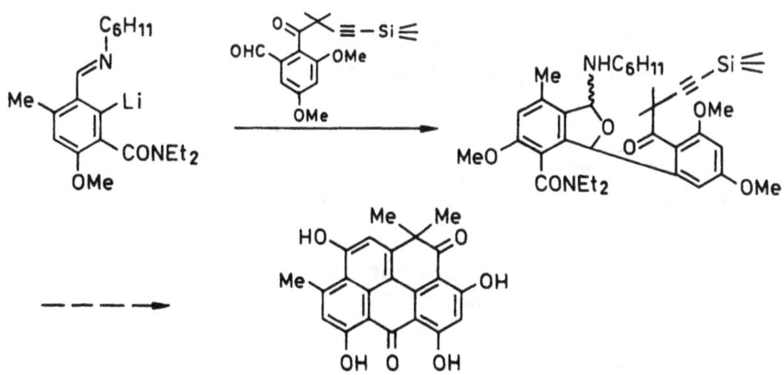 Ref. 105

Monomethyl resistomycine

26

Phathalans as Synthons

Synthesis of monomethyl resistomycine *26*, which requires 1-aminophthalan, is shown above. The complement inhibitor K-76, *28* has also been synthesised in racemic form, from phthalan *27*. The steps involved are as shown below [106].

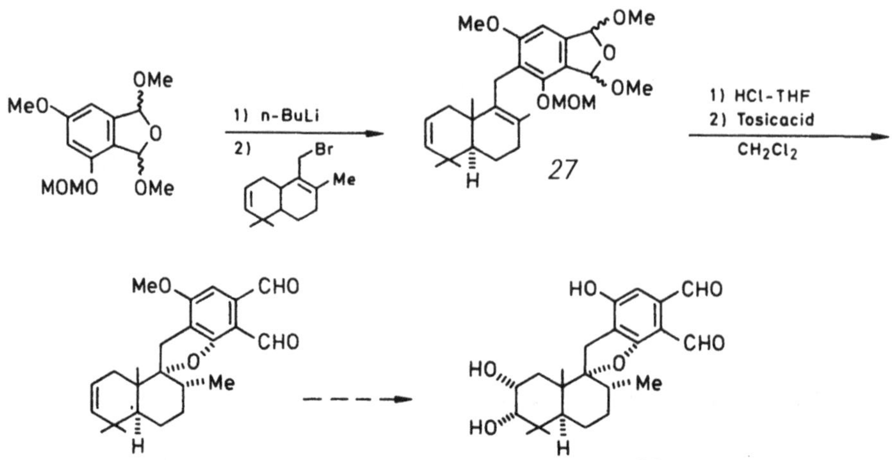

28

4 Phthalic Anhydrides

Phthalic anhydrides can be prepared from 1-hydroxyphthalans by oxidation. The direct synthesis of phthalic anhydride can be achieved by lithiation of a carboxamide and introducing a carboxyl group at *ortho* position using carbon dioxide [57].

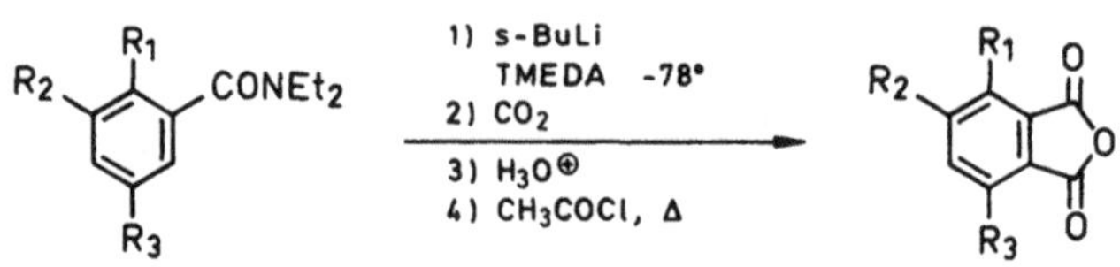

The thiophene analogues have been prepared through a lithiation reaction, using N,N-dimethylaminomethyl group for directing lithiation [101].

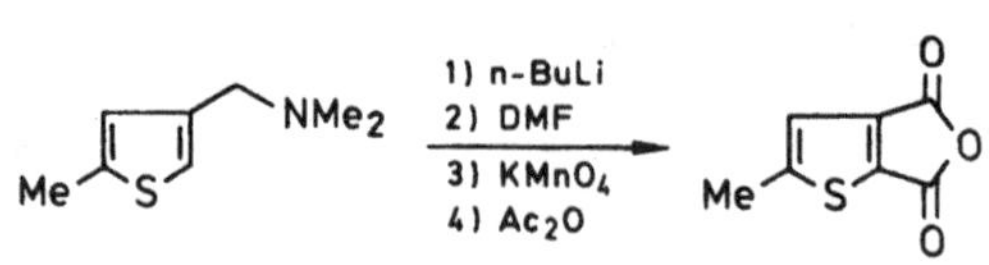

Recently phthalic anhydride has been synthesised by thallation followed by carbonation of the organothalium intermediate [45].

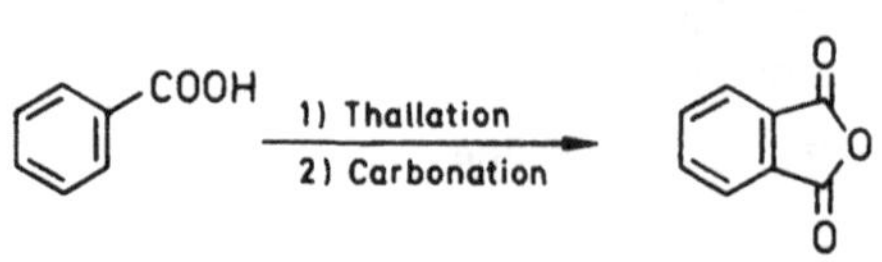

Phathalic Anhydrides as Synthons

Phthalic anhydrides can be easily converted to phthalimides, simple phthalides, alkylidene and benzylidine phthalides, and anthraquinones. A few interesting examples are depicted below.

1)

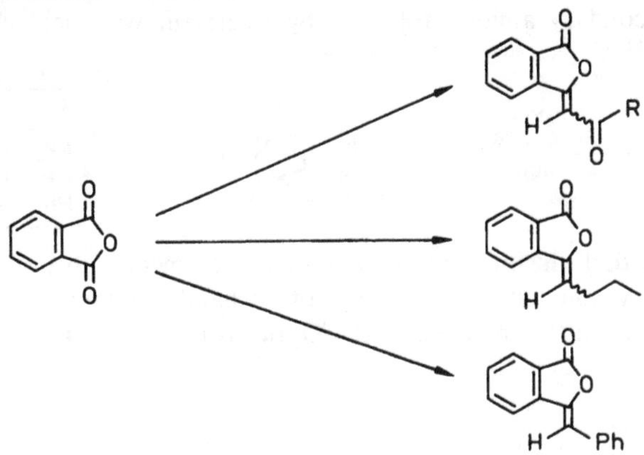

Ref. 107

Ref. 108, 109

Ref. 110

2)

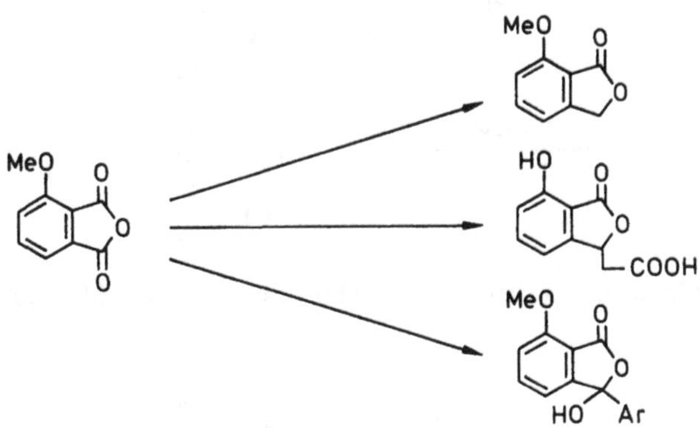

Ref. 111

Ref. 62

Ref. 112

3)

Ref. 113

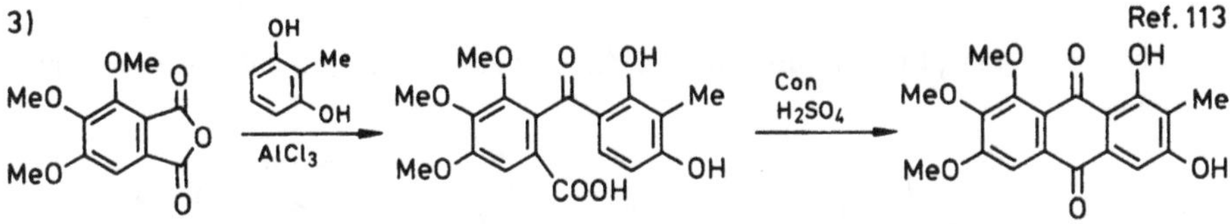

Occurs naturally

5 2,3-Dihydro-isoindole-1-ones

Several of the phthalimidines are antifungal agents [114].

2,3-Dihydroisoindole-1-ones can be obtained by directed metalation of secondary-benzamides or benzyl amines as shown below [53].

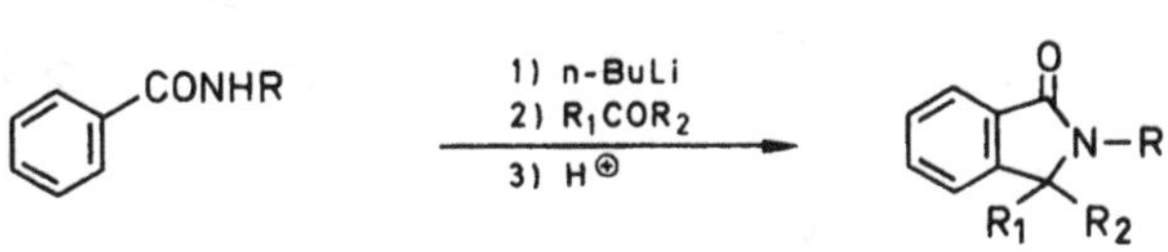

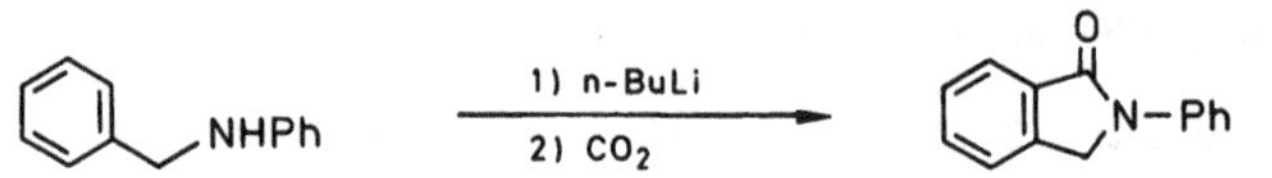

Ref. 116

3-Hydroxy and 3-amino, 3-aryl(alkyl)-2,3-dihydroisoindole-1-ones are obtained by lithiation of secondary amides, followed by treatment with acid chloride [117, 118] or benzonitrile [119].

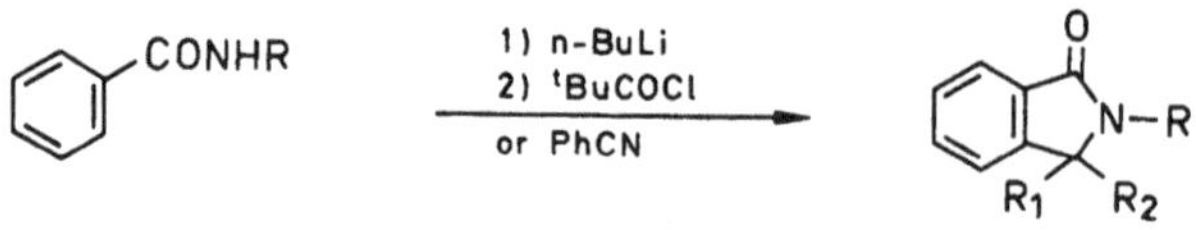

R	R_1	R_2
Me	OH	${}^tBu$
Me	NH_2	Ph
Ph	NH_2	Ph

m-Methoxy-N-methylbenzamide on lithiation and treatment with DMF furnishes 4-methoxy-3-hydroxy compound *29*. On the other hand *meta*-methoxy N-phenyl benzylamine on lithiation and treatment with CO_2 furnishes the 7-methoxy compound *30*.

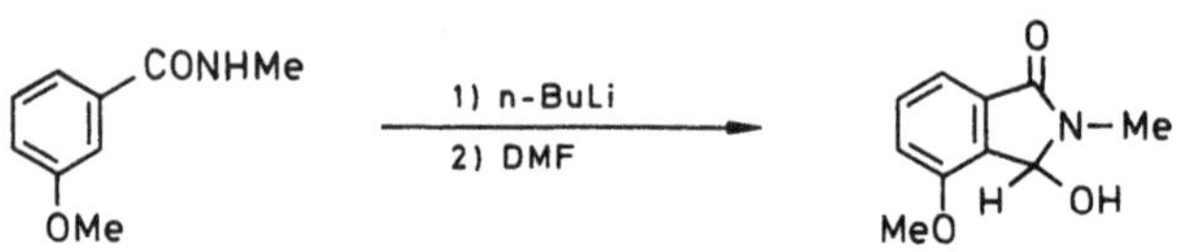

29

Ref. 71, 96

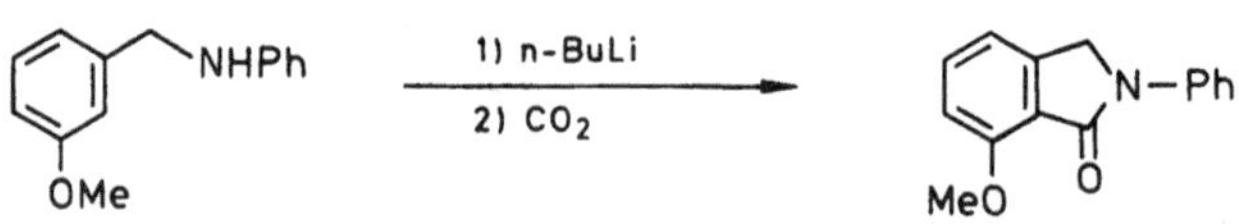

30

Ref. 116

The pyridine analogue is reported from pyridine-2-carboxylic acid amide.

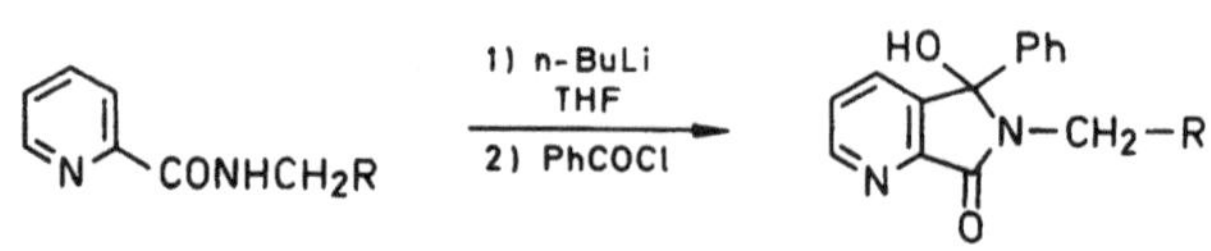

Ref. 120

R = H, Ph

3-Hydroxy-3-arylphthalimidine can be directly converted into substituted anthraquinone *31* [121].

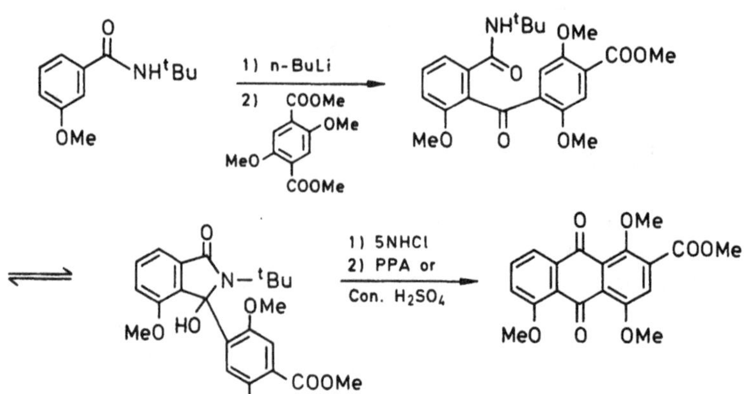

31

1,3-Dihydroisoindoles (Isoindolines)

Isoindolines can be obtained through directed metalation reaction from N-methyl or N-phenyl benzylamines, benzamides or benzanilides.

1) 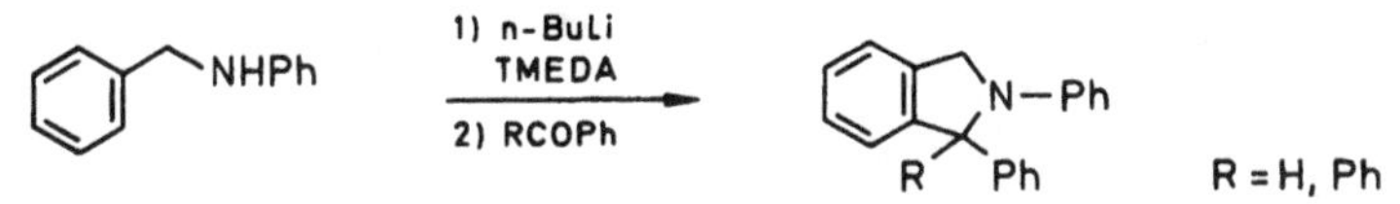 Ref. 116

2) 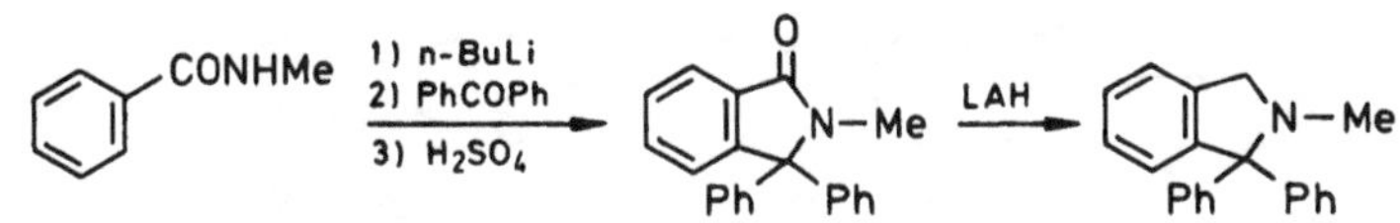 Ref. 122

3) 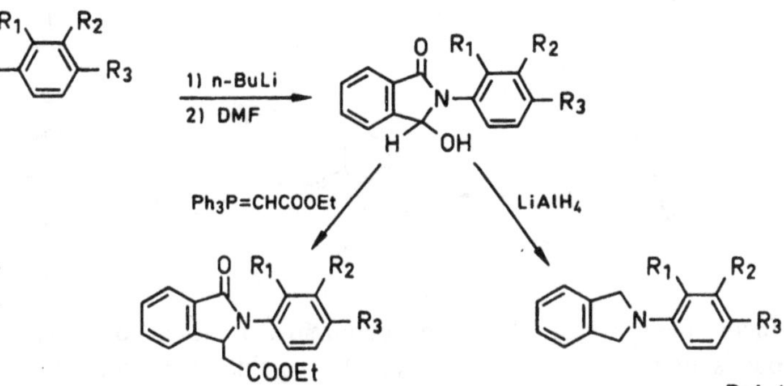

R_1	R_2	R_3
OMe	H	H
H	OMe	H
H	H	OMe

Ref. 61

Ref. 123

6 Isocoumarins and Dihydroisocoumarins

The isocoumarin and the dihydroisocoumarin ring systems occur widely in nature [124]. Their synthesis by lithiation method is according to type C and involves *ortho* metalation of N-methyl arylcarboxamides. The organometallic compounds on treatment with electrophilic reagents such as an epoxide or allylbromide followed by hydrolysis and cyclisation furnish the dihydroisocoumarins. Bromination at benzylic position followed by dehydrobromination then gives the isocoumarins themselves.

The oxazole derivative of an aromatic acid can also be used instead of the amides. Some interesting syntheses of dihydroisocoumarin are given below.

Ref. 32, 125

1) 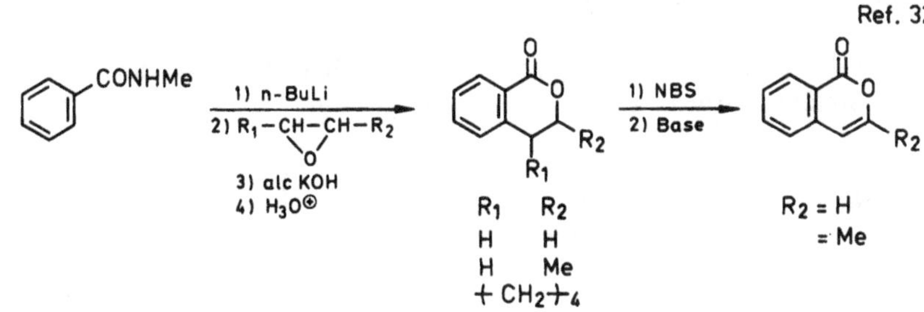

2) 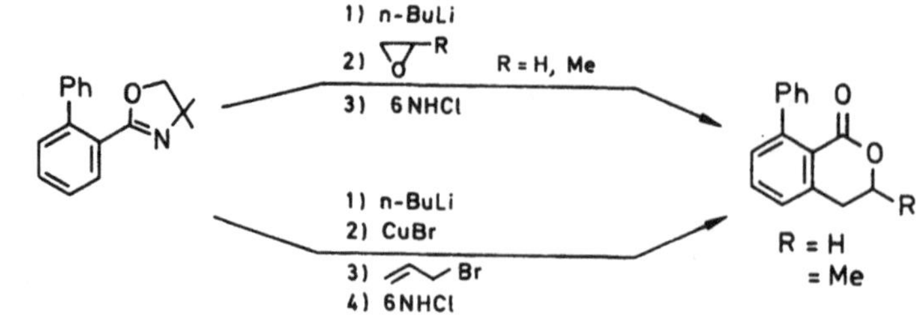 Ref. 126

3) 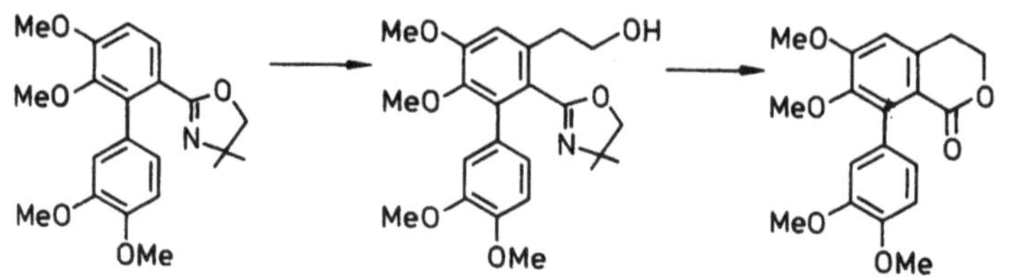 Ref. 127

4) 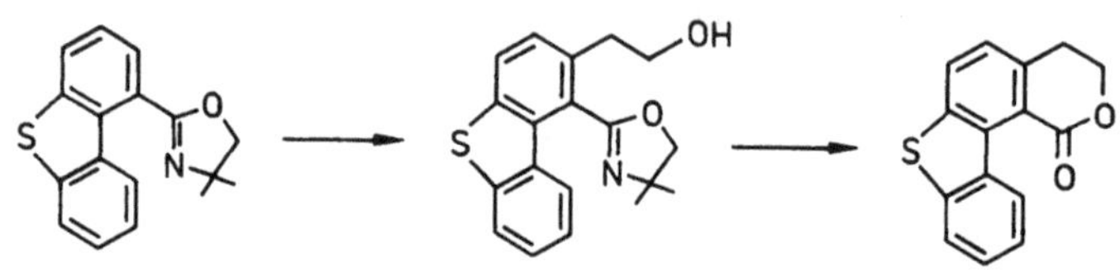 Ref. 128

In another synthesis of dihydroisocoumarin a one carbon fragment such as —CHO group is introduced in the first instance. The aldehyde, which is actually a 1-hydroxyphthalide (mentioned earlier) gives ethylenic compound by Wittig reaction which gets cyclised to give 3-methoxy-3,4-dihydroisocoumarin. The latter on treatment with conc. H_2SO_4 gives isocoumarin [29].

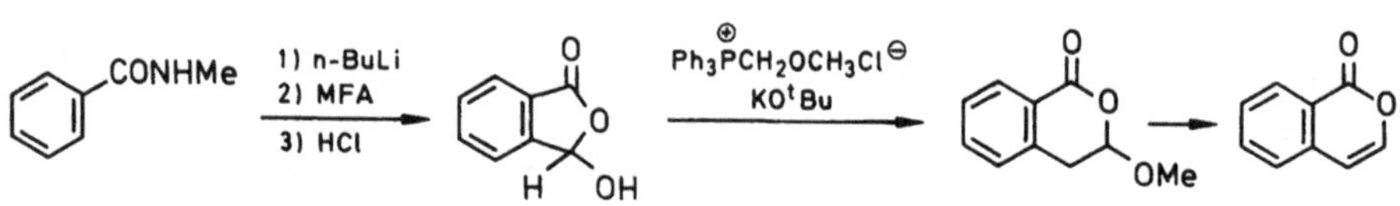

By using this method 5-methoxy and 6-methoxy isocoumarin have been synthesised [96].

Substituted Dihydroisocoumarins

5-Methoxy dihydroisocoumarin, which is difficult to obtain by acid catalysed methods, is readily obtained through aromatic lithiation reaction. This is shown below [51,129].

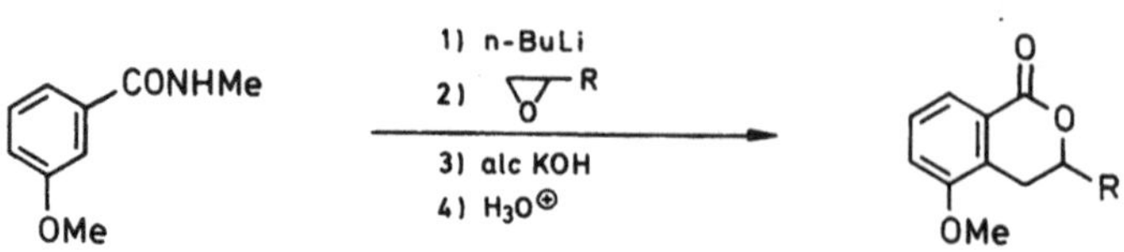

R = H, Me

8-Methoxy isocoumarin can also be obtained by aromatic lithiation reaction. However, an important side reaction, which occurs during lithiation, is a demethylation reaction. The demethylated compound is not lithiated further, presumably because it is precipitated from the reaction medium [51].

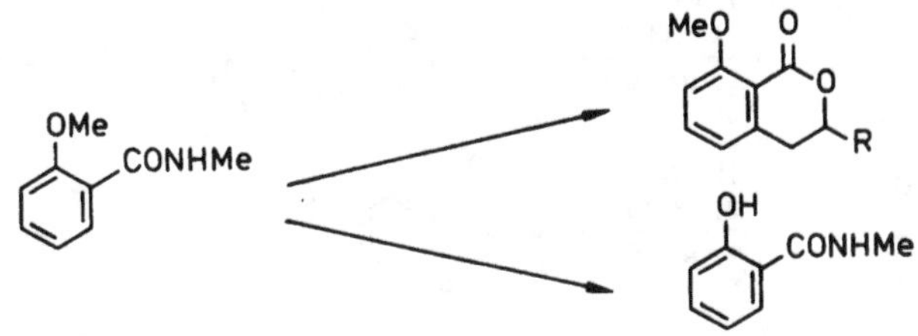

The lithiation of isophthalic acid can be achieved either at 2- or 4- position after conversion of the carboxylic acid groups to oxazole derivatives or N,N'-butylamide derivatives respectively. Conversion to N-methyl amide derivative leads to the precipitation of the N-lithio salt on treatment with BuLi which fails to lithiate further. The N-lithio salt of the N-butylamide, however, remains in solution under the experimental conditions. It may be noted that the two lithiation reactions furnish two different carboxydihydroisocoumarins. The 5-carboxy dihydroisocoumarin can be converted to erythrocentaurine, which is naturally occurring [124].

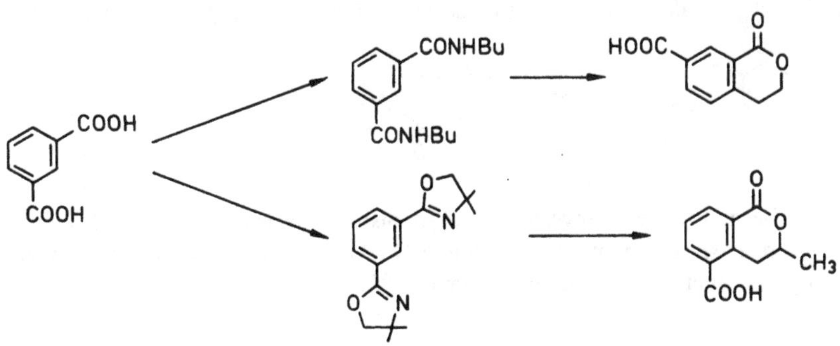

The angular and linear benzodihydroisocoumarin can be synthesised through organolithiation reaction as shown [132, 133].

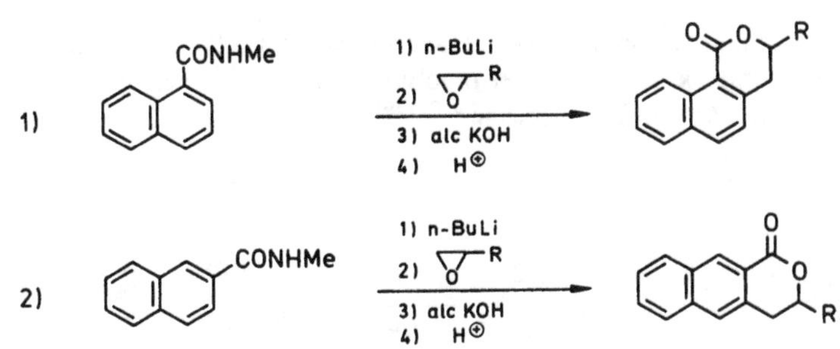

The dihydroisocoumarin can also be formed from N-methyl-o-toluamides or esters of o-toluic acids on treatment with BuLi followed by an aldehyde or ketone. It should be noted that the syntheses do not involve aromatic lithiation reaction.

1)
2)

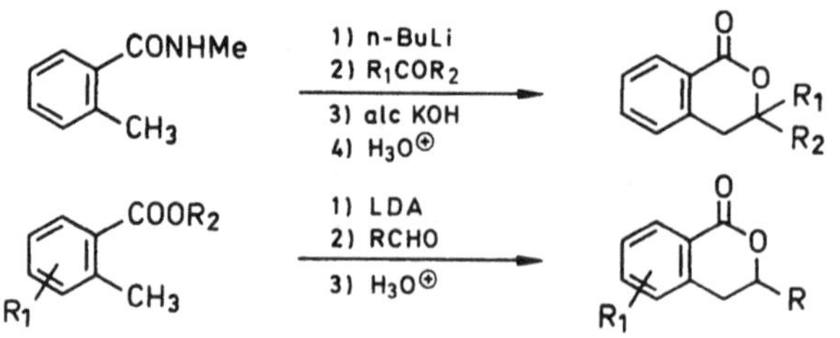

Ref. 134

Ref. 135

However, since o-toluamides can be synthesised by lithiation of N-methylcarboxamide followed by reaction with CH_3I, the above synthesis can be rendered into a one pot [136, 137] reaction, in which the ortho-methyl compound, formed *in situ*, is further lithiated and reacted with an appropriate electrophile.

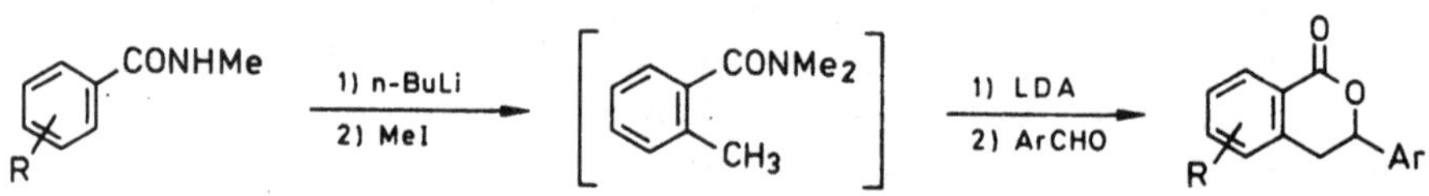

In principle it should be also possible to synthesise a dihydroisocoumarin by lithiation of a phenylethyl alcohol. Such a synthesis however is not reported. Phenylethyl

alcohols have been, however, converted to dihydroisocoumarins by thallation followed by carbonation. It should be noted that when $R=OCH_3$, thallation occurs at *para*-position to give 6-methoxy dihydroisocoumarin [45].

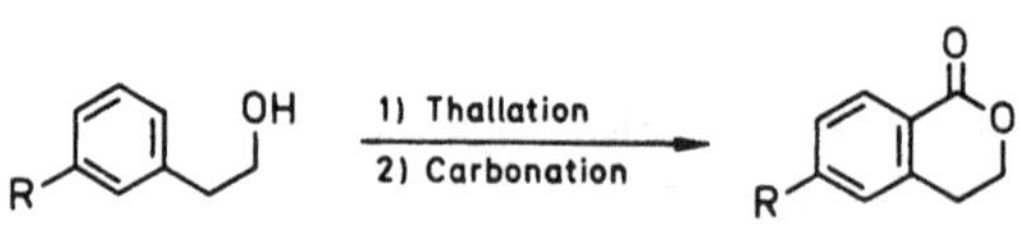

An interesting synthesis of isocoumarins uses the π-allyl nickel halides and π-olefin palladium complex, while another involves an aromatic nucleophilic substitution by the carbanion derived from acetone [138, 139].

Ref. 138

1)

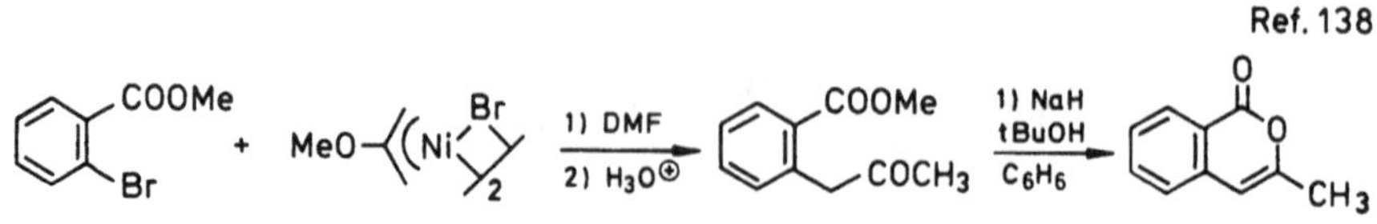

Ref. 138

2) 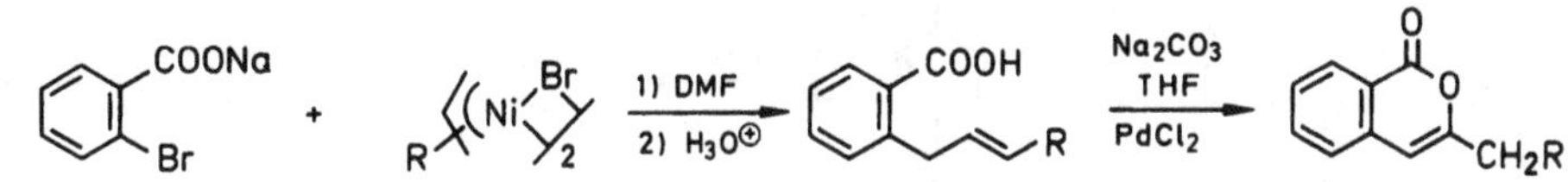

Ref. 139

3) 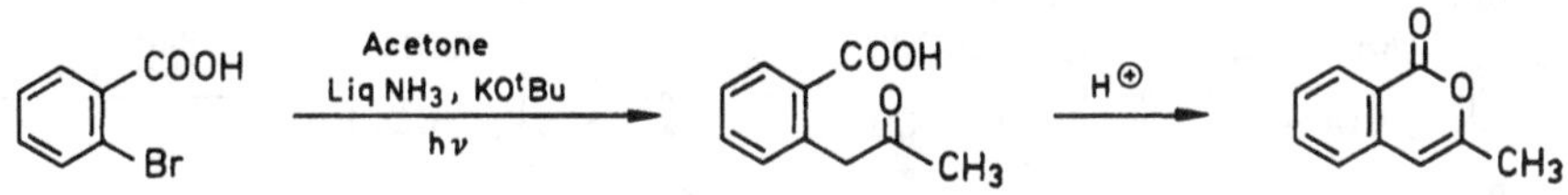

It may be noted that in both these methods the starting compound is *ortho* bromobenzoic acid, not readily available.

Natural Isocoumarins

Mellein 32

This metabolite was first isolated from *Aspergillus mellus* [140]. It has been synthesised using aromatic lithiation reaction as shown below [51, 141].

Ref. 51

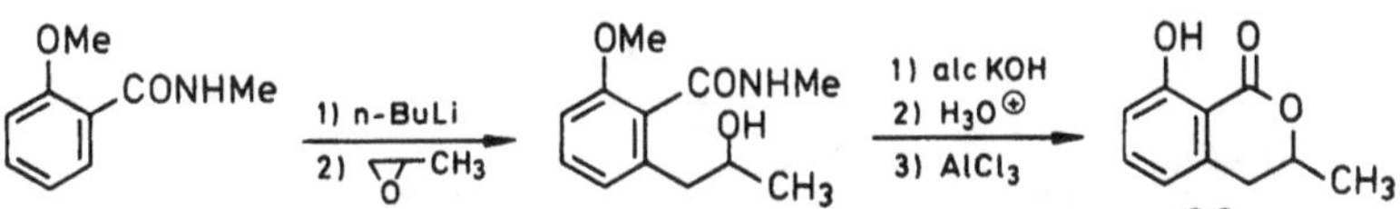

32

Ref. 141

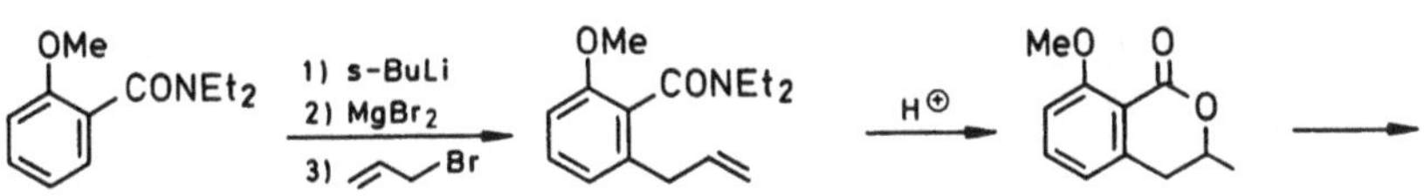

32

The earlier syntheses involved several steps. These are given below. The second method involves the formation of an indanone in the first instance. Bromine atom is used as a blocking group to achieve the desired cyclisation to obtain the starting indanone [142, 143].

1) 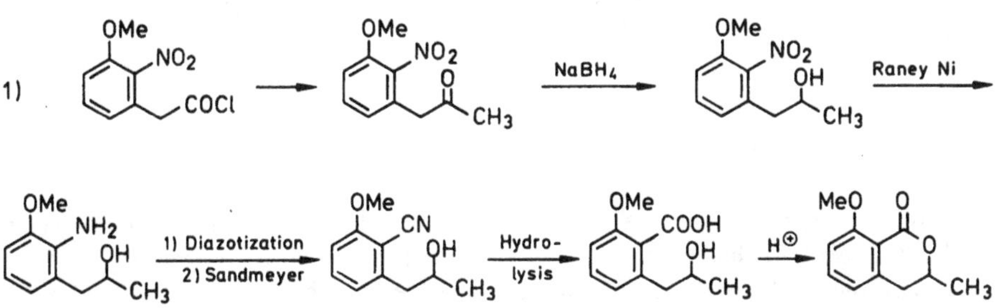

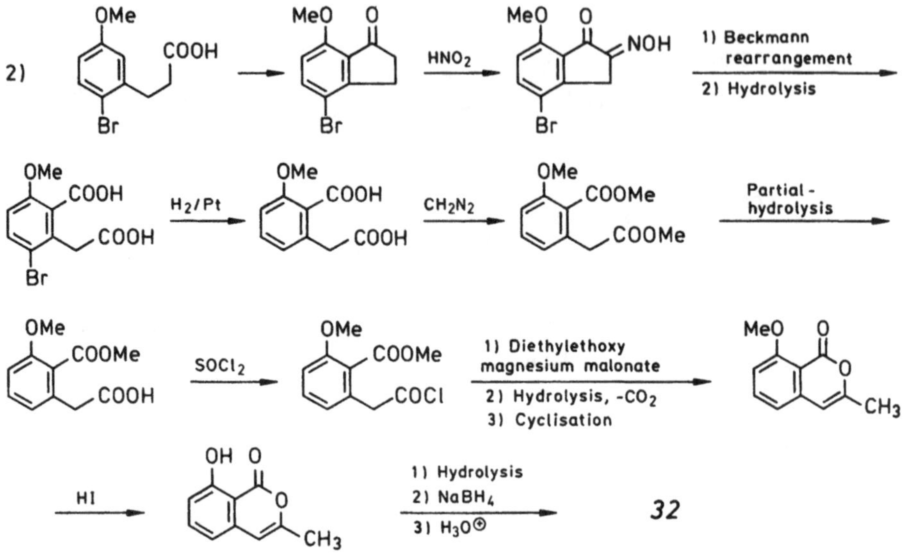

A recent synthesis of mellein involves a Diels-Alder reaction, which gives the expected 1,2,3-trisubstituted benzene derivative [144].

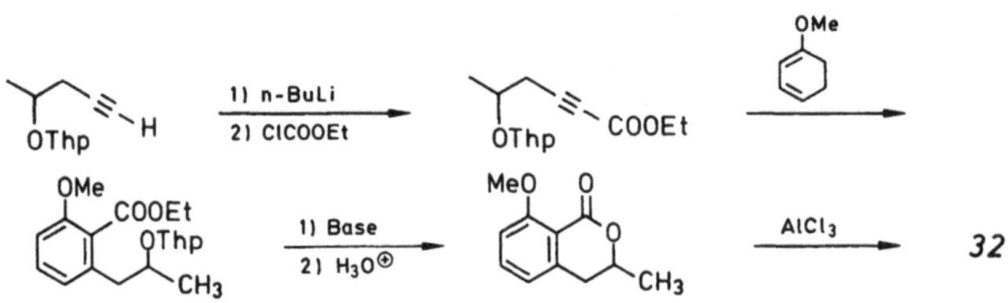

Trifluoroacetic acid catalysed claisen rearrangement also provides another route to mellein *32* as shown below. [145].

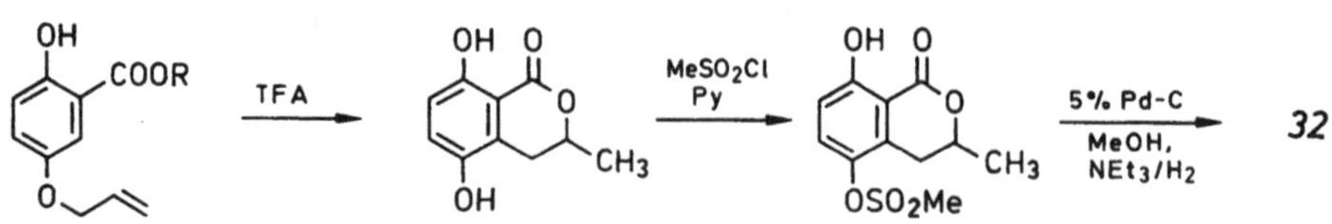

(—)-5-Methylmellein 33

This phototoxic metabolite was isolated first from the fungus *Fusicoccum amygdali* Del [146]. Recently it has been isolated from different *Semecarpus* species [147].

The synthesis of (±) *33*, involving acid catalysed route requires several steps [148].

The recent synthesis through directed metallation requires fewer steps. This is presented below [149].

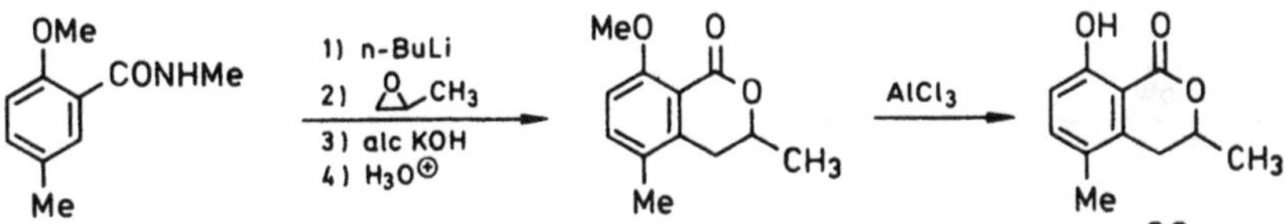

33

(—)-6-Methoxymellein 34

This dihydroisocoumarin was first isolated from mouldy carrots in 1957 [150]. Its synthesis, involving acid catalysed methods, are lengthy [151]. The lithiation route, presented below, needs only two steps [152].

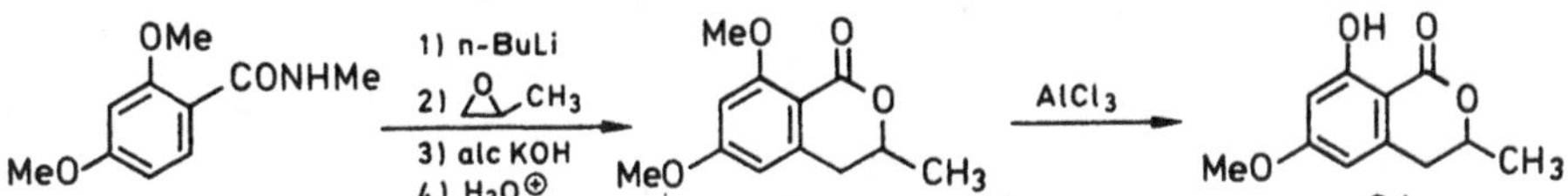

34

Kigelin 35

This dihydroisocoumarin was isolated from *Kigelia pinnata* DC [153]. Several methods are reported for its synthesis.

Aromatic lithiation reaction provides an efficient synthesis as shown below [154, 141].

Ref. 154

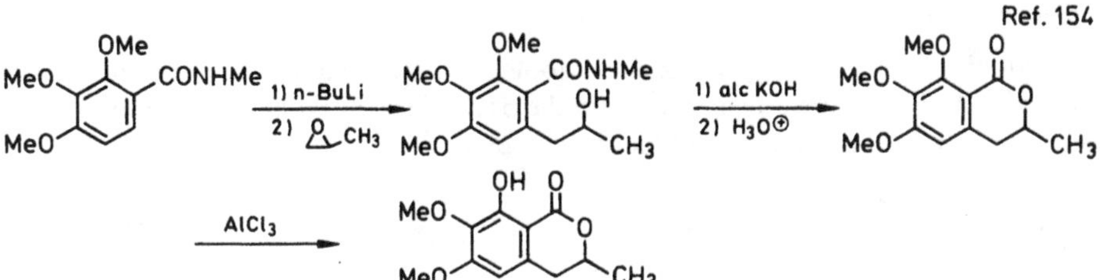

35

Ref. 141

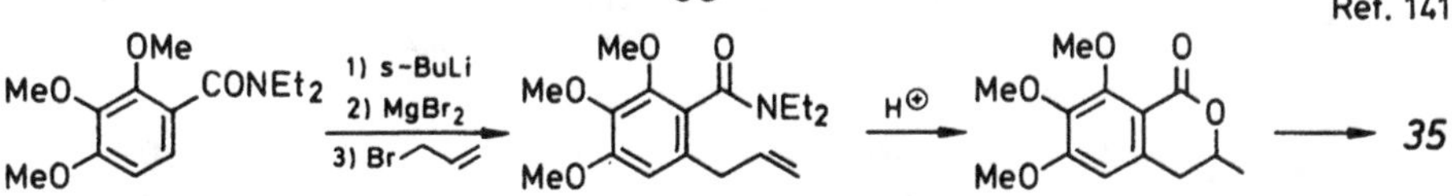

The other syntheses of kigelin are depicted below. The starting compound for the third synthesis is the naturally occurring elemicine.

1)

Ref. 155

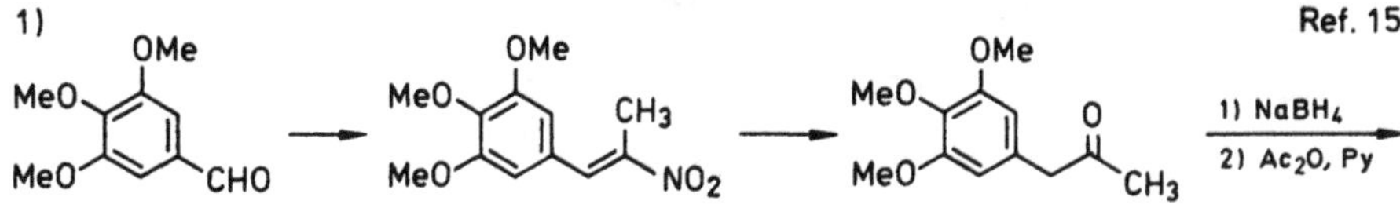

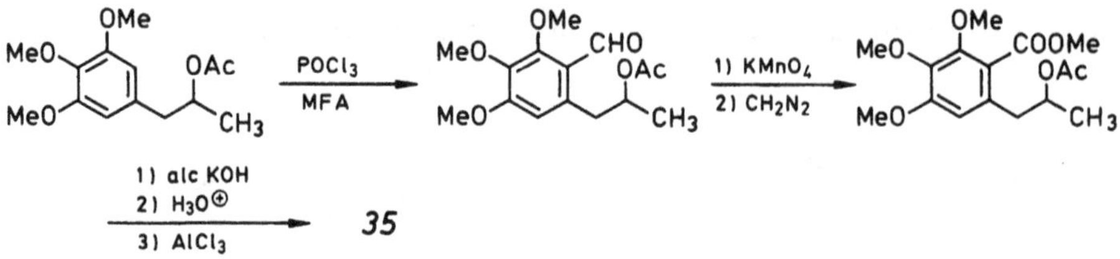

2) Ref. 156

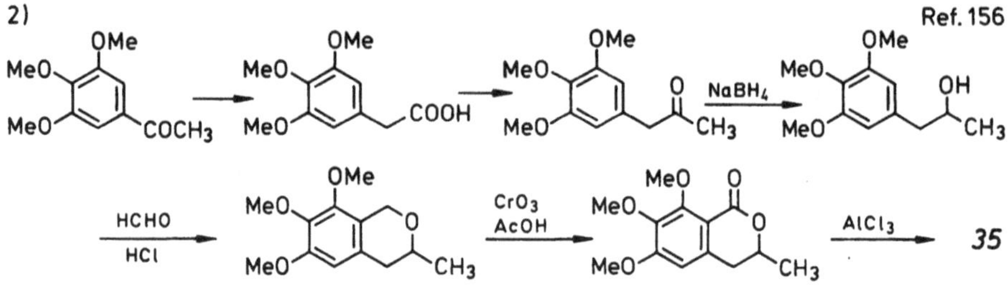

3) Ref. 157

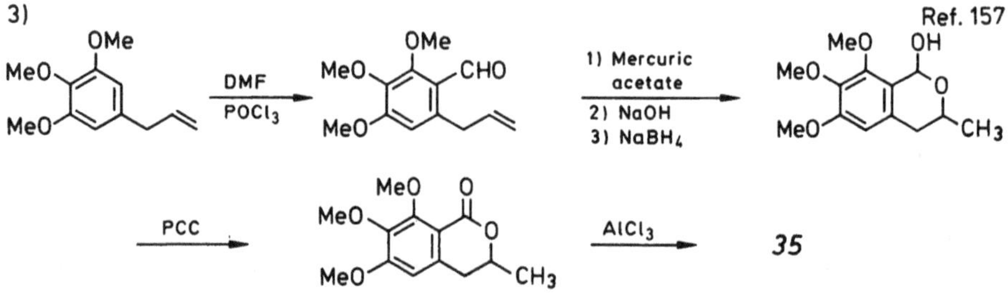

Ochratoxin A 37

Ochratoxin A *37* is a highly toxic metabolite isolated from *Aspergillus ochraceus* Wilh [158]. The structure assigned to it on the basis of spectral data, was confirmed by an anambiguous synthesis, which is shown below [159].

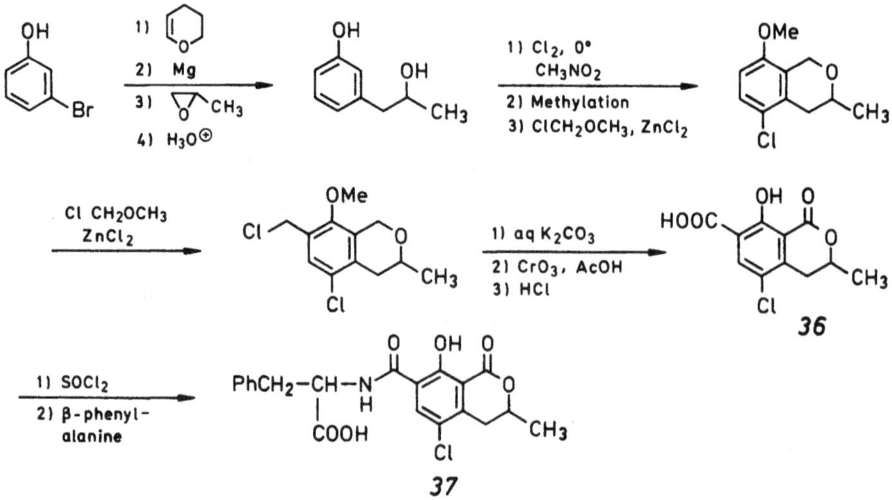

An alternate synthesis from chloroindanone is also known. The key intermediate here is 5-chloro-8-hydroxy-3-methyl-3,4-dihydroisocoumarin *38* [160]. The key intermediate *38* has been conveniently prepared via aromatic lithiation reaction as follows [161]:

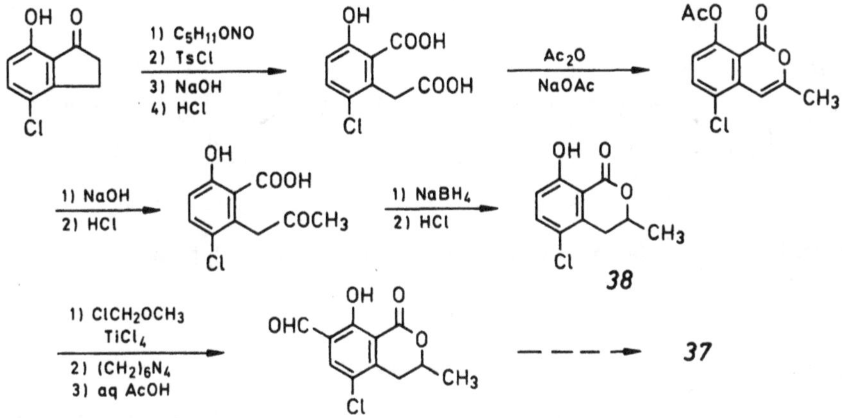

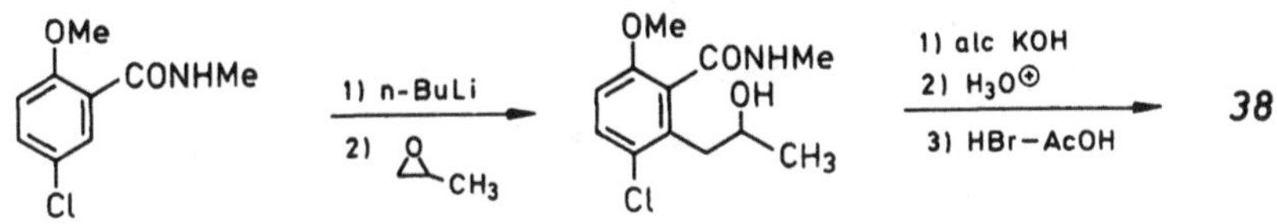

The intermediate *36* involved in the earlier synthesis can also be prepared by lithiation at the aromatic methyl group as indicated below [135].

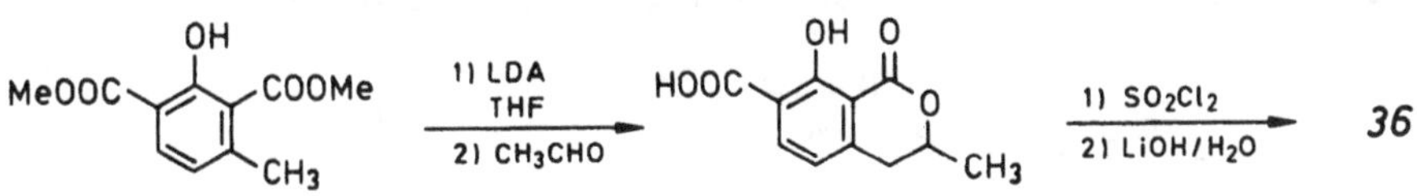

Phyllodulcin 39 and Hydranginol 40

The dihydroisocoumarins *phyllodulcin 39* and *hydranginol 40* have been isolated from *Hydrangea macrophylla* leaves [162]. They have been synthesised as shown below [136].

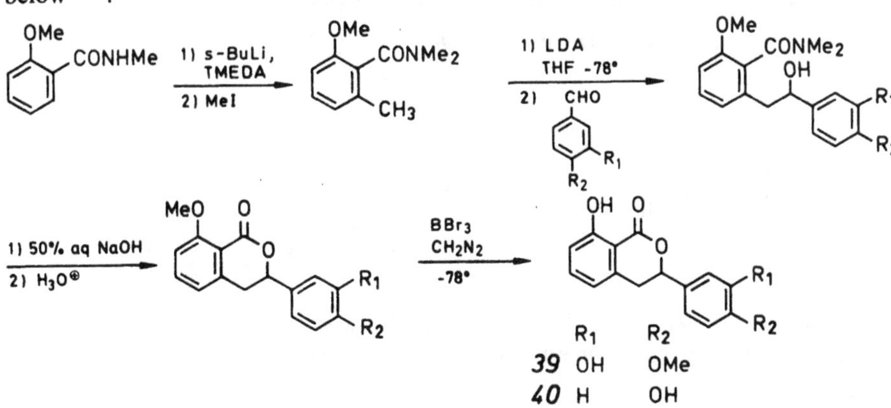

Homalicin 41 and (—)-Dihydrohemalicine 42

These glycosides have been isolated from the roots of *Homalium zeylanicum* Benth [163].

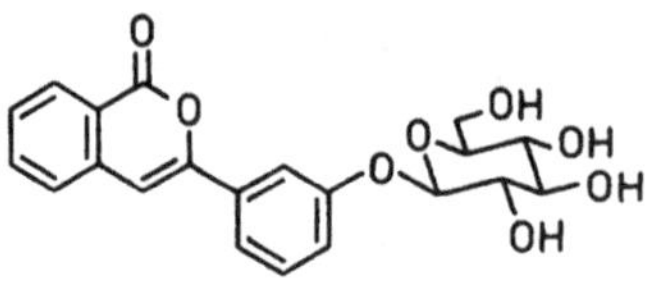

41

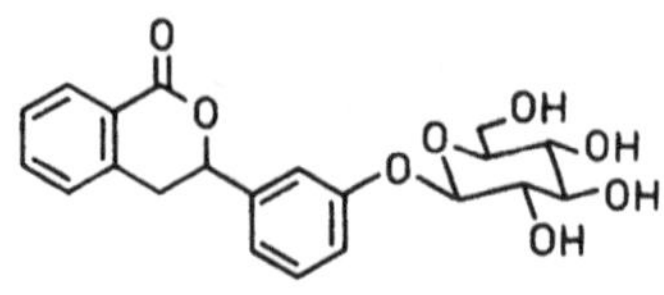

42

The synthesis of the aglycone of (±)-dihydrohomalicine by lithiation method is presented below [115].

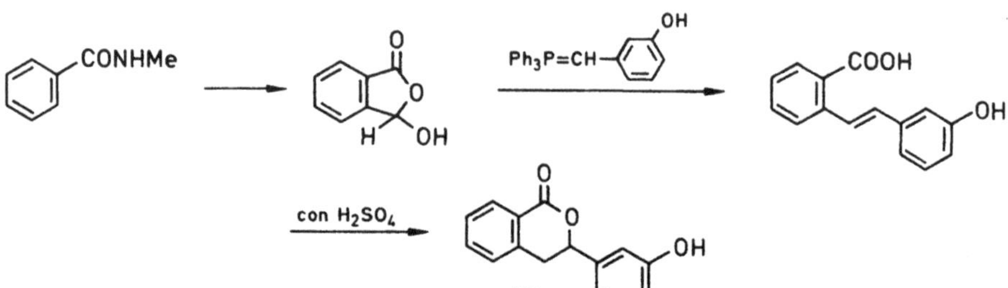

43

The methyl ether of the aglycone of (±)-dihydrohomalicine *43*, has also been synthesised by lithiation at an aromatic methyl group, which is *ortho* to the lithiation-directing N-methyl carboxamide function. The synthesis is shown below [163].

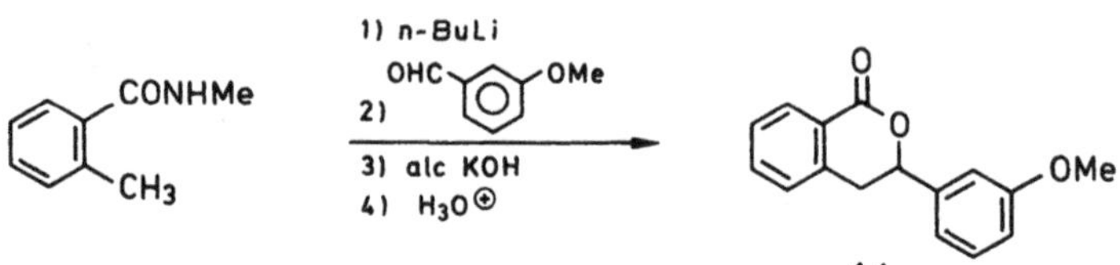

44

Artemidine 45, Artemidinal 46 and the Related Compounds

These are isolated from *Artemisia dracunculus* [164, 165].

Artemidinal *46* can be readily synthesised from homophthalic anhydride (vide infra). Its syntheses from 3-methyl dihydroisocoumarin and 3-methyl isocoumarin are given below [125, 166].

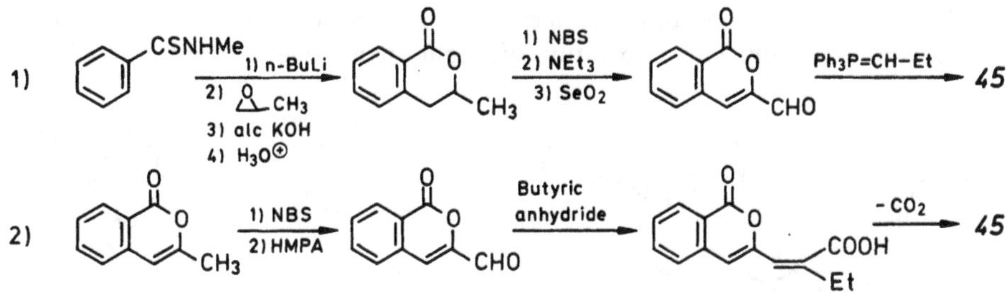

Another method makes use of the potassium salt of phthalaldehydic acid [167].

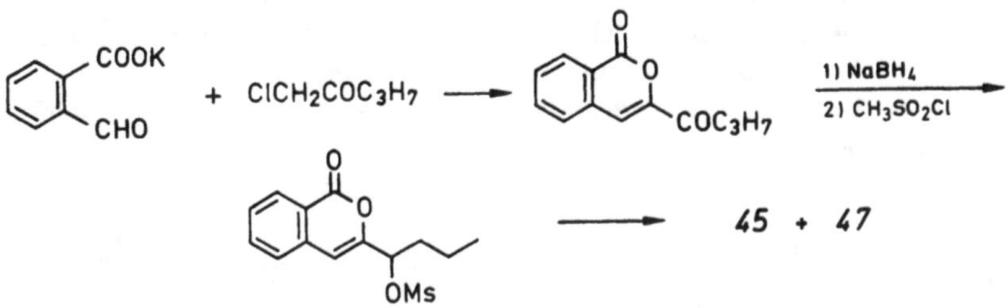

Copper acetylides have also been used, for the synthesis of 3-alkyl isocoumarins [168].

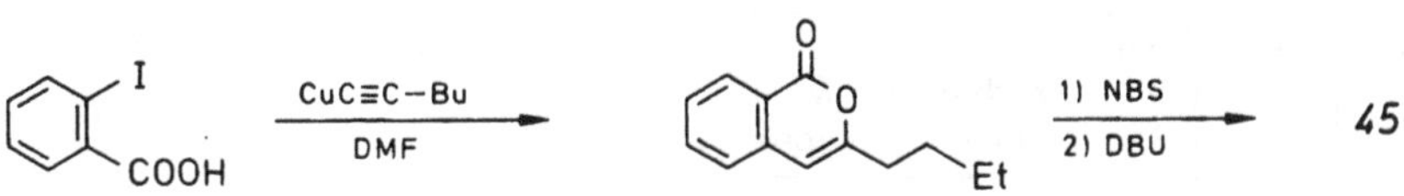

7 Isochroman-3-ones

A few isochroman-3-ones occurs in nature [169]. The importance of isochroman-3-ones however is because they are valuable intermediates for the synthesis of isoquinolines. A most interesting use of isochroman-3-ones is that they provide benzocyclobutenes which are valuable synthons for berbines, spiro-benzylisoquinolines, 3-aryl isoquinoline, benzophenanthridines, benzocarbazoles, yohimbin, tetracycline, steroids etc.

The lithiation reaction for the synthesis of isochroman-3-ones is of interest because the usual methods i.e. hydroxymethylation of phenylacetic acid, do not furnish certain

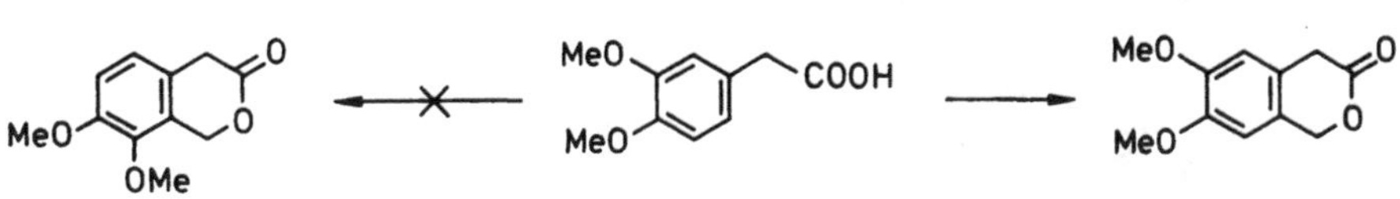

methoxyl substituted isochroman-3-ones. Such compounds can be readily prepared by lithiation reaction. The preparation of isochroman-3-one is shown below [170].

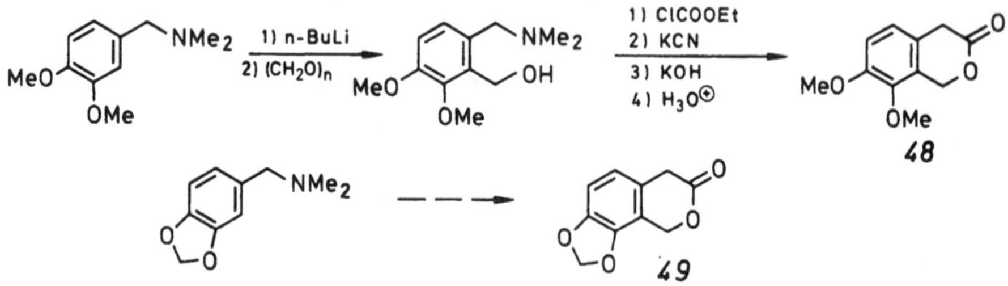

48

49

The lithiation occurs specifically at the position which is *ortho* to both the aminomethyl and the $-OCH_3$ groups [18]. The hitherto not reported 7,8-methylenedioxy isochroman-3-one 49 has also been synthesised by this method [170].

Several 1-aryl-7,8-dialkoxyisochroman-3-ones-have been synthesised by lithiation method as shown below [171]:

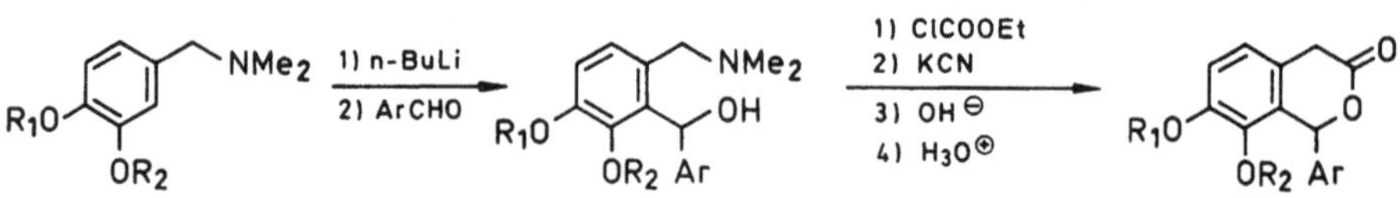

R_1	R_2	Ar
Me	Me	phenyl or 3,4-dimethoxyphenyl or 3,4-methylenedioxyphenyl
$-CH_2-$		

Recently 7,8-dimethoxy isochroman-3-one *48* has been synthesised by an interesting route which involves hydroxymethylation of *meta*-hydroxy phenyl acetic acid in presence of phenylboronic acid. The presence of the latter in the reaction medium specifically brings about hydroxymethylation *ortho* to the phenolic group [172].

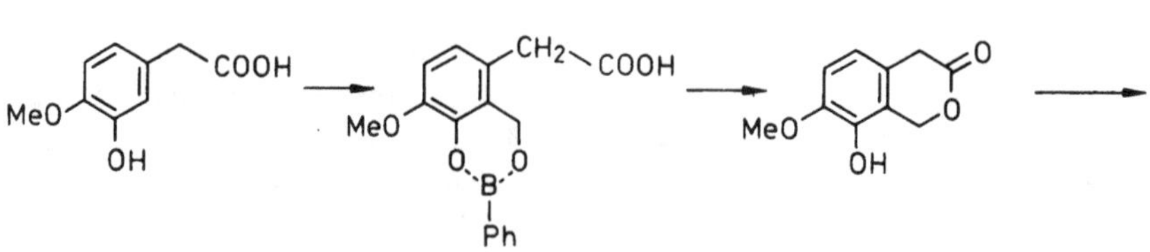

48

Another interesting synthesis of the isochromanone ring system is through metallation of an aromatic methyl group which is *ortho* to a hydroxymethyl group in the aromatic ring [173].

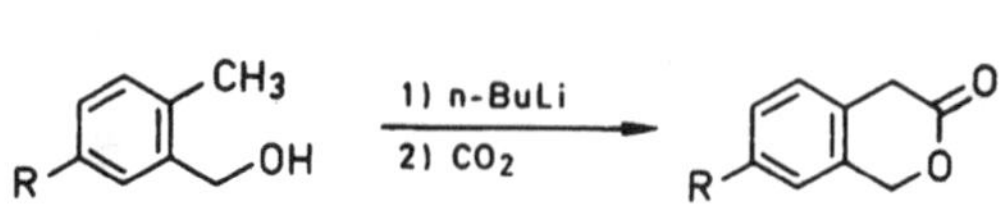

Isochromanones as Synthons

A large number heterocyclic compounds have been synthesised from isochromanones.

Isoquinolines

Isochroman-3-ones on reaction with HBr/EtOH furnish the bromoesters which on treatment with amines provide isoquinolones. The latter on reduction with $LiAlH_4$ give 1,2,3,4-tetrahydro isoquinolines [174–176].

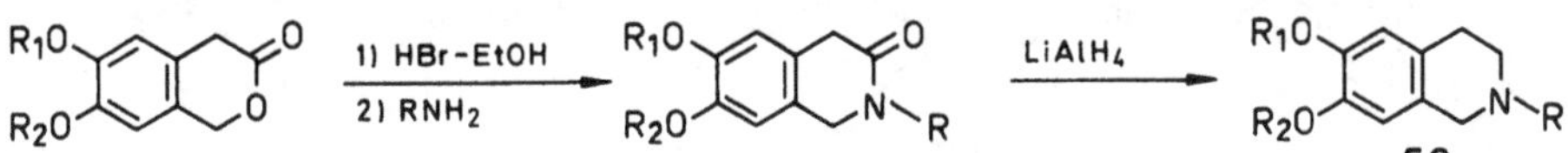

50

50	R_1	R_2	R_3	
a	Me	Me	Me	O-methylcorypalline
b	Me	H	Me	Corypalline
c	Me	H	$-CH_2-C_6H_4-OMe$	Sendavarine

Berbine Alkaloids

Isochromanones on condensation with β-phenylethyl amines, followed by Bischler-Napieralski cyclisation and $NaBH_4$ reduction furnish berbine alkaloids in good to excellent yields. The 7,8-dimethoxy and methylenedioxy isochromanones prepared by lithiation methods provide the 9,10-oxygenated berbine alkaloids, such as tetrahydropalmatine *51a* [170], canadine *51b* [170], stylopine *51c* [170], sinactine *51d* [170], corypalmine *51h* [177], tetrahydrogroenlandicine *51i* and cheilanthifoline *51j*. The benzyloxy compounds *51e*, *51f* and *51g* on debenzylation furnish the phenolic berbine alkaloids (±) corypalmine [177], (±) tetrahydrogroenlandicine [178] and (±) cheilanthifoline [178], respectively.

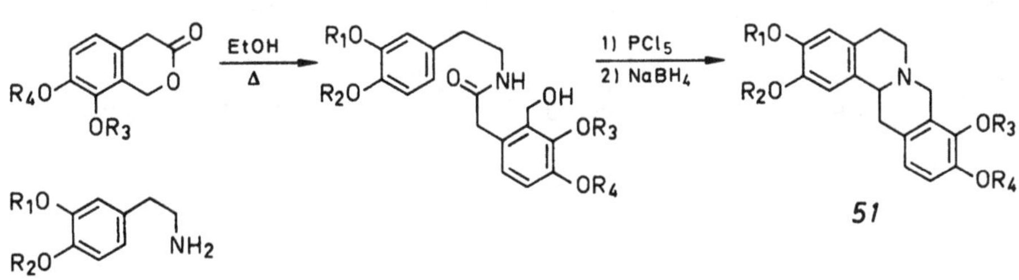

51

51	R_1	R_2	R_3	R_4	
a	Me	Me	Me	Me	tetrahydropalmatine
b	$-CH_2-$		Me	Me	canadine
c	$-CH_2-$		$-CH_2-$		stylopine
d	Me	Me	$-CH_2-$		sinactine
e	Bz	Me	Me	Me	—

f	Bz	Me	$-CH_2-$		—
g	Me	Bz	$-CH_2-$		—
h	OH	Me	Me	Me	corypalmine
i	OH	Me	$-CH_2-$		tetrahydrogroenlandicine
j	Me	OH	$-CH_2-$		cheilanthifoline

The recently [179] isolated (±) bharatamine has also been synthesised by making use of this approach.

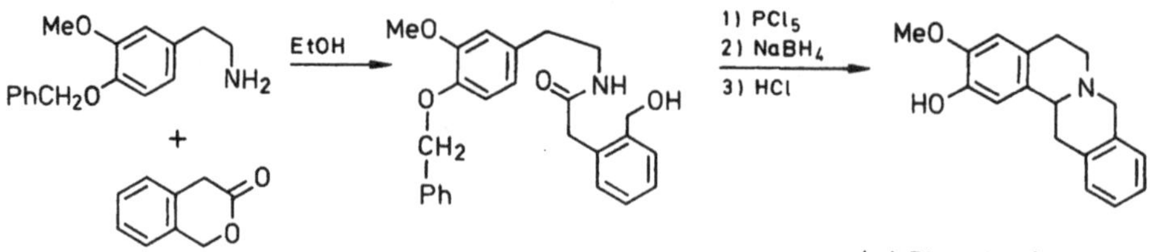

(±) Bharatamine

1-Phenyl-7,8-dimethoxy-isochroman-3-one has been converted into 8-phenyl-tetrahydropalmatines as shown below [180].

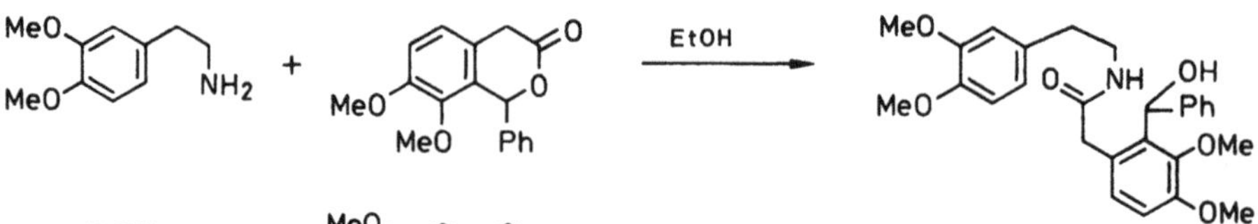

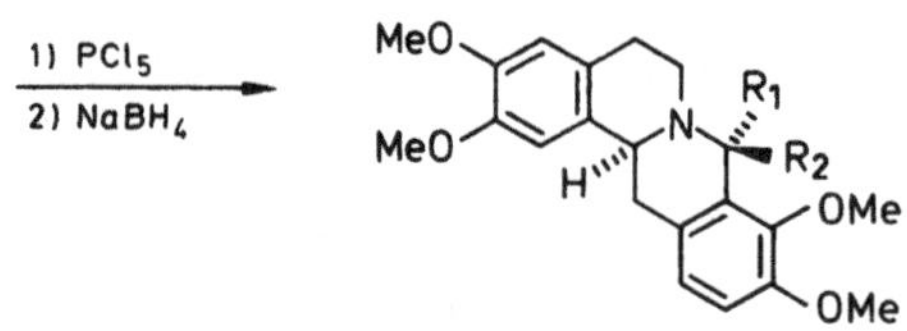

R_1	R_2
Ph	H
H	Ph

Yohimbane Skeleton

Condensation of isochromanones with tryptamine followed by Bischler-Napieralski cyclisation and $NaBH_4$ reduction provide the Yohimbane skeleton [181].

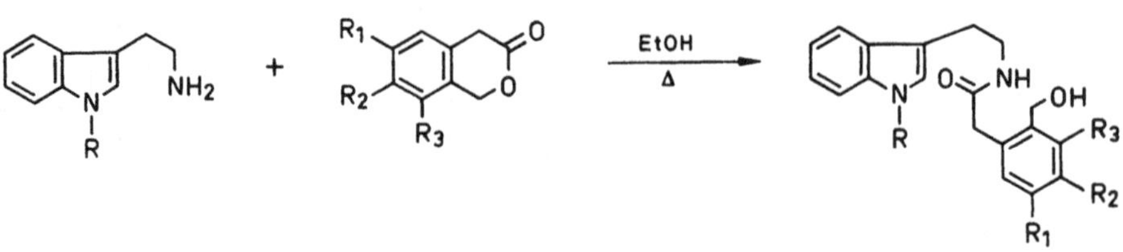

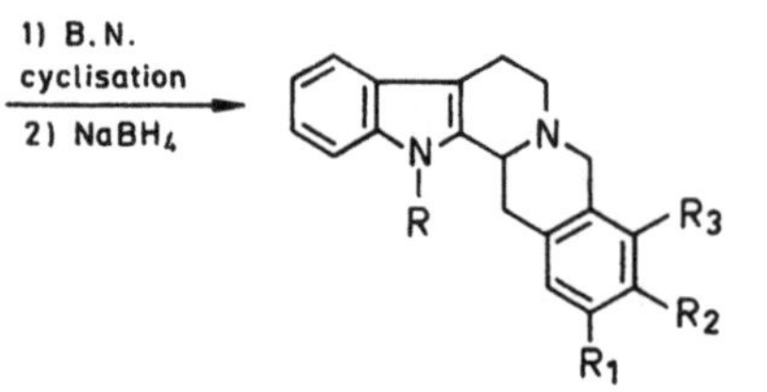

R	R_1	R_2	R_3
H	H	H	H
Me	H	H	H
H	OMe	OMe	H
Me	OMe	OMe	H
H	H	OMe	H
H	H	OMe	OMe

Similarly isochromanones have been used for the synthesis of selenium analogues [182].

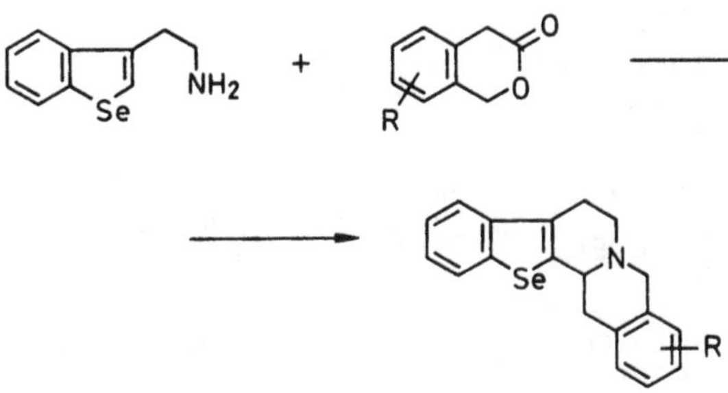

Benzocyclobutene

Pyrolysis of isochromanones gives benzocyclobutene in good yield [183]. The latter have been converted to various heterocyclic compounds and steroids [183].

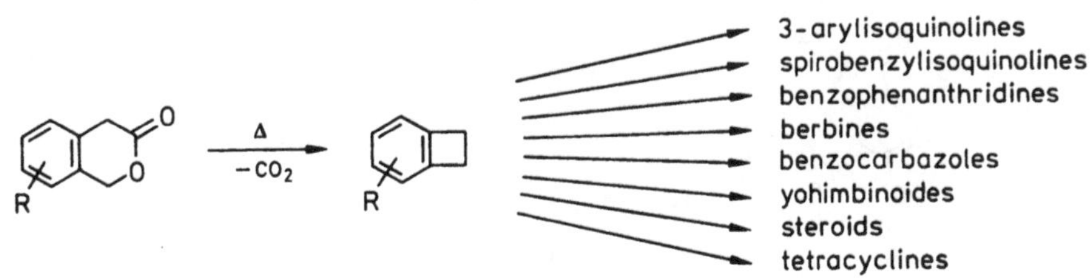

8 Homophthalic Anhydrides

Homophthalic anhydrides are valuable synthons for the synthesis of isocoumarins, isoquinolones and particularly for several berbine alkaloids.

Homophthalic anhydrides can be prepared by oxidation of isochroman-3-ones which in turn can be also obtained by organolithiation reactions, as mentioned earlier.

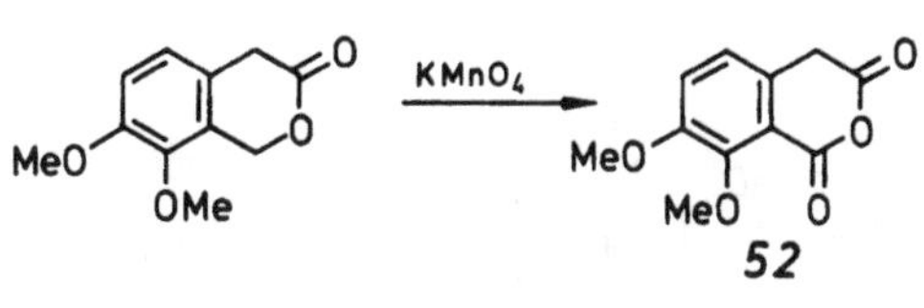

Ref. 184

A direct synthesis using lithiation of aromatic benzylamine is shown below [47, 185].

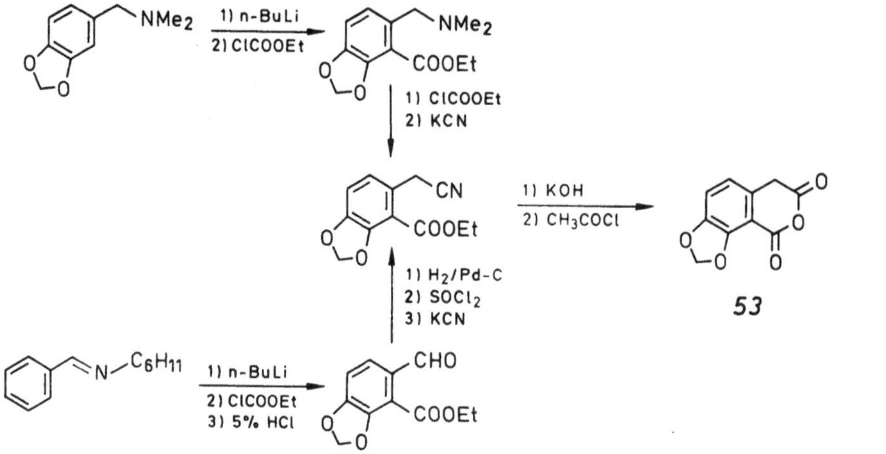

53

It is interesting to compare this synthesis with another synthesis through an indanone using acid catalysed reaction. In the latter, to prevent the formation of the indanone *54*, the more reactive *para*-position is blocked by Br. Towards the end, the bromine is removed by hydrogenolysis [186].

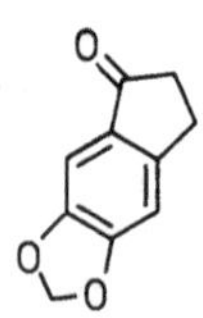

54

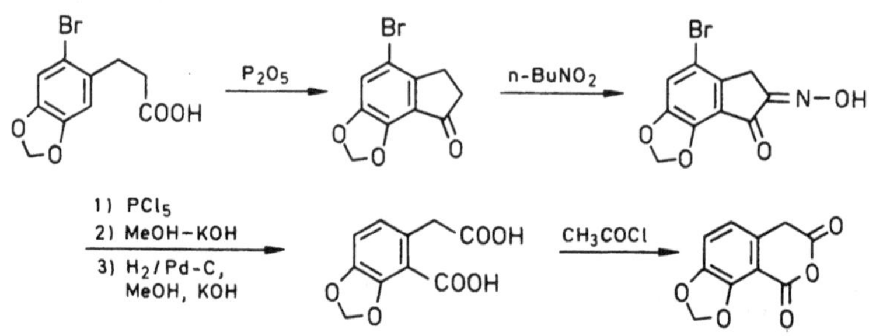

53

Another interesting synthesis of the homophthalic anhydride involves lithiation of an amide and first introducing an *ortho* methyl group. Further lithiation leads to metalation at the aromatic methyl group. The organo metallic compound reacts with CO_2 to furnish the anhydride. The synthetic sequence is shown below [187]. The whole sequence may be rendered into a one pot reaction, without isolating the aromatic methyl compound.

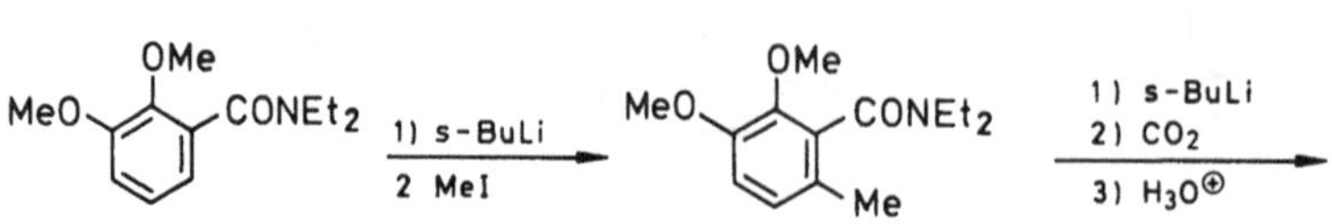

52

A recent synthesis of homophthalic anhydride is by thallation of phenylacetic acid followed by carbonation[45].

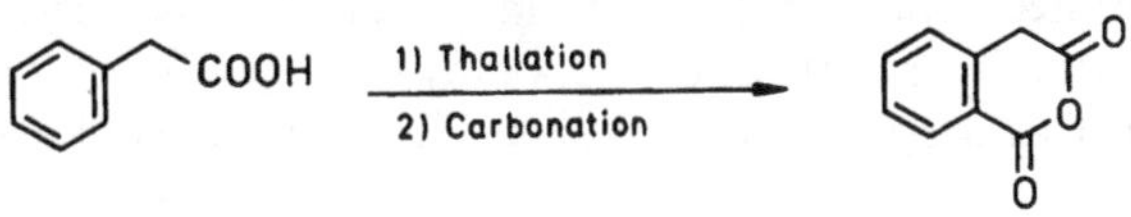

Homophthalic Anhydrides as Synthons

Homophthalic anhydrides have been used for the synthesis of isocoumarins, isoquinolones, berbines and benzophenanthridine alkaloids. Some of the recent examples are given below.

Isocoumarins

Artemidinal *46*, the naturally occurring 3-formyl isocoumarin has been synthesised in good yield, which was then converted to the biologically active isocoumarin-3-carboxylic acid *55*[188]. The natural isocoumarin *56* was also synthesised from the parent homophthalic anhydride as shown below[189].

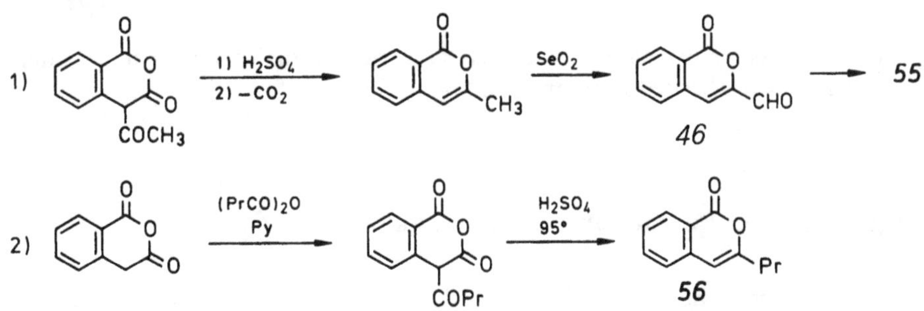

Isoquinolones

3-Substituted isoquinolones and 4-substituted isoquinolones have been obtained from homophthalic anhydrides.

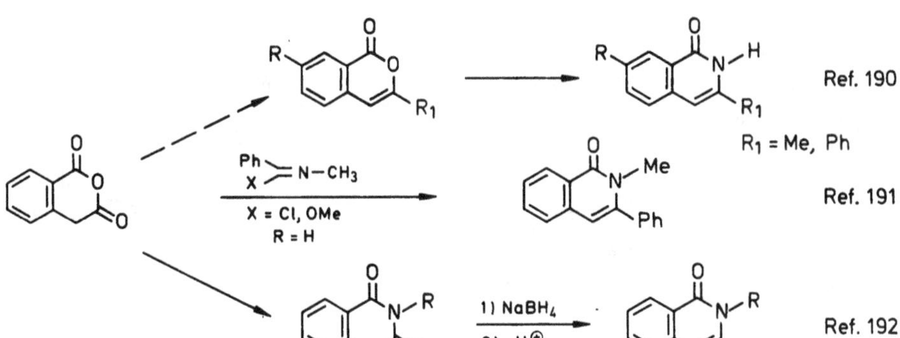

Berbines

Several 13-unsubstituted and 13-methyl-berbine alkaloids have been synthesised by reacting homophthalic anhydride with 3,4-dihydroisoquinolines followed by a sequence of reaction. Synthesis of (±) corydaline (*57*) involves the following steps [193].

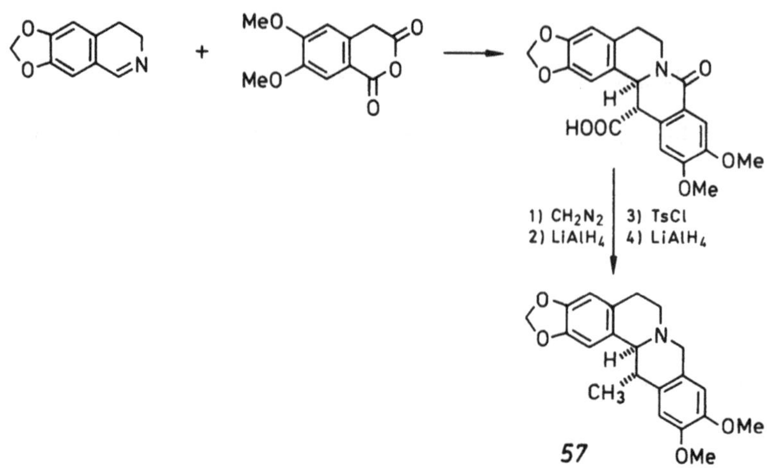

By using a similar sequence of reactions, thalictricavin, thalictrifoline and cavidine have been also synthesised in racemic and optically active forms [186, 194].

Isoquinoline and Benzophenanthridine Alkaloids

The 7,8-dimethoxy homophthalic anhydride *52* prepared via lithiation of 2,3-dimethoxy N,N-diethylbenzamide has been used to obtain some phthalide isoquinolines, while the 7,8-methylenedioxy homophthalic anhydride *53* has been used to prepare (±) chelidonine (*58*) a benzophenthridine alkaloid [47, 185].

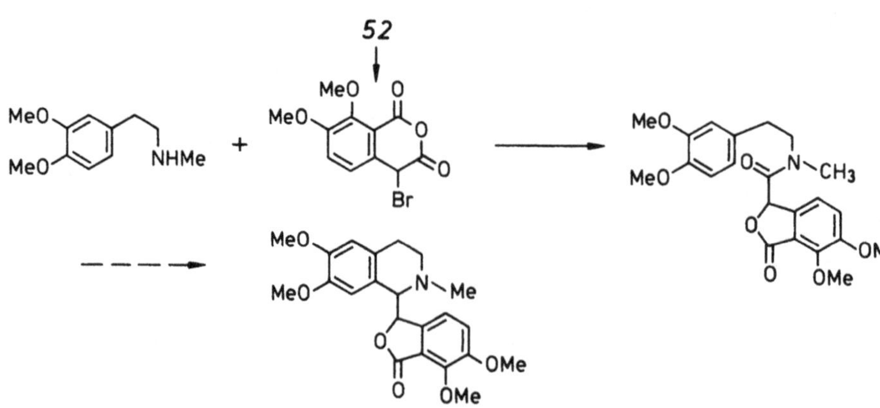

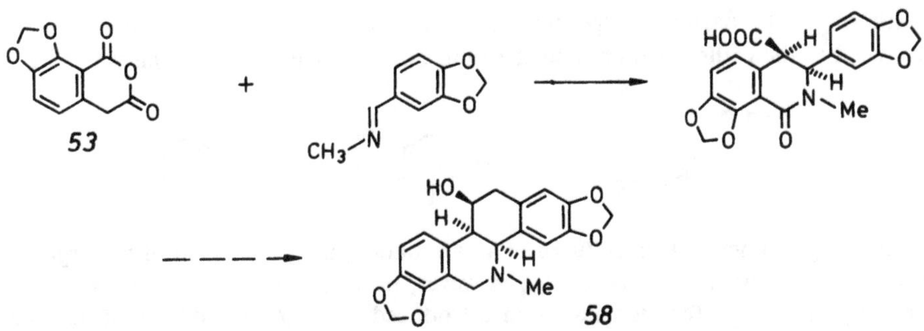

53 58

Similarly 7,8-methylenedioxy-4-methylhomophthalic anhydride has been converted to (±) corydalic acid methyl ester [195].

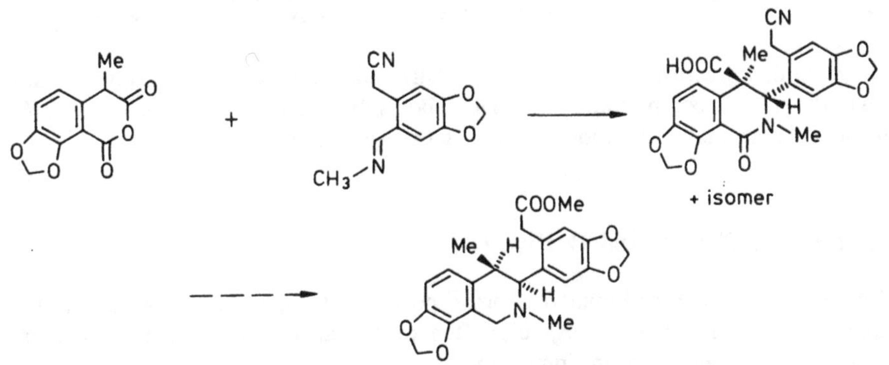

Corydalic acid methyl ester

9 Coumarins

Coumarin is one of the earliest ring systems to be studied. In view of the interesting physiological activities exhibited by the synthetic and the naturally occurring coumarins, there is still continued interest in this field [196].

The syntheses of coumarins by organolithiation reaction are of interest because they provide compounds with methoxyl substitution pattern not available by the usual acid catalyzed methods such as Pechman condensation.

Coumarin belongs to type 13 mentioned in page 8. Two approaches to its synthesis are possible. One is according to type B and has an oxygen substituted aromatic ring (e.g. PhOH) as the starting compound. A carbon substituent such as —CHO could be introduced at the ortho position and the coumarin synthesis completed by standard methods [197]. In the other method which belongs to type A a carbon substituted aro-

Type B

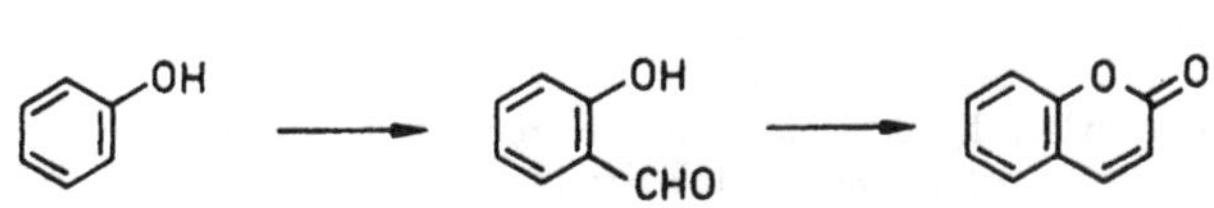

matic ring is the starting compound. An oxygen substituent then will have to be introduced at the ortho position, and the synthesis completed in the usual way [198].

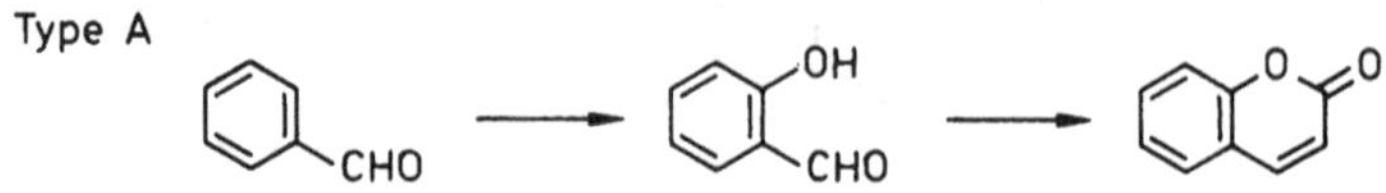

The oxygen function is introduced in an aromatic ring by treatment of the organolithium reagent with O_2 or peroxide. However, these reactions are not satisfactory. Recently an oxygen function has been introduced *ortho* to an aldehyde group via the nitrostyrene. The reaction sequence is shown below [198, 199].

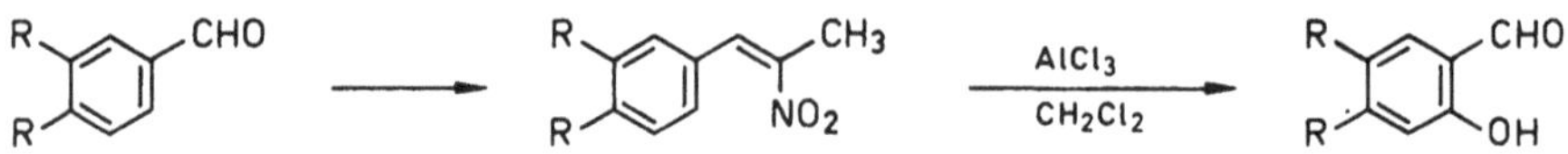

In general, a practical synthesis of a coumarin has an oxygen substituted aromatic ring as the starting compound, in which a carbon substituent is introduced at the *ortho* position. An aromatic lithiation reaction is valuable for this.

Lithiation of Phenols and Derivatives

Although phenols can be lithiated at ortho position, the yields are poor. One of the reasons is that lithium phenolates, quite often, precipitate out of the reaction medium. The other one, which is more important, is that, since the lithiation in these cases would actually occur on the phenolate ion and since the ortho hydrogen in the phenolate ion is considerably less acidic, it is not replaced by lithium in any significant yield.

To overcome the above, phenols are converted to their methyl ethers and lithiation is carried out on these. Lithiation on the methyl ethers occur in improved yield, and through this method ortho carbon substituents can be introduced in better yields [4].

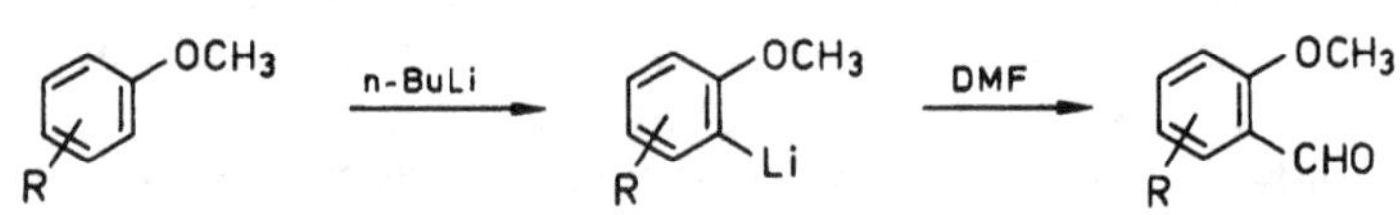

In further steps of the synthesis it is necessary to demethylate the methyl ether and get the phenol back. The usual demethylating agents are HBr, $AlCl_3$, etc., which being strong, are not satisfactory in many cases.

A readily hydrolysable ether of a phenol is its methoxymethyl derivatives. The methoxymethyl ethers are hydrolysed in dil. HCl or by traces of pTSA in benzene. Besides ready removability, CH_2OCH_3 is also readily introduced. The phenols then are preferably converted into their methoxymethyl derivatives[5] before a lithiation reaction is carried out [4, 18].

5 Chloromethyl methyl ether, used for the preparation of methoxymethyl ether is a carcinogen. It is desirable to convert the phenol into a tetrahydropyranyl ether using dihydropyran.

An additional and most useful feature of lithiation of a methoxymethyl ether of a phenol is that yields of lithiation are very high (90% or more). This is due to better complexation of R—Li with $-OCH_2OCH_3$ group which then confers better basicity to the complexed BuLi. The *ortho* carbon substitution is then achieved in excellent yield and further reaction with appropriate electrophiles give *ortho* hydroxyaromatic aldehydes which are starting compounds for coumarins.

Further steps in the synthesis of coumarin via the ortho-hydroxy benzaldehydes involve a Perkin reaction. Here, in the first instance, a cinnamic acid is obtained. The cinnamic acids are obtained more conveniently by the Wittig reaction of the aldehydes with the stable phosphoranes, $Ph_3P{=}\underset{\displaystyle R}{\underset{|}{C}}{-}COOEt$[18c]. Indeed lithiation reaction, coupled with the Wittig reaction constitutes one of the best synthesis of the coumarin ring system. The reaction proceeds under neutral condition and in high overall yield.

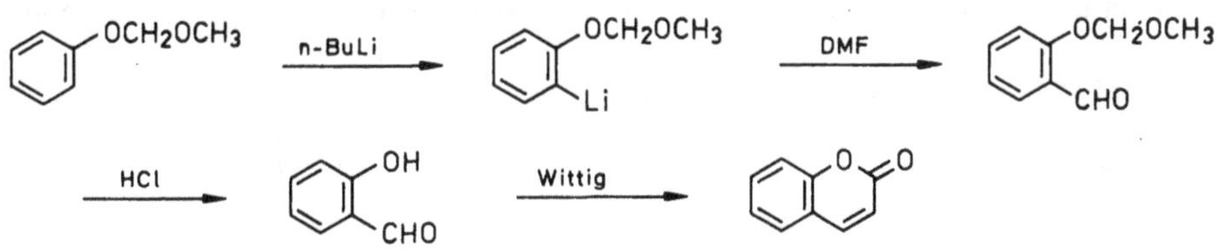

Substituted Coumarins

The synthetic utility of the methods described above is even more if it is realised that the lithiation reaction, being subject to orientation rules mentioned earlier, can provide aromatic hydroxyaldehydes which are not readily available by other acid catalysed methods. These are illustrated by the following examples [18a].

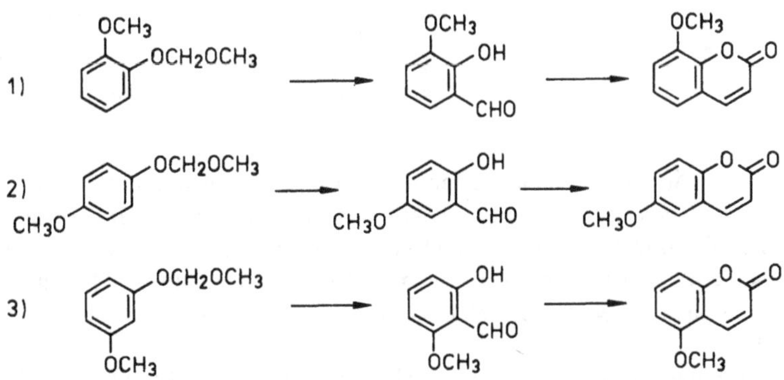

Lithiation occurs *ortho* to $-OCH_2OCH_3$ because the lithiating reagent complexes better with this group rather than with OCH_3. It may be noted that acid catalysed electrophilic substitution on the resorcinol ethers occurs only at 4-position [6].

Lithiation of Naphthol Ethers

The hydroxynaphthaldehydes can be synthesised by lithiation of naphthol ethers and later converted to benzocoumarins. The linear benzocoumarin *59* is of special interest, since its synthesis by other methods is not readily achieved [7].

Lithiation of 1-methoxy naphthalene and 2-methoxy naphthalene needs some comments. In both cases lithiation occurs at two alternate positions which are *ortho* or sterically close to the lithiation directing group ($-OCH_3$) [4].

In 1-methoxy naphthalene, lithiation occurs predominantly at 2-position [16]. This is because the hydrogen at this position is more acidic than that at 8-position. The greater acidity for the 2-H is due to the presence of an oxygen atom at the closer 1-position.

In 2-methoxy naphthalene lithiation occurs, to larger extent, at 3-position than at 1-position [7]. This is unexpected. Both the positions are ortho to the OCH_3 group. The acidity of 1-H is expected to be more because of the presence of 1,2-double bond in naphthalene (this structure is more important than the one with 2,3-double bond).

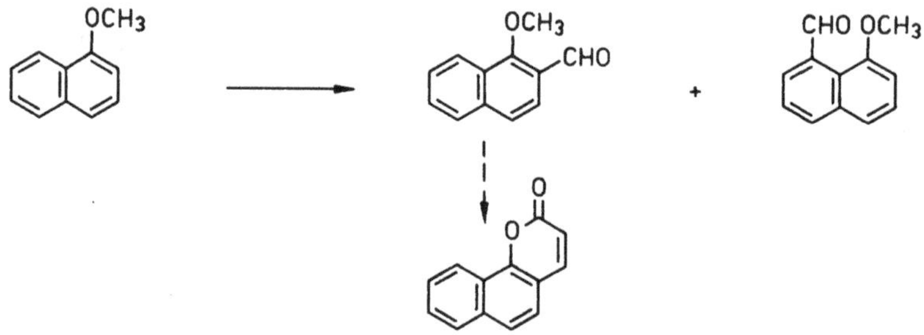

Ref. 7

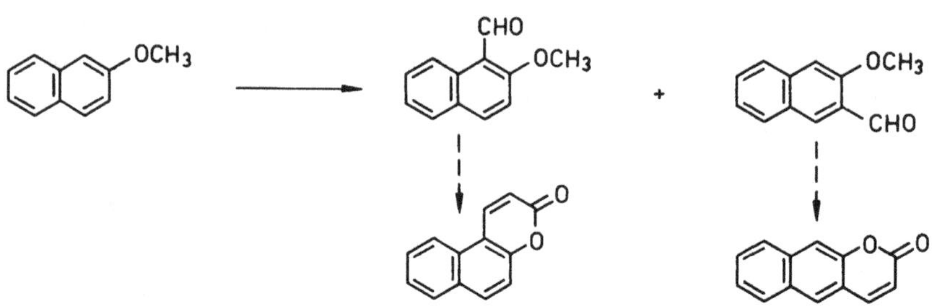

59

Ref. 7

Gilman invoked resonance factors to explain why lithiation occurs to a smaller extent at 1-position in 2-methoxynaphthalene. According to him, the electron density at 1-position would be more due to resonance interaction of the ring with the OCH_3 substituent. This electron density would decrease the acidity of H at that position. The H at 3-position would be relatively more acidic and lithiation would be favoured there.

It is however now known that π electron density is not important when the acidity of an aromatic hydrogen is considered [200]. This is because the π orbital is orthogonal to the σ orbital to which the aromatic H is attached. These orbitals do not interact.

That π electron density does not affect the acidity of aromatic H is evident from the following points.

(i) Between ferrocene and benzene the former has higher π electron density on the individual carbon atoms. However in lithiation reactions, the ferrocene ring is more readily lithiated than the benzene ring [36].

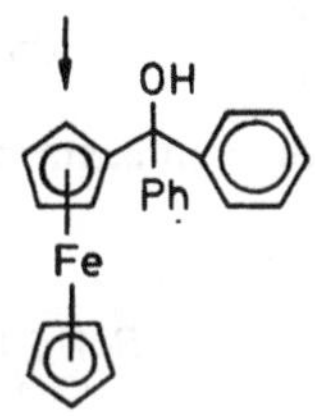

(ii) The benzyne *60* reacts with the amide anion to give *61* and not *62* [200]. Here, the π electron density due to the OCH_3 would be more at 2-position than at 3-position and should have destabilised the anion, which is, however, not the case. The orientation observed in the reaction of the benzyne is essentially due to the greater stability of the carbanion *61* due to the inductive effect of the adjacent OCH_3 group.

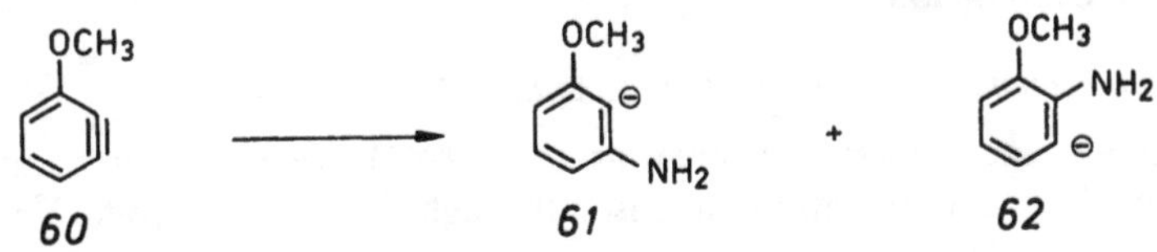

The predominant lithiation at 3-position in 2-methoxynaphthalene may be due to steric factors. When one compares the two transition states corresponding to lithiation at 1-position and at 3-position, one may note that the alkyl group of the lithiating agent is hindered by the peri-H in *63*, while no such hindrance is present in *64*.

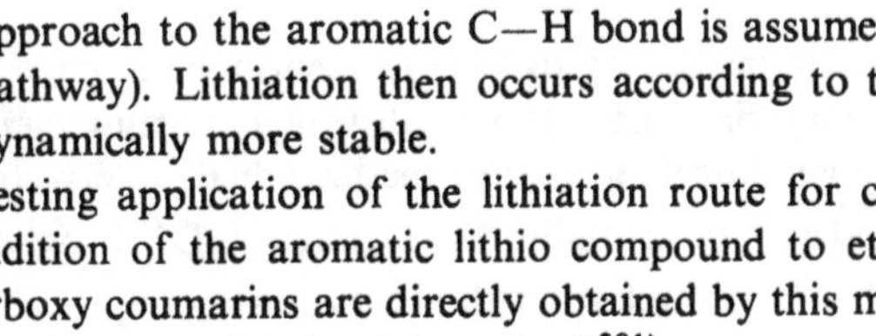

(A linear approach to the aromatic C—H bond is assumed since this would be the favoured pathway). Lithiation then occurs according to transition state *64*, which is thermodynamically more stable.

An interesting application of the lithiation route for coumarin synthesis is the Michael addition of the aromatic lithio compound to ethoxy methylene malonic ester. 3-Carboxy coumarins are directly obtained by this method. The linear benzo-coumarin is also accessible by this method [201].

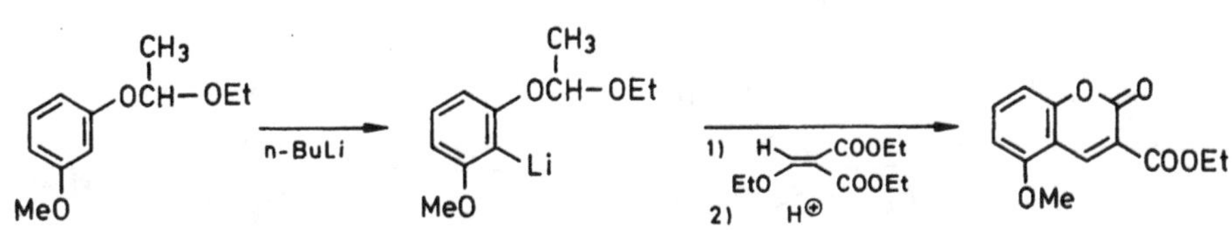

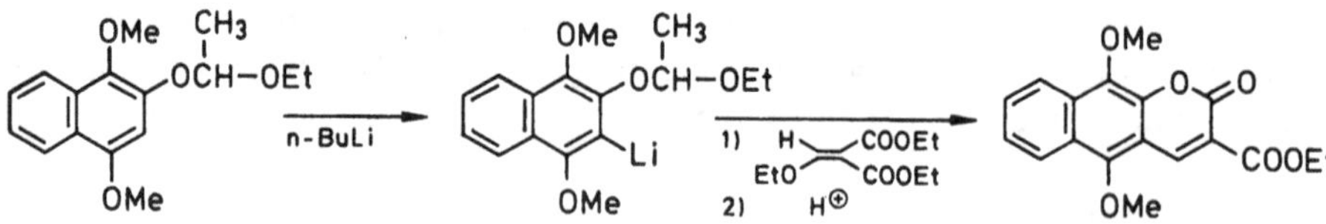

Two recent methods, involving some novel reagents for coumarin synthesis, are the following [202, 203].

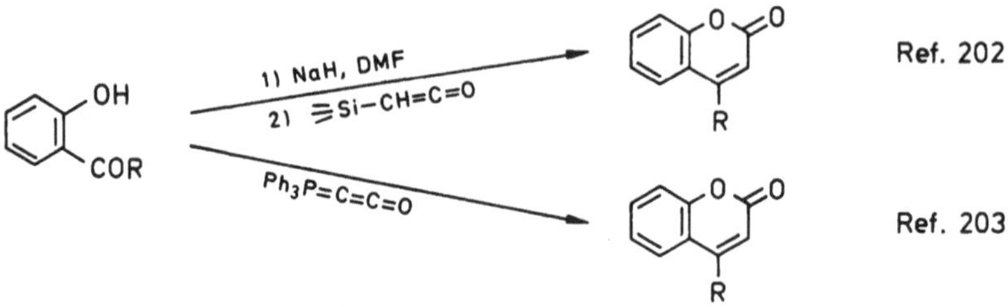

Natural Coumarins

5-Methoxycoumarin 8 and 8-methoxycoumarin 65

5-Methoxy- and 8-methoxycoumarins (*8* and *65*) are naturally occurring [204, 205] and have been conveniently synthesised through the lithiation route [18a].

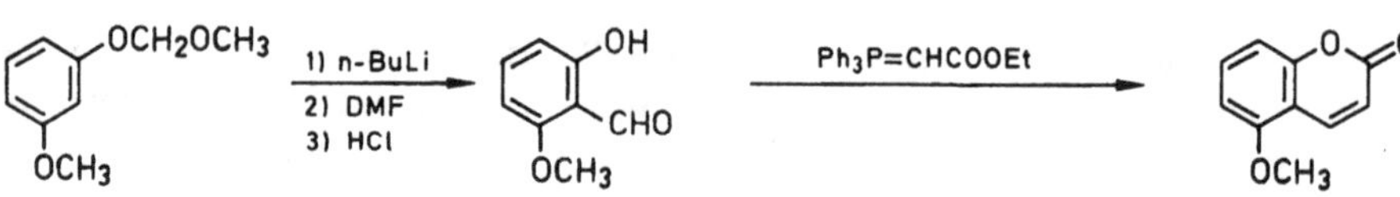

8

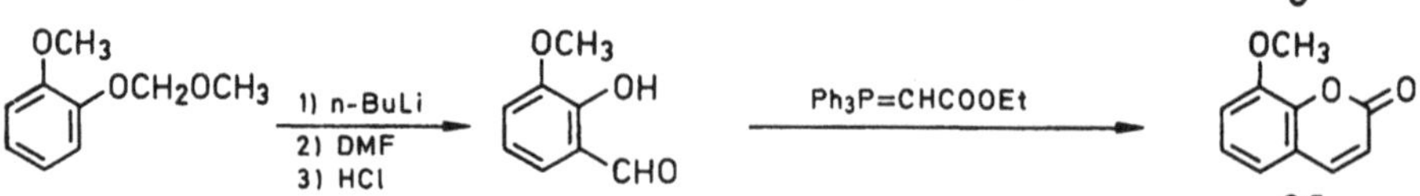

65

10 Condensed Furans

Condensed furans like coumarins belong to type 13 mentioned in page 8. As in the case of the synthesis of coumarins, the ortho disubstituted starting compounds are obtained by lithiation of phenol or its methyl or methoxymethyl derivatives (Type B synthesis). These are reacted with electrophilic reagents such as DMF, , 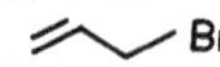 and the synthesis is completed by standard methods. Typical examples are the following, which also indicate the versatility of the method to provide condensed furans not available by usual methods.

1) 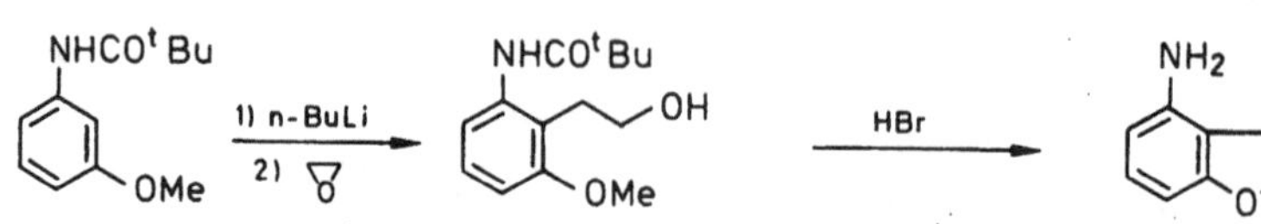 Ref. 206

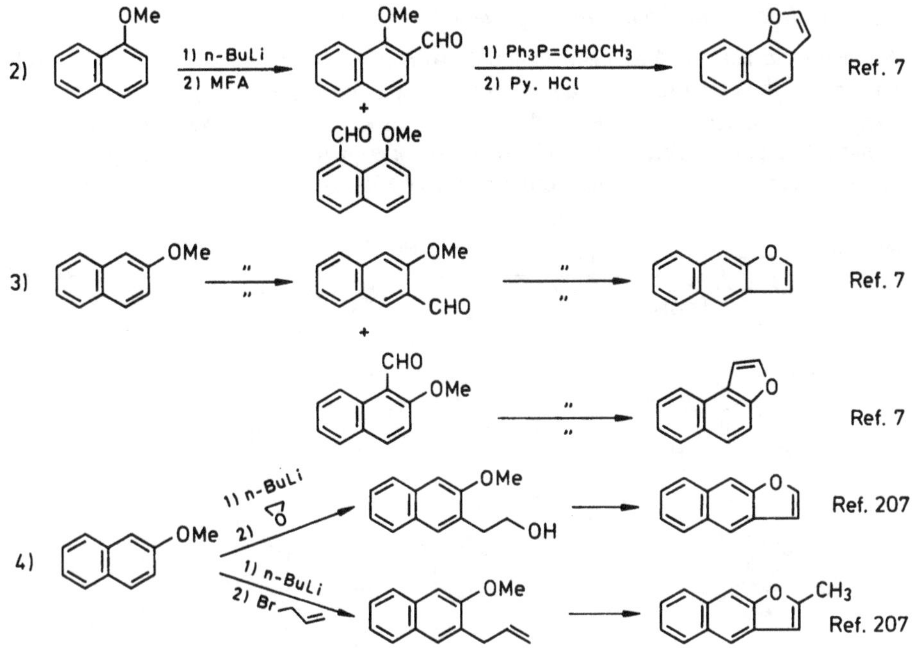

Naturally occurring Benzofurans

Several benzo(2,3-b)furans and dihydro benzofurans have been recently reported from various sources [208]. The structures of these compounds are determined on the basis of their analytical and spectral data.

Fomannoxin 66

(±) Fomannoxin *66* obtained from *Fomes annosus* [209] has been synthesised using copper acetylide as shown below [210].

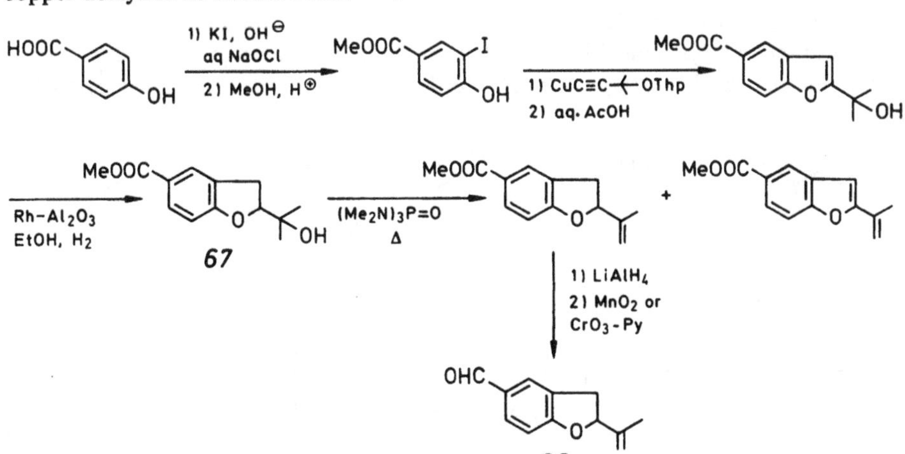

66

(±) Anodendroic acid 68, Euparin 69 and Pterofuran 70

Anodendroic acid *68* has been isolated from *Anodendron affine* Durce [211], while euparin *69* and pterofuran *70* have been obtained from *Eupatorium purpureum* [212] and *Pterocarpus indicus* [213] respectively. The synthesis of (±) anodendroic acid *68*, euparin *69* and pterofuran *70* also involves use of copper acetylide. The dihydrobenzofuran, intermediate *67* has been converted to (±) anodendroic acid *68* [210].

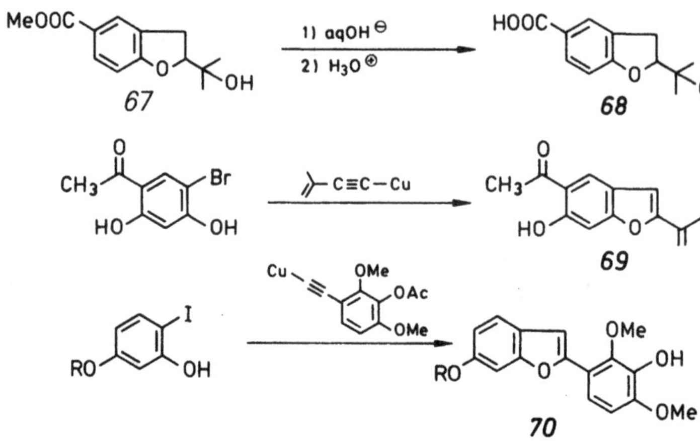

Ref. 210

Ref. 214

Ref. 215

Dehydrotremetone 71

Dehydrotremetone *71* isolated [216] from *Eupatorium urticaefolium* has been synthesised by an intramolecular Wittig reaction approach as shown below [217].

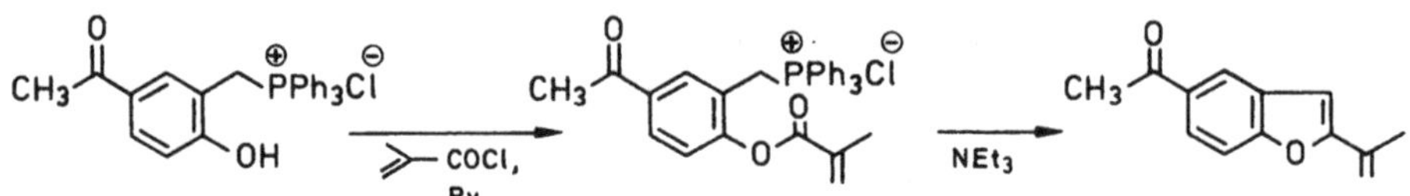

71

Seco-Furanoeremophilane 72

Seco-Furanoeremophilane *72*, isolated from the aerial parts of the South African composite *Ehryops hebecarpus* DC B. Nord [218], is synthesised as shown below [219].

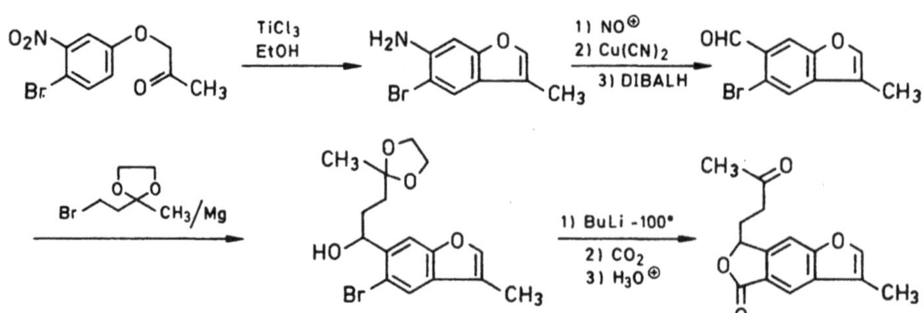

72

K-76 (28)

The complement inhibitor K-76 (*28*) contains a 2,3-dihydrobenzofuran moiety. The key step is the formation of *73*, which is obtained by an aromatic lithiation reactions. This was then converted to K-76 [106].

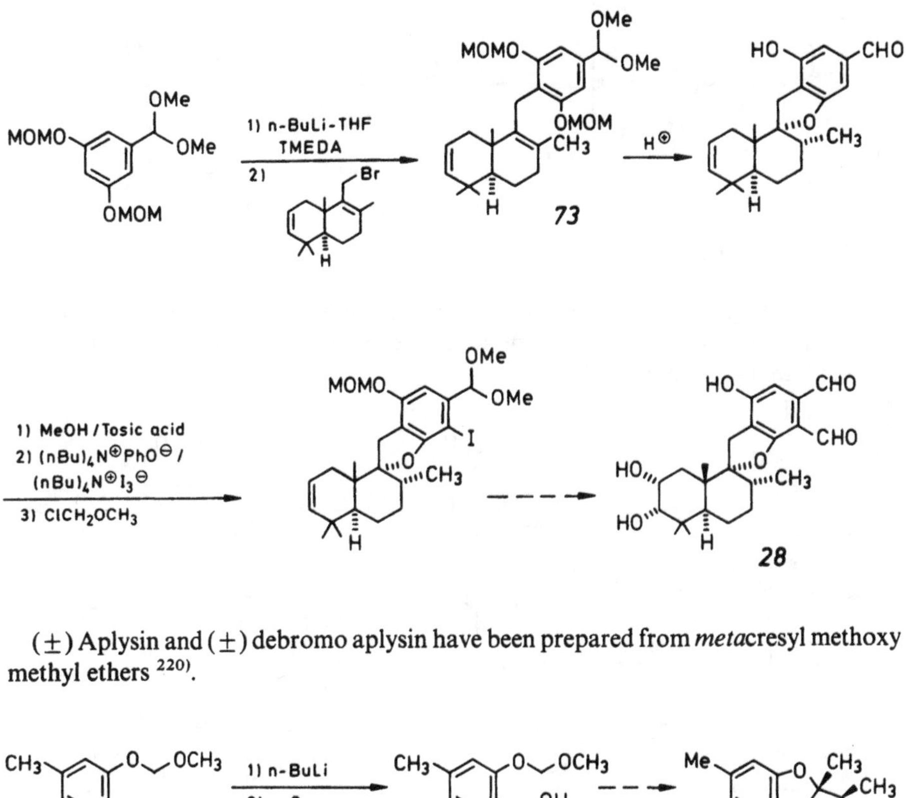

(±) Aplysin and (±) debromo aplysin have been prepared from *meta*cresyl methoxy methyl ethers [220].

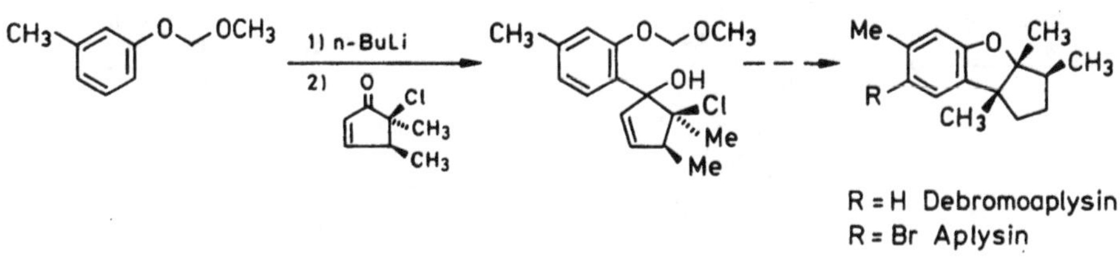

(—) Aplysin and (—) debromoaplysin were prepared [221] analogously via lithiation (t-BuLi) of the following optically active ether.

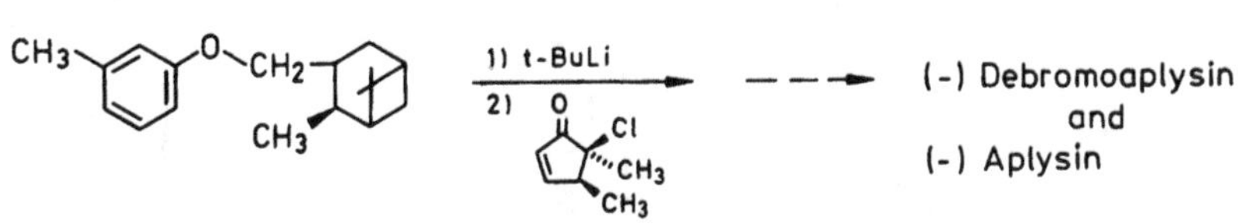

Benzopyrans [222)]

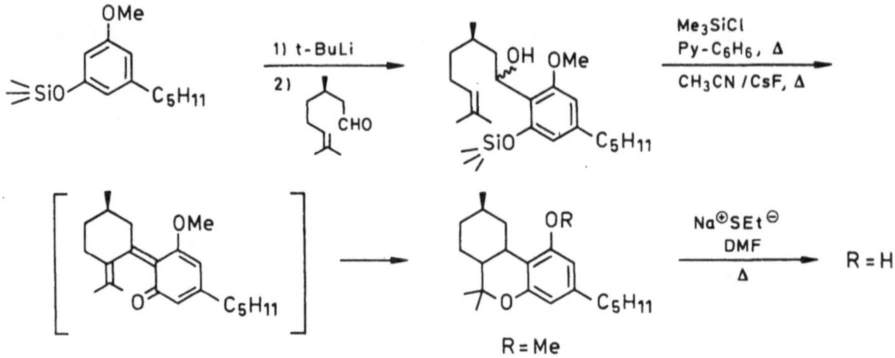

11 Oxaphenalenes *74*

Oxaphenalene [naphtho(1,8-b,c)-pyran] *74* is interesting as it contains 1,8-disubstituted naphthalene structure. Two syntheses are reported for this compound. The steps involved are depicted below [223)].

Scheme I

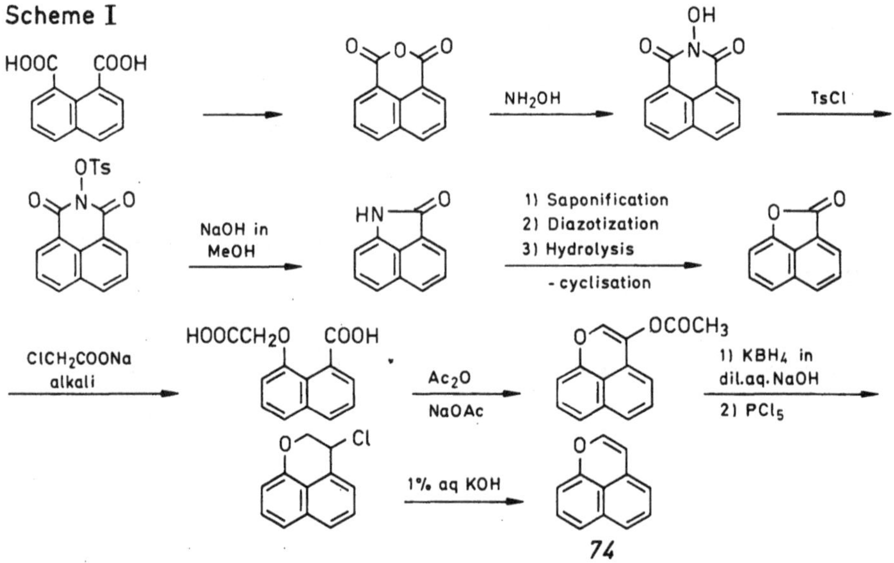

74

Scheme II

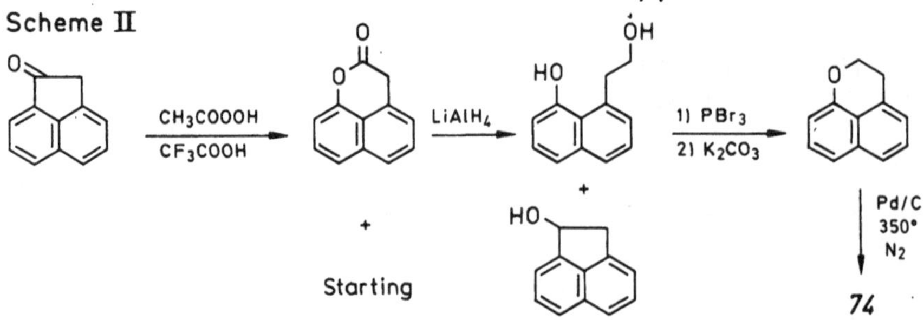

Xanthorrhoein *75*, isolated from *Xanthorrhoea resins* [224], has also been synthesised [225] by *Scheme I*.

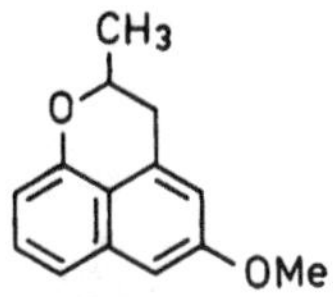

75

Both methods start with 1,8-disubstituted naphthalene, which are not readily available. Further the syntheses involve several steps.

The synthesis of the oxaphenalene ring system can be achieved by lithiation method readily. The starting compound is also readily available. The synthesis is complete in three steps [226].

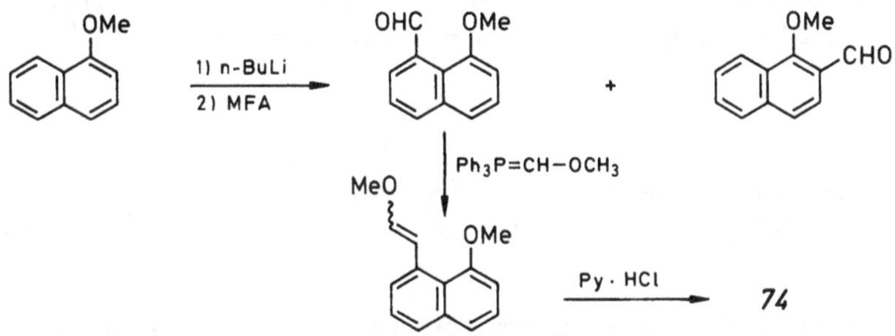

In the method given above 1-methoxynaphthalene, on lithiation, gives a mixture of 2-lithio and 8-lithio compounds. 8-Lithio compound, is however, obtained only in minor amount. In contrast when 1-hydroxy naphthalene is lithiated in presence of TMEDA 8-lithio compound is the exclusive product. Further treatment of the organometallic compound with DMF furnish 1-hydroxy-8-naphthaldehyde [27].

12 Isoquinolines

The isoquinoline ring system is readily accessible by well known methods such as Pictet-Spengler [227], Pomaranz-Fritz [228], Bischler-Napierlaski [5] and their modified procedures. Despite this, however, certain specific methoxy isoquinolines are not readily synthesised by the usual methods as mentioned in the beginning of the chapter. This is because the above methods are acid catalyzed and lead only to certain specific orientation in the cyclisation reaction. Isoquinoline ring system can be synthesised through aromatic lithiation reaction, which obviates the difficulties mentioned above.

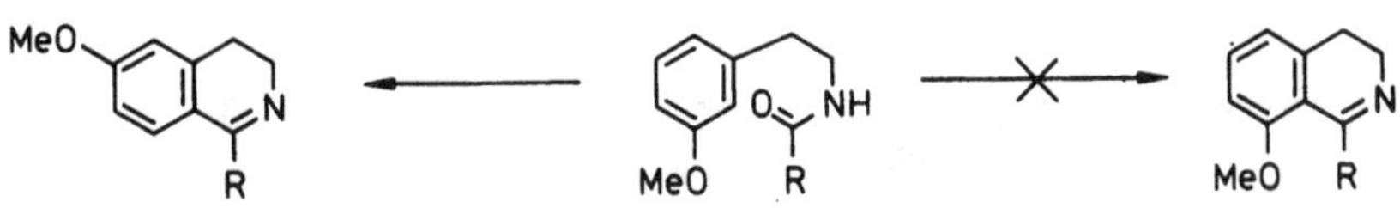

The synthesis of isoquinoline by lithiation route is according to type C. Thus aromatic benzylamines or β-phenylethylamines are lithiated at *ortho* position. The organo-metallic compounds on treatment with epoxide or an aldehyde furnish *ortho* substituted compounds which are converted to isoquinolines [229].

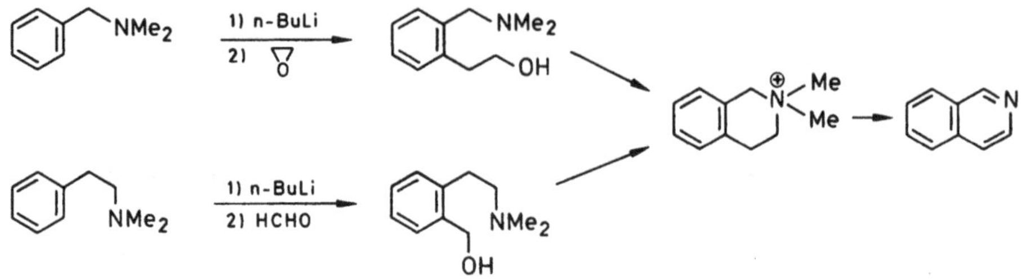

The yield of lithiation is good only in the case of the benzylamine. In the case of the β-phenylethylamine the chief reaction is the formation of an elimination product, presumably through the competing benzylic lithiation reaction [229].

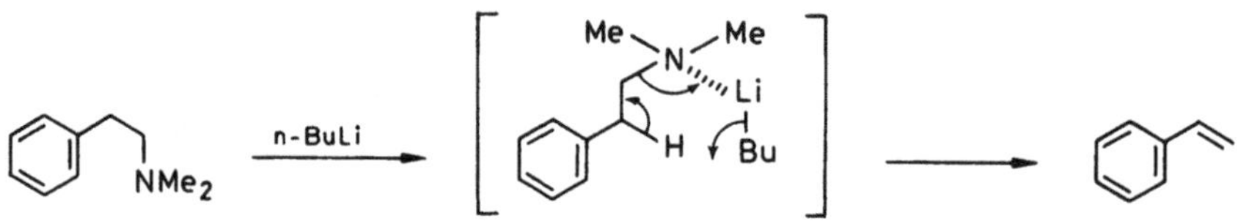

The yield of the lithiation, in case of benzylamine, is again good only when the amine is tertiary. This then leads to the quaternary salt in the cyclisation reaction. The dehydrogenation of quaternary salt does not proceed in synthetically useful yield. For this reason the method is not suitable for the synthesis of isoquinolines themselves. However, when the desired product is a tetrahydroisoquinoline the method can be of some value as in a synthesis of tetrahydropalmatine (vide infra).

In a modification of the above method, after introduction of an appropriate substituent at the *ortho* position, the NMe_2 group can be replaced successively by Cl and CN and finally converted to isoquinoline. This is illustrated by the synthesis of 8-methoxyisoquinoline (vide supra) which is difficultly synthesised by other methods.

The RLi complexes better with amines than with others. Hence considerable specificity in the lithiation reaction can be observed. Thus methoxy substituted isoquinoline can be obtained by lithiation reaction as shown below.

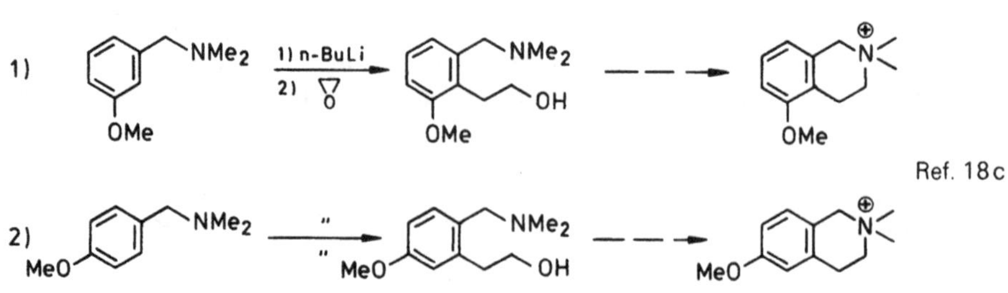

Ref. 230

3) 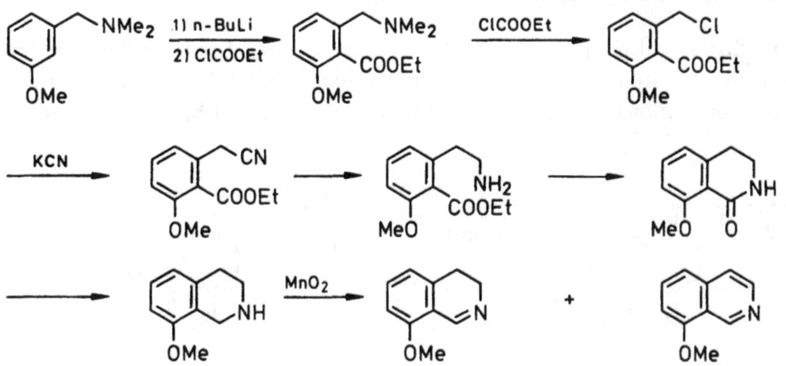

Lithiation of benzanilide derivative followed by treatment with benzonitrile, also furnishes substituted isoquinoline as shown below [231].

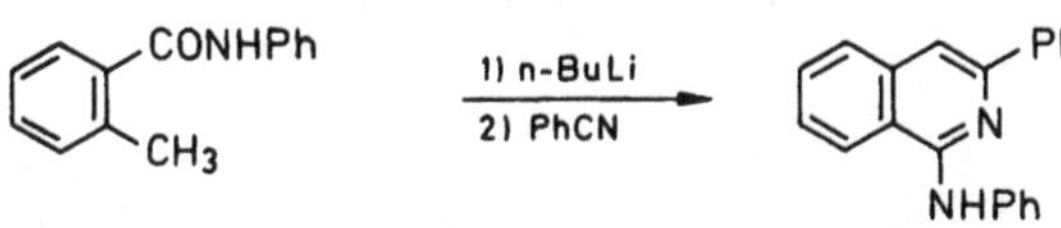

Thiophene analogues of isoquinolines have also been synthesised through lithiation reaction [232].

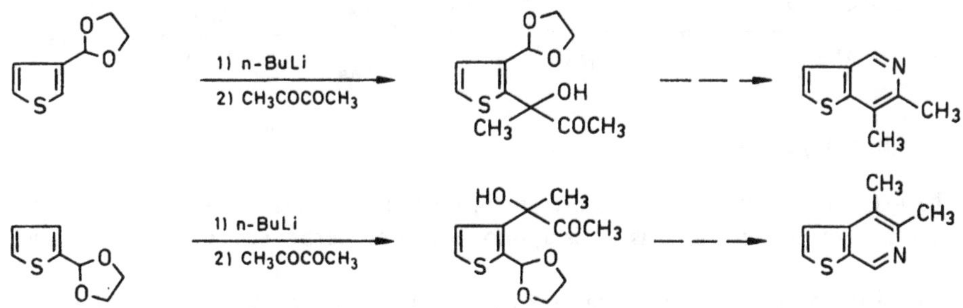

It has already been mentioned that specific isocoumarin and isochroman-3-ones can be obtained by aromatic lithiation reaction. These can then be converted to isoquinolines.

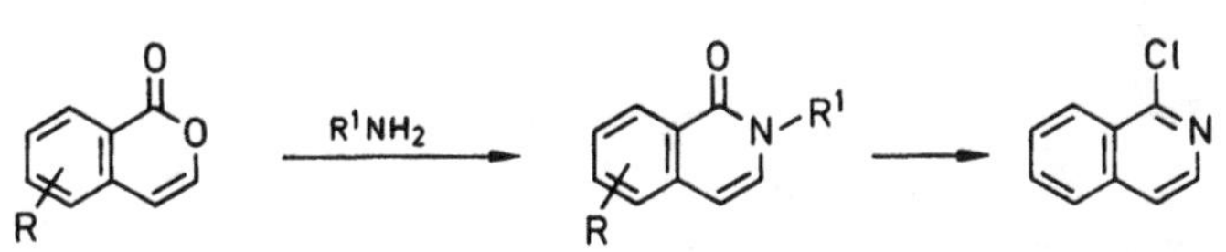

Ref. 190, 233

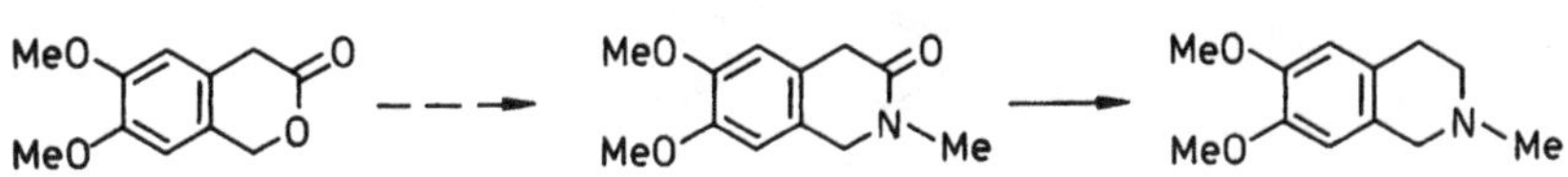

Ref. 174

Isoquinolines as Synthon

Tetrahydropalmatine (*51a*), a berbine alkaloid, has been synthesised from isoquinolines. The steps involved are presented here [229, 234]. It may be noted that in the first the benzoquinolizine ring is directly obtained in the lithiation reaction.

1) Ref. 229

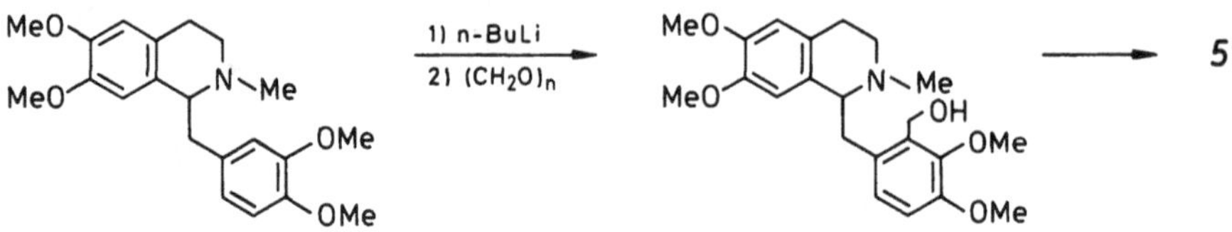

2) Ref. 234

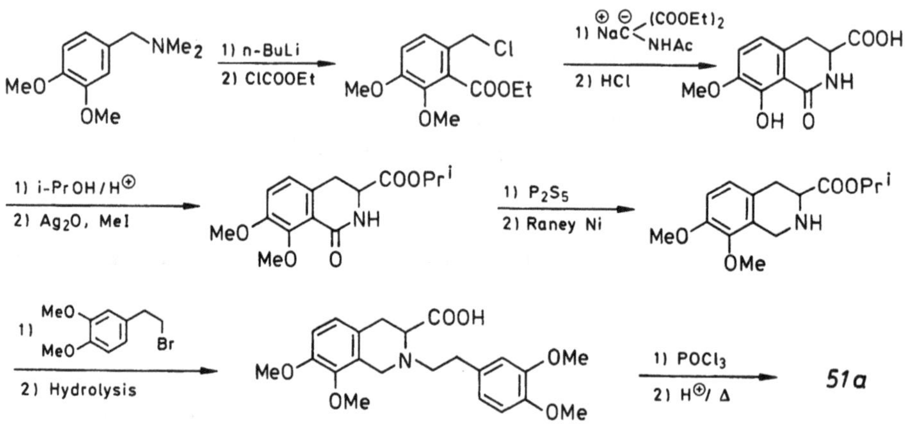

13 Isoquinolones

The isoquinolone ring system is present in several alkaloids [237]. It is also a useful intermediate in the synthesis of isoquinolines, protoberberines, dibenzoquinolines etc.

Aromatic lithiation reaction can provide this ring system conveniently. These are shown in the following examples.

1)

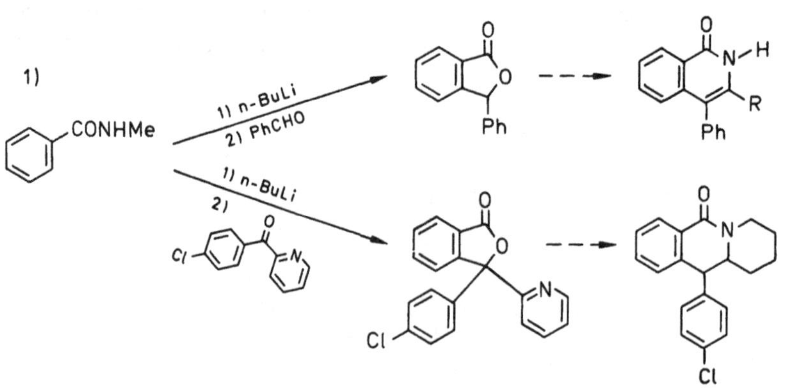

Ref. 235

Ref. 236

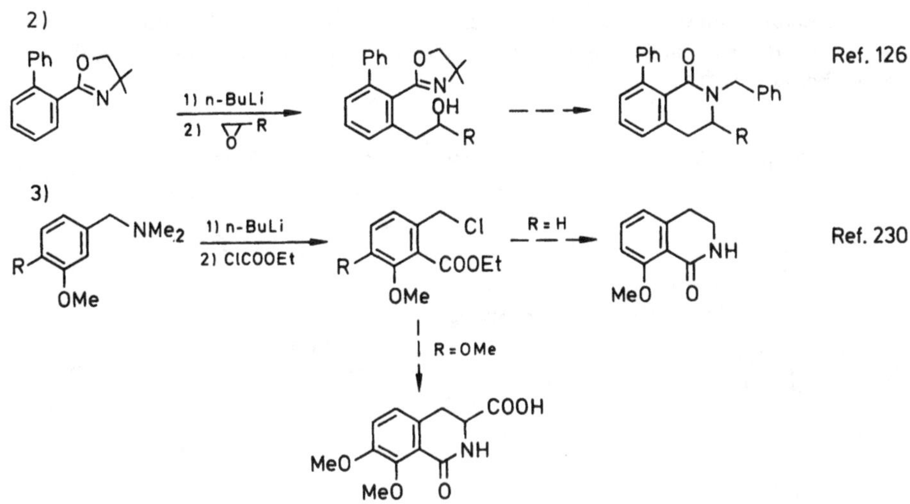

In addition to directed aromatic lithiation reactions, lithiation of aromatic methyl groups can also furnish the isoquinolones. Examples of these are the following:

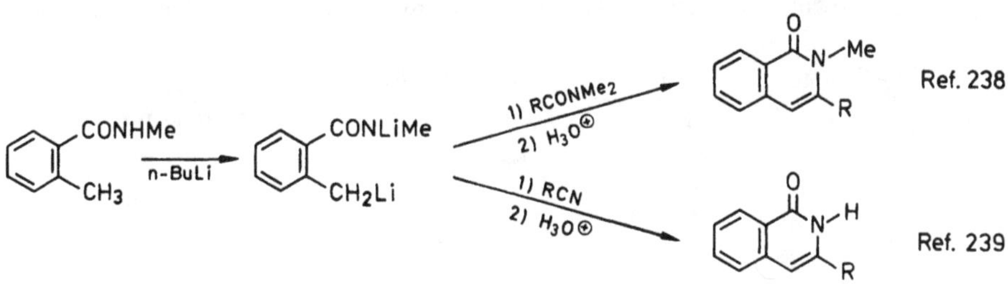

Similarly 3,4-disubstituted-N-methylisoquinolone has been synthesised as shown in the following scheme [240].

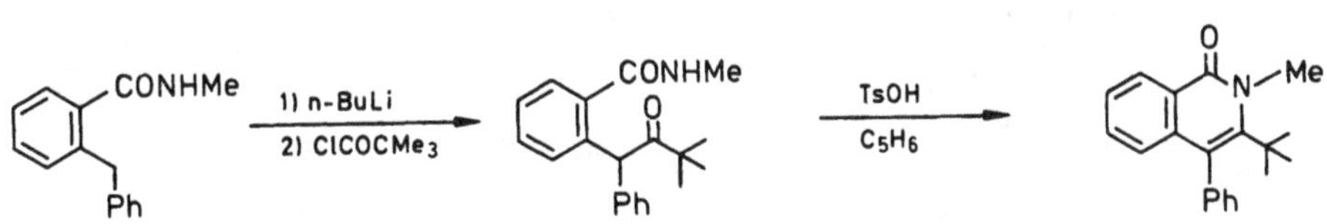

All the examples mentioned above belong to 1-oxoisoquinoline ring system. 3-Oxo-isoquinolones *76* are also found in nature. They possess various biological activities [241]. On reduction they furnish tetrahydroisoquinolines.

76

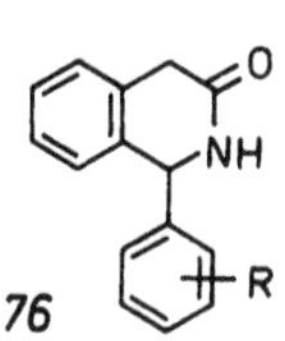

The 3-oxoisoquinolines have been recently synthesised by condensation of phenylacetylchloride with shiffs bases or phenylacetic acids with various benzamides as shown below [241, 242].

1) 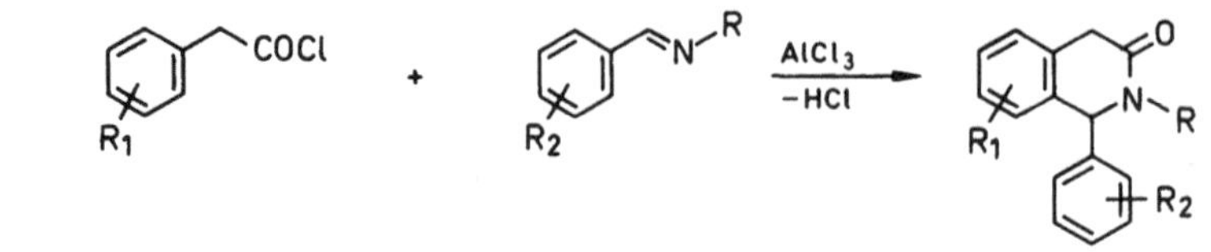 Ref. 241

2) 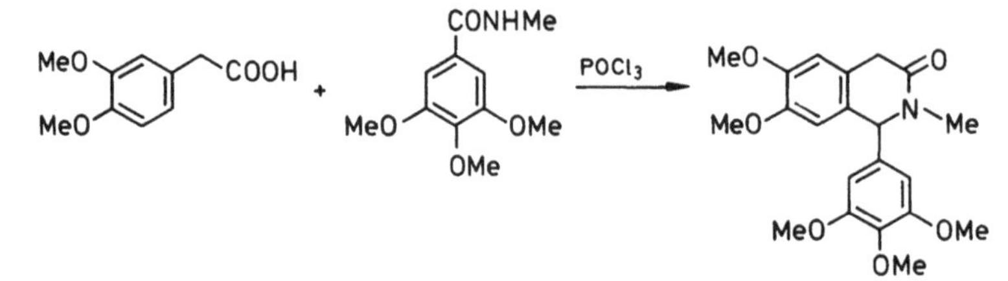 Ref. 242

Both the methods being acid catalysed provide only 6,7-dialkoxy isoquinolones from 3,4-dialkoxyphenyl acetic acid or 3,4-dialkoxyphenyl acetyl chloride.

The 7,8-dimethoxy isoquinolones can be synthesised by lithiation reaction according to the following sequence [243]. The feature of particular interest is that 7,8-dialkoxy

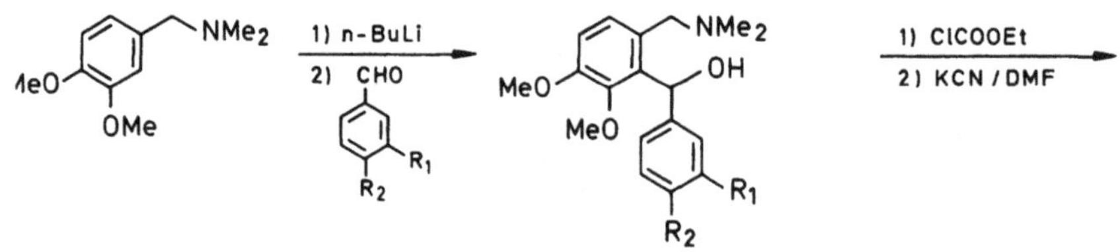

77

	R_1	R_2
a	H	H
b	OCH_3	OCH_3
c	$O-CH_2-O$	

substituents are present in the final compound 77, a pattern not readily available by other methods. The final cyclisation step in the above is presumably an intramolecular Ritter reaction.

The 7,8-dialkoxy isoquinolones can be also obtained from the 7,8-dialkoxy isochroman-3-ones.

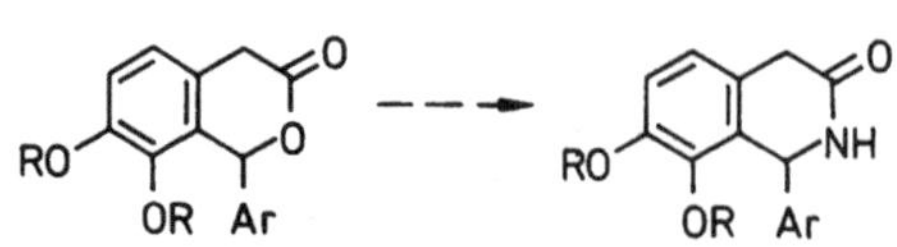

Isoquinolones as Synthones

Various isoquinolines and isoquinoline alkaloids are easily available from isoquinolones [192].

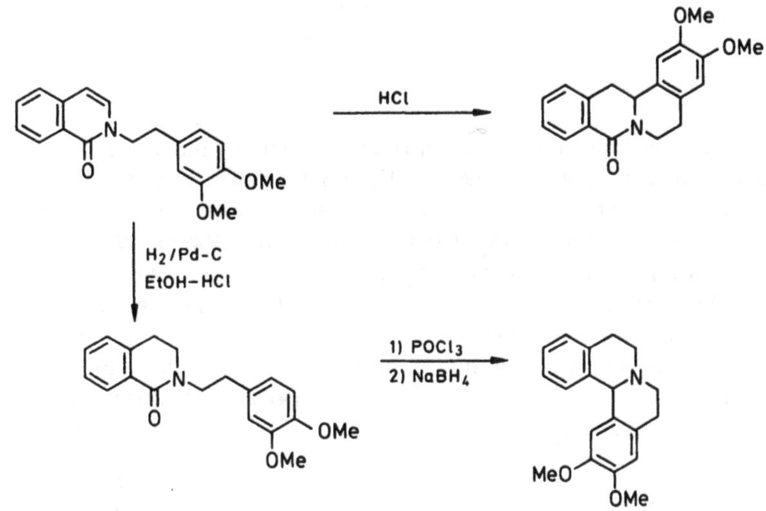

Berbine and yohimbane skeleton is also available from isoquinolone [244].

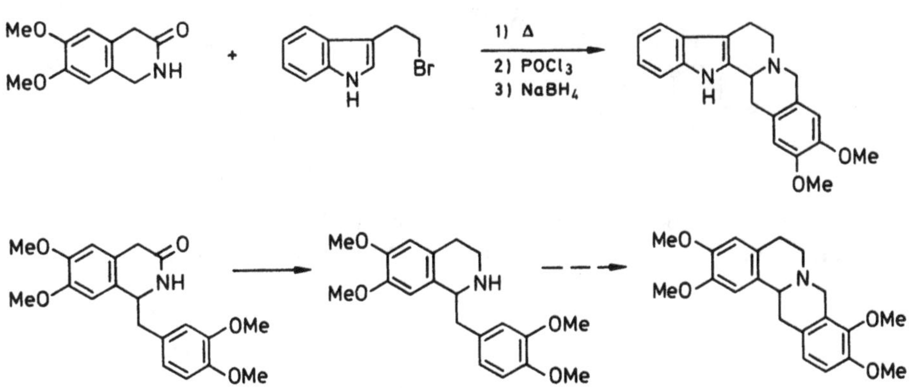

14 Phenanthridine

The general synthesis of phenanthridines, which are benzoisoquinolines, are patterned on isoquinolines. The lithiation route is similar to the one, where, for an isoquinoline synthesis, a β-phenylethyl amine (*78*) is the starting compound. The starting compound now is a 2-amino biphenyl (*79*). Interestingly, it is no more necessary to convert the primary amine to its dimethyl derivative.

The 2-aminobiphenyl resembles β-phenylethyl amine in the skeletal structure.

The amino group can then direct lithiation at 2′ position and lead to a carbon substituent (like COOH or CHO) and lead to a phenanthridine synthesis.

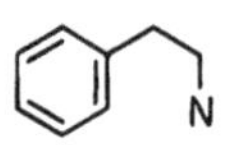

78

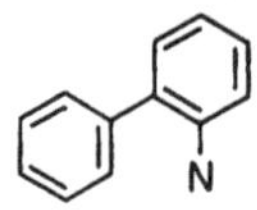

79

A possible competing lithiation reaction would be the one at 3-position i.e. *ortho* to the amino group in the same ring. Fortunately the factors governing lithiation reaction are such that this does not happen. Thus in the lithiation of 2-aminobiphenyl by R—Li reagents, the first event that would occur is the formation of N-lithio salts. In the N-lithio salts the hydrogen *ortho* to the nitrogen group would be less acidic than the hydrogen at the more distant 2′-position. (This situation is similar to what is observed in the lithiation of 1-hydroxy-naphthalene). Lithiation at 2′-position is then uniquely favoured and indeed occurs exclusively there. Further treatment with CO_2 or DMF introduces a functionalised one carbon substituent which can be readily converted to phenanthridine [15a, 15b].

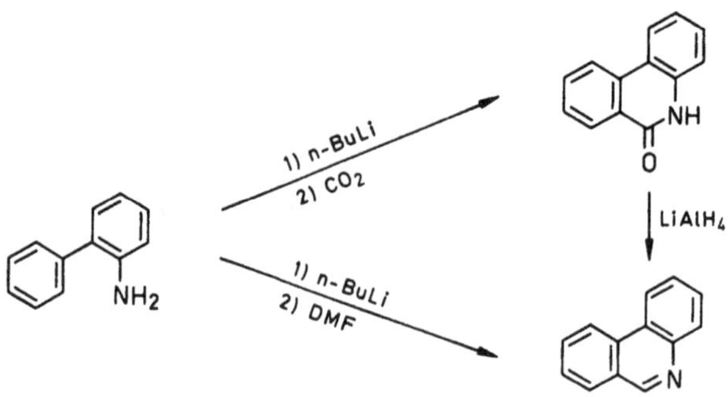

The utility of the above method is again evident in the synthesis of 7,8-dimethoxyphenanthridine [15b], which is not obtained by Morgan-Walls reaction.

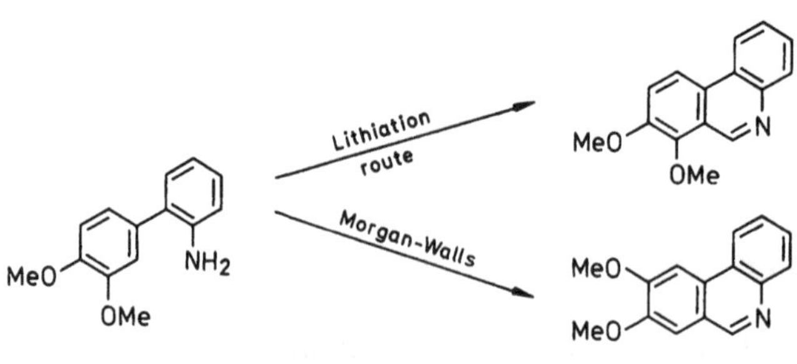

15 Dibenzothio-, ox- and di-azepines

Dibenzo(b,e)azepine(morphananthridines) have some importance since they manifest spasmolytic, antihistamine, anticonculsive and other psychotropic properties [245].

In view of the interesting pharmacological properties of the dibenzoazepines, the synthesis of the ring system has been exhaustively investigated. Some of the methods used for their synthesis are the following [246]:

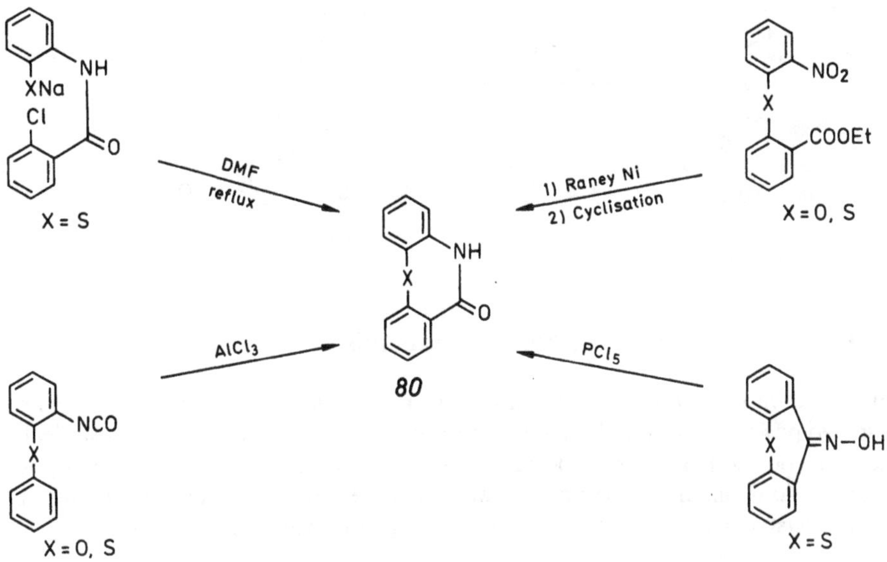

The compound *80* (X = O, S, NH) have also been obtained by directed metallation of amino compounds as shown below [247].

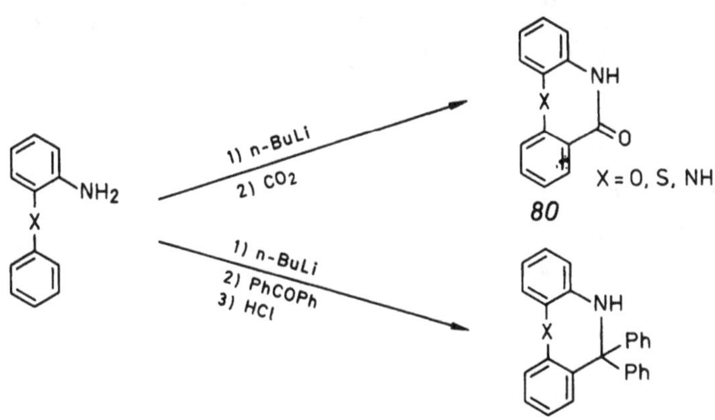

The compounds dibenzothiazepine *82*, dibenzooxazepine *83* and dibenzodiazepine *84* are elegantly prepared via the organolithium compound as depicted below [248].

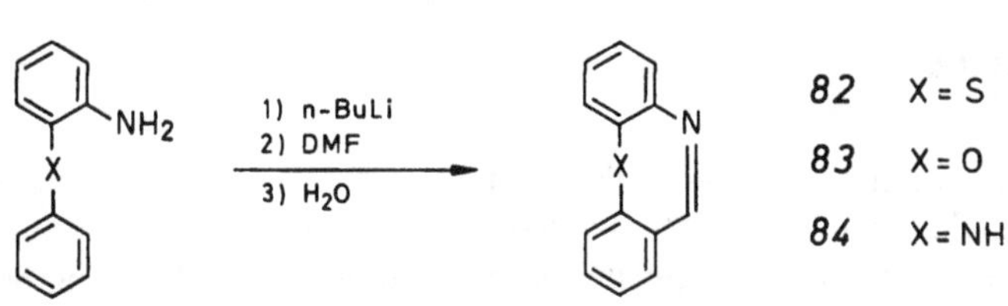

The dimethoxy derivatives of *82* and *83*, which once again cannot be prepared by normal acid catalyzed methods, are prepared by using the lithiation reaction [276].

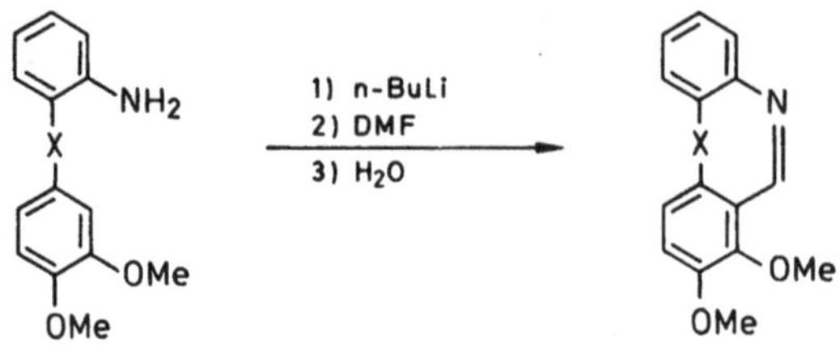

X = O, S.

16 Naphthastyrils and Related Compounds

The lithiation of an aromatic primary or secondary amino compound, where more than two positions for lithiation are present, occurs at the more distant but sterically accessible position. This principle can be used to introduce a lithium atom at 8-position in a 1-aminonaphthalene ring system. This route then gives a convenient synthesis of naphthastyrils and related compounds, examples of which are given below.

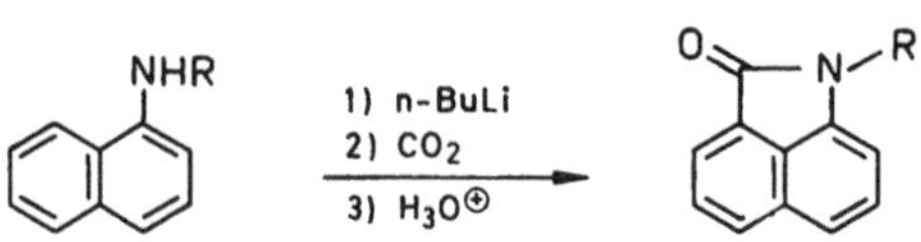

R = H Ref. 26
= Ph Ref. 15c

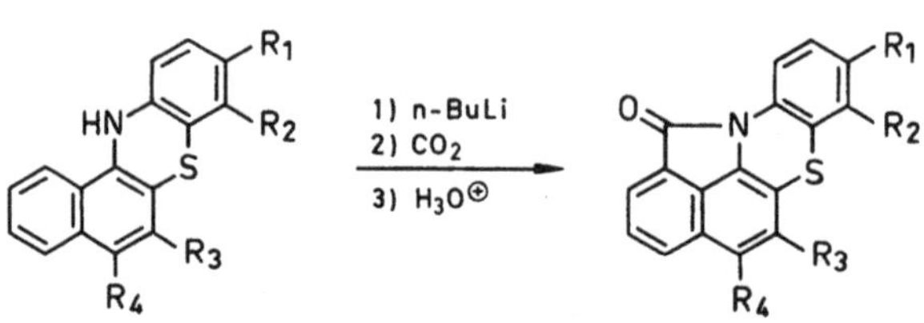

R_1	R_2	R_3	R_4
$-(CH=CH)_2-$		H	H
H	H	$-(CH=CH)_2-$	

Ref. 249

17 Acridones

Although lithiation *ortho* to an amino group is not achieved in any significant yield, with N-phenyl derivatives lithiation indeed can be achieved at *ortho* position. Further treatment with CO_2 gives N-phenyl anthranilic acids which are important intermediates. The synthesis of acridones through aromatic lithiation reaction is illustrated by the following examples.

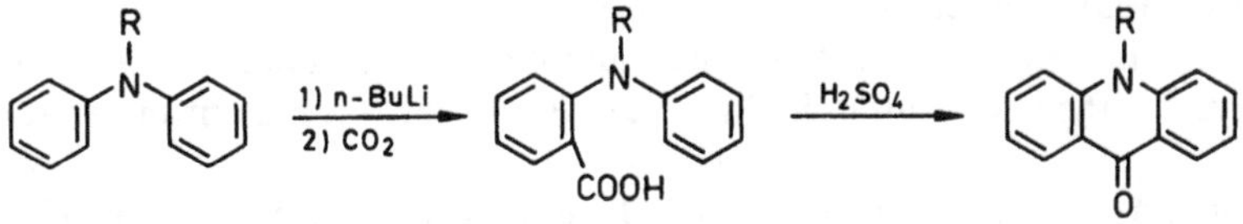

Ref. 250

Ref. 249

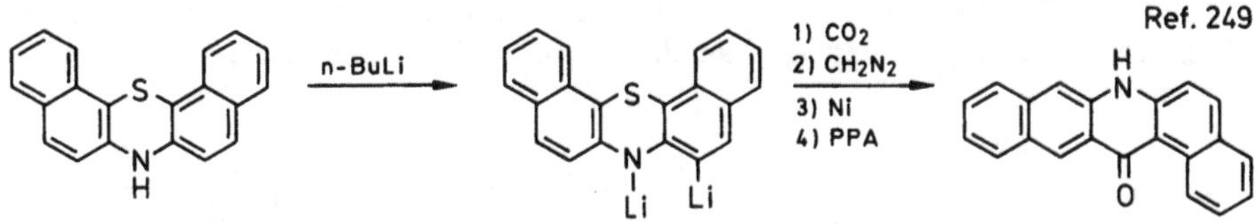

Recently N-acyl substituents have been introduced *ortho* to amide, to obtain anthranilic acids, which have been converted to acridones [251].

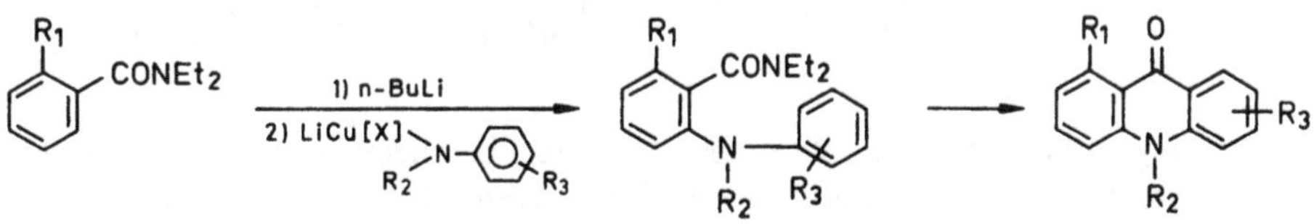

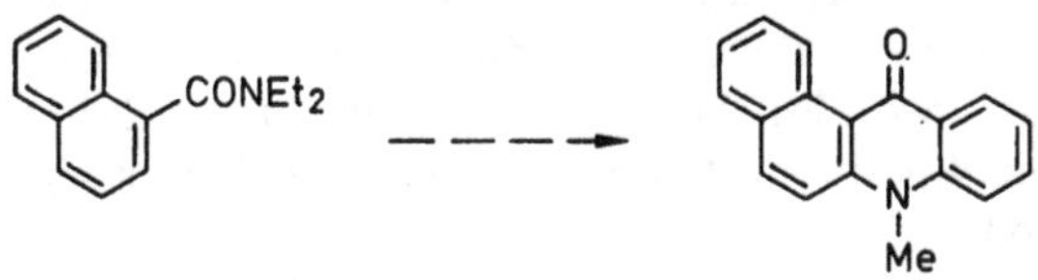

N,N-Dimethylbenzamide, on similar reactions gave the acridones [251].

Naturally occurring acridones

Various acridones are found to occur in nature [252]. Des-N-methylacronycine *85* has been synthesised *via* organolithiation reaction as shown in the following scheme [253].

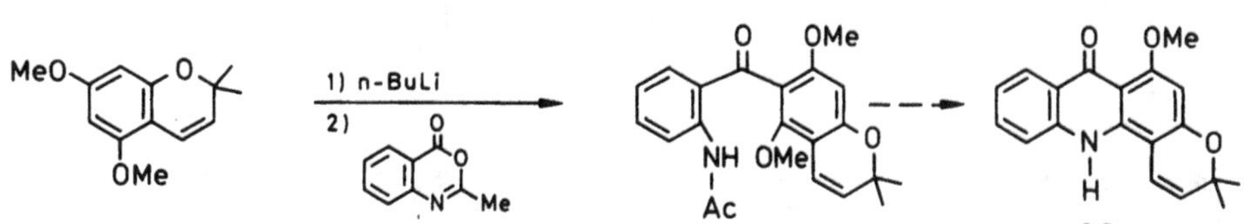

85

18 Furoquinolines

Lithiation of aromatic methyl ethers as mentioned earliar gives a new route to condensed furans. When the aromatic substrate is a quinoline, furoquinolines result. Of particular interest is the linear furoquinoline ring system which is present in several alkaloids.

2-Ethoxyquinoline is lithiated at 3-position. The lithio derivative on reaction with ethylene oxide or allylbromide furnishes the 3-hydroxyethyl or 3-allyl derivatives. These can be readily converted to the dihydrofuroquinolines as shown [254].

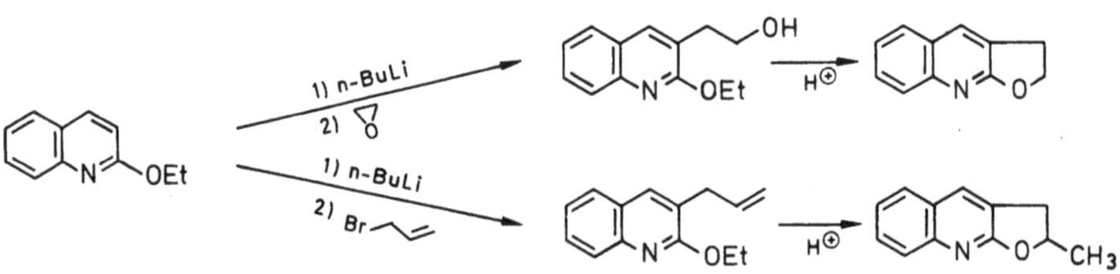

The fully aromatic furoquinolines themselves can be directly synthesised by treatment of the organometallic intermediate with DMF to obtain a 3-formyl derivative. Condensation with $Ph_3P{=}CHOCH_3$ furnishes the enolether which on hydrolysis gives the 3-formylmethyl derivative. Further hydrolysis of the 2-OCH_3 group, followed by cyclisation then provides the linear furoquinolines. When methoxyl groups are present at 2- and 4-positions it is possible to bring about a selective demethylation of 2-OCH_3 group [255].

The 3-formylmethyl derivative can also be obtained from the 3-lithio compound by treatment with an allylbromide followed by ozonolysis and hydrolysis [256].

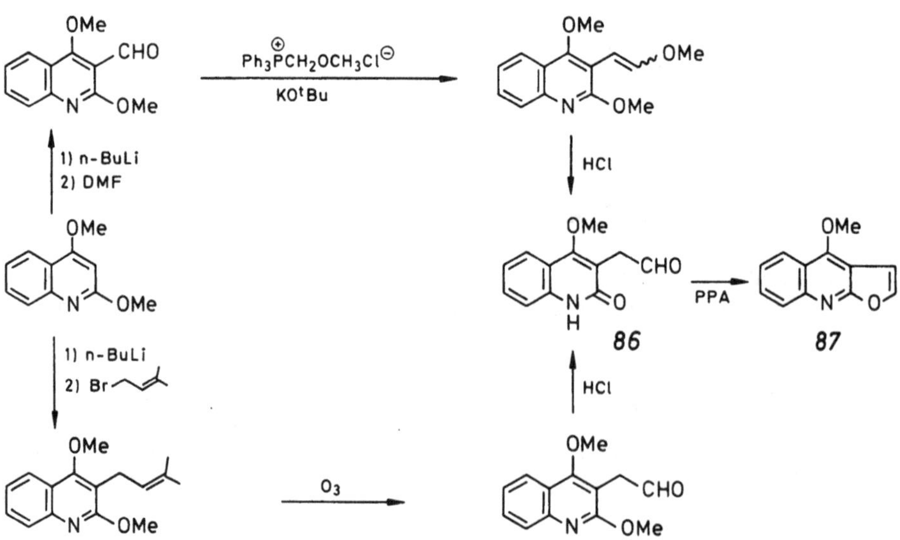

Furoquinoline Alkaloids

The 4-methoxy linear furoquinoline *87* is actually the naturally occurring alkaloid [257], dictamine. The other furoquinoline alkaloids have also been synthesised in analogus way [255, 258, 259].

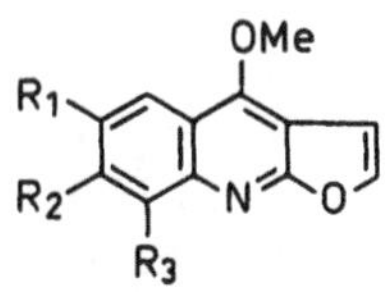

87...92

	R_1	R_2	R_3	
87	H	H	H	Dictamnine
88	OMe	H	H	Pteleine
89	H	OMe	H	Evolitrine
90	H	H	OMe	γ-fagarine
91	OMe	OMe	H	Kokusaginine
92	-OCH$_2$-O-		H	Maculine

In the above reactions when the lithio compound is reacted with ethylene oxide, dihydrofuroquinolines are obtained as shown below [254].

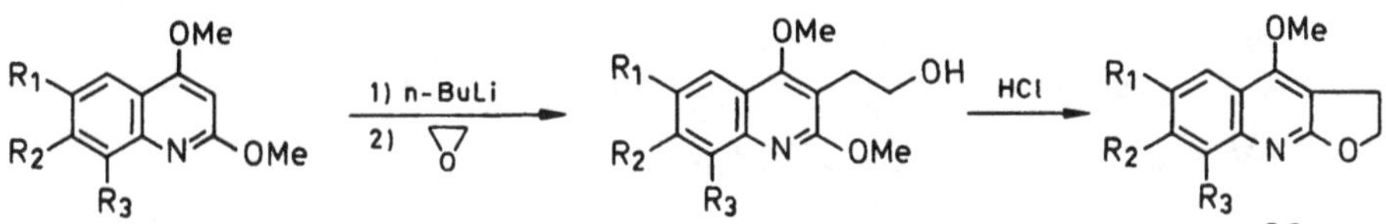

93

93	R_1	R_2	R_3	
a	H	H	H	Dihydrodictamnine
b	OMe	H	H	Dihydropteleine
c	H	H	OMe	Dihydro-γ-fagarine

The compound *93a* is then oxidised by successive treatment with NBS followed by refluxing with collidine to dictamnine *87*.

19 Pyranoquinolines

The pyranoquinoline ring system is of two types, the linear and angular. Both the systems can be synthesised by lithiation route through the same intermediate, useful in the synthesis of the furoquinoline. This is shown below [260].

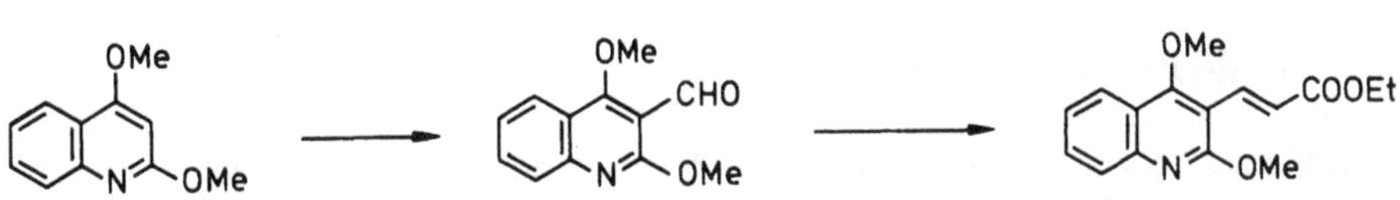

94

The intermediate *94* undergoes thermal cyclisation to give the angular pyranoquinoline *95*.

94 95

To obtain the linear pyranoquinoline the 2-OCH_3 of *94* has to be demethylated to obtain *96*. Interestingly *96* also gives only the angular compound *97* either by thermal or acid catalysed cyclisation. However acid catalysed cyclisation, *in the cold*, in presence of UV radiation (to convert the *trans* to *cis* olefin) gives the linear compound *98* [260].

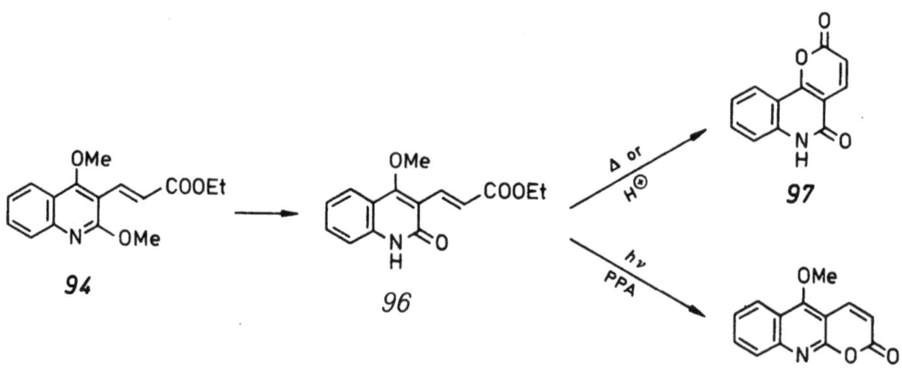

94 96 97 98

Natural Pyranoquinolines

The angular isomer on treatment with CH_3MgI gives flindersine *99*, a pyranoquinoline alkaloid [261].

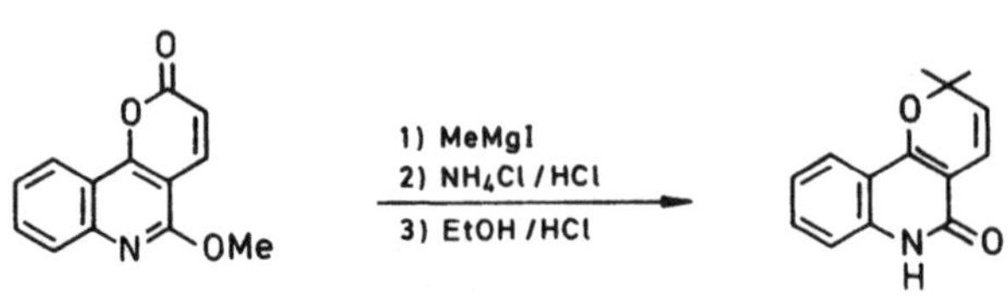

95 99

20 Condensed Pyrimidines

Pyrimidines, condensed to various carbocyclic and heterocyclic compounds like benzene, naphthalene, indole, quinoline etc., have been synthesised through the organo lithiation reactions where the electrophilic reagent is an aromatic nitrile. The synthetic schemes used are presented below.

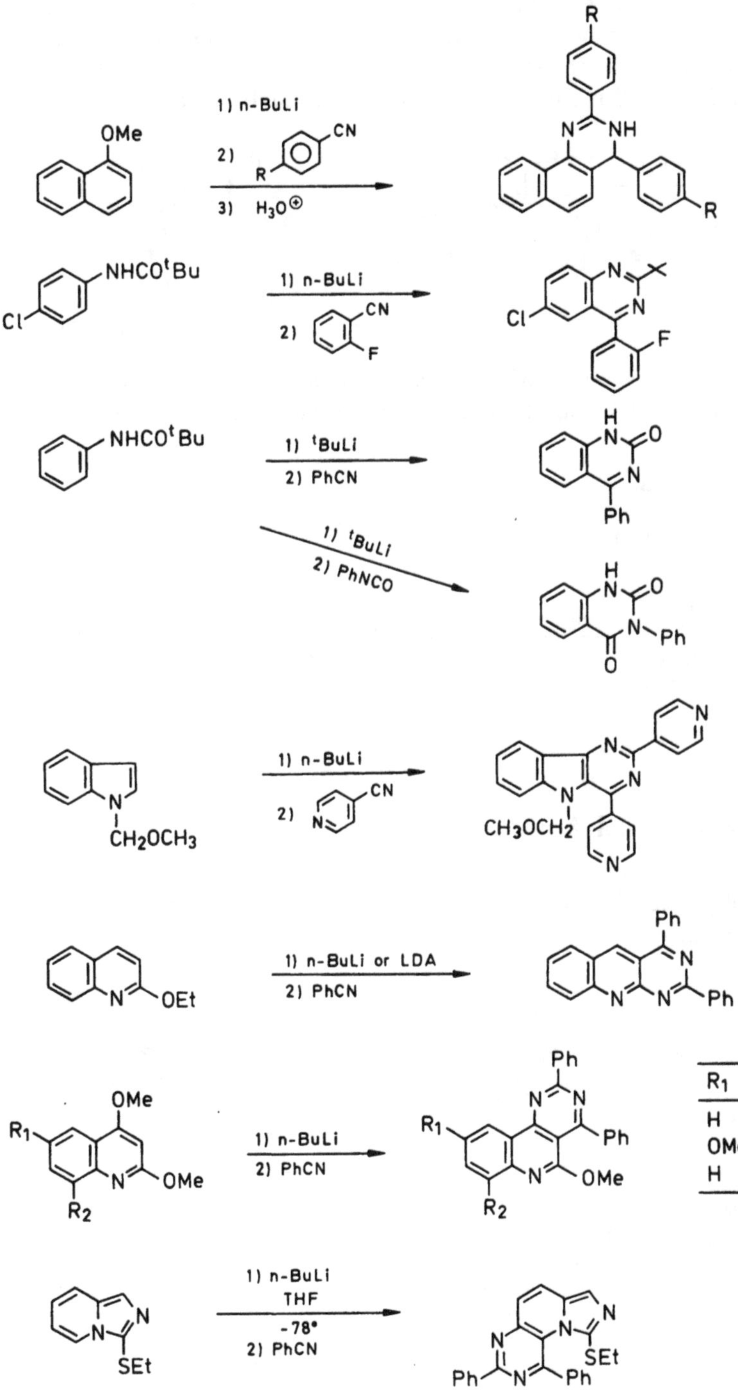

Ref. 262 R = H, Cl

Ref. 25

Ref. 263

Ref. 263

Ref. 264

Ref. 265

R_1	R_2	
H	H	Ref. 266
OMe	H	Ref. 267
H	OMe	— „ —

Ref. 268

21 Indoles

No synthesis of indoles via aromatic lithiation reaction is known. However aromatic methyl group *ortho* to a protected amino group have been lithiated. The benzylcarbanion further reacts with the amide carbonyl to give 2-substituted indoles [25, 269]. This intramolecular attack is similar to the Modelung indole synthesis.

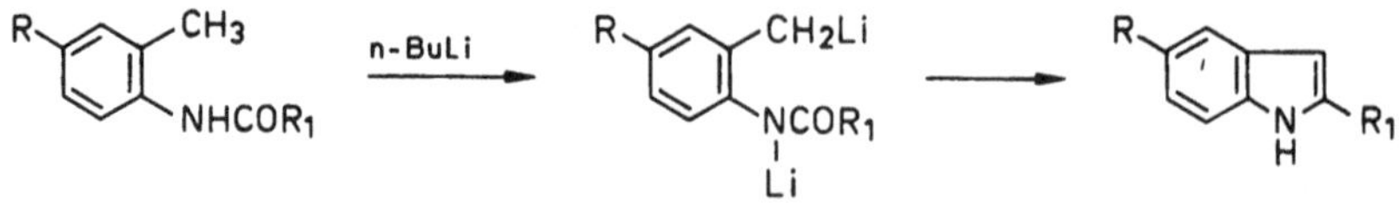

R = H, Cl

R_1 = Ph, t-Bu etc.

Other methods for the synthesis of indole involve an intramolecular Wittig reaction or an aromatic nucleophilic substitution by the carbanion derived from acetone.

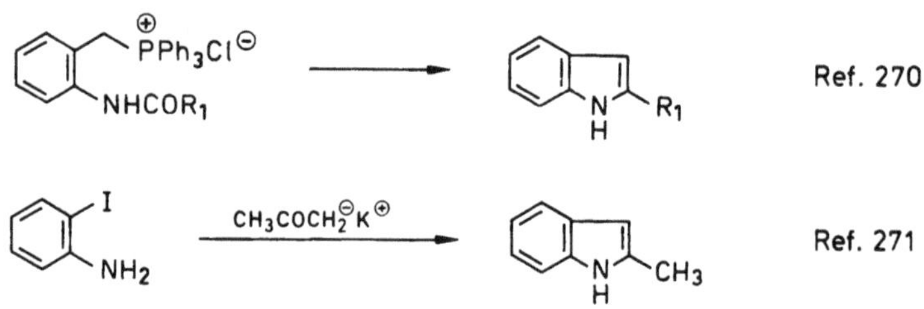

22 Azacarbostyrils and Anthramycin

A careful use of protected primary amino group in regiospecific lithiation reaction has been demonstrated in the following examples [272]:

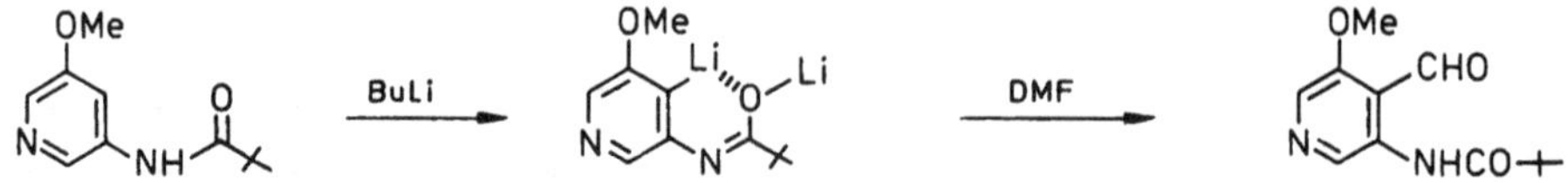

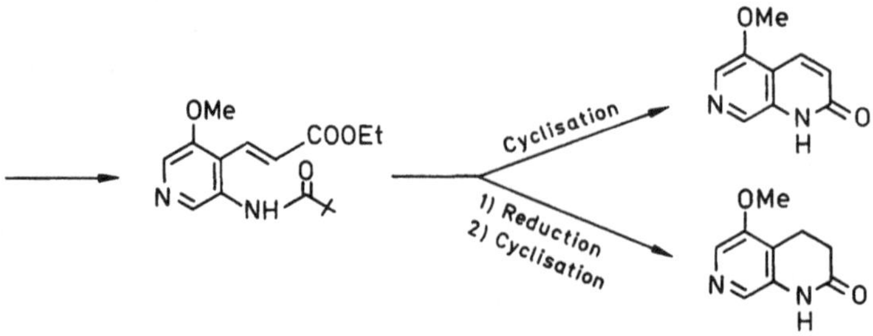

Sequential directed *ortho*-metallation reaction has been utilised in the synthesis of anthramycin [273].

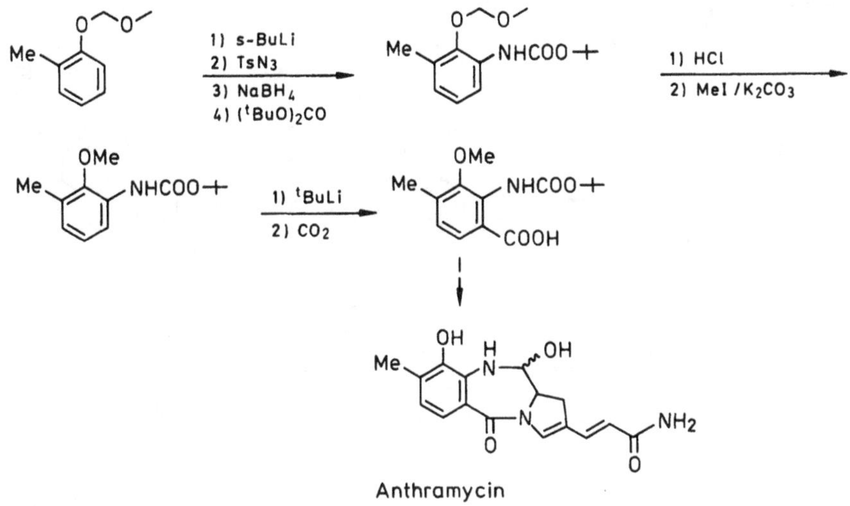

23 Miscellaneous Heterocycles

The organometallic compounds, obtained by aromatic lithiation reaction have been used for the synthesis of various heterocyclic compounds containing more than one heteroatom in a ring or metal atoms in the ring. These types of compounds available by the method are included in a recent review [53].

Some recently reported examples are presented below.

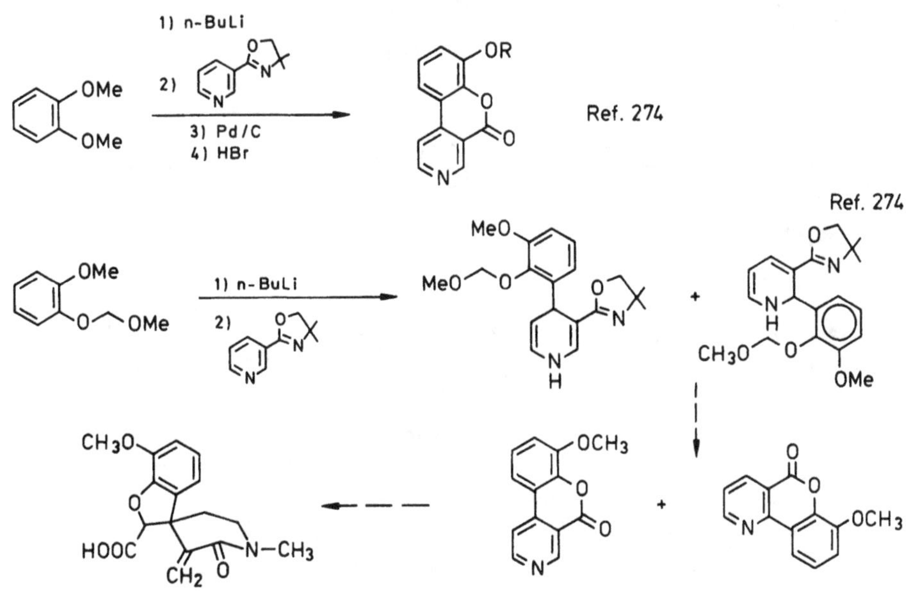

Ref. 92

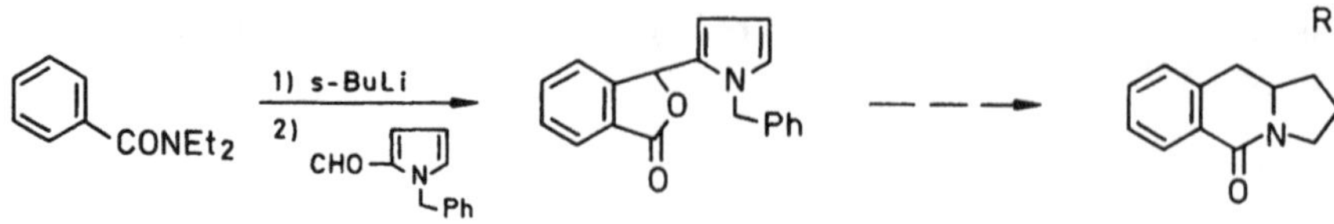

Ref. 92

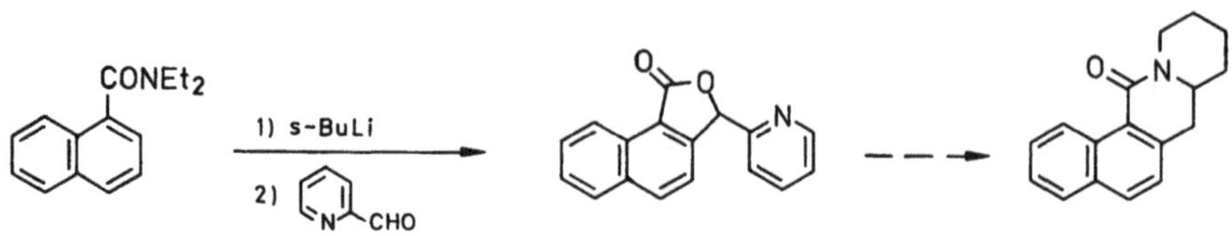

Ref. 92

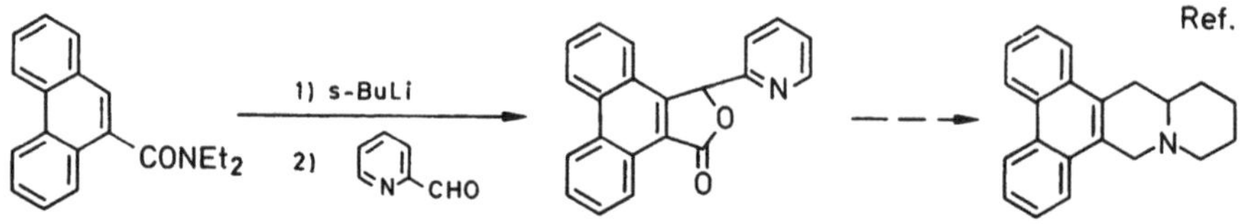

Ref. 275

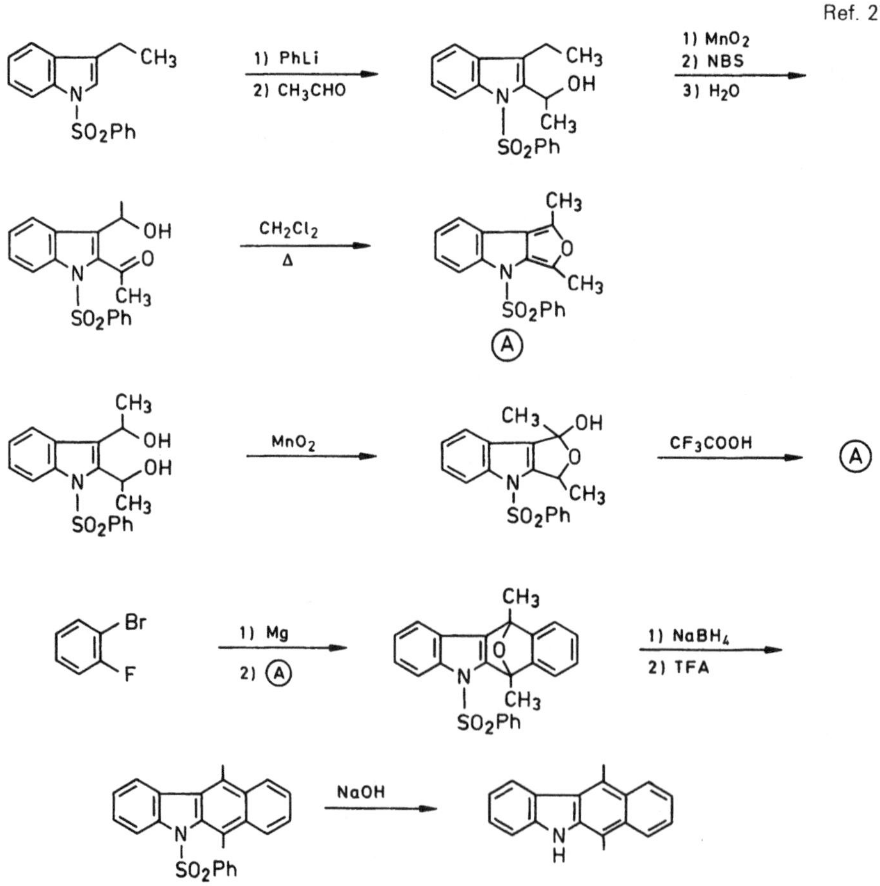

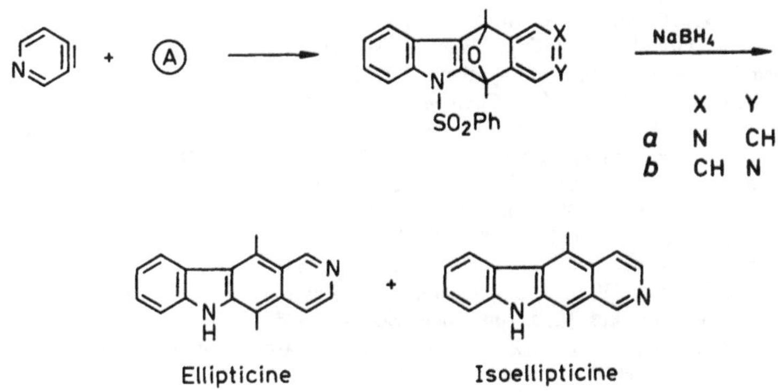

24 References

1. Gilman, H., Morton Jr., J. W.: Org. Reactions *8*, 258 (1954);
   Jones, R. G., Gilman, H.: ibid. *6*, 339 (1951)
2. Mallan, J. M., Bebb, R. L.: Chem. Review *69*, 693 (1969)
3. Wakefield, B. J.: The Chemistry of Organolithium Compounds, Pergamon Press 1974
4. Gschwend, H. W., Rodriguez, H. R.: Org. Reactions *26*, 1 (1979)
5. Whaley, W. M., Govindachari, T. R.: ibid *6*, 74 (1951)
6. Truce, W. E.: ibid. *9*, 37 (1957);
   Mali, R. S., Yadav, V. J., Zaware, R. N.: Indian J. Chem. *21B*, 759 (1982)
7. Narasimhan, N. S., Mali, R. S.: Tetrahedron Lett. 843 (1973); idem, Tetrahedron *31*, 1005 (1975)
8. Kametani, T. and Ihara, M.: J. Chem. Soc. (C), 530 (1967)
9. Ahluwalia, V. K., Chandra, Prakash: Indian J. Chem. *15B*, 423 (1977)
10. Norman, R. O. C.: Principles of Organic Synthesis, 2nd Ed., p. 398, Chapman and Hall, London 1978
11. Gould, E. S.: Mechanism and Structure in Organic Chemistry, Holt, Rinehart and Winston, New York 1959, p. 438
12. Wittig, G., Meyer, F. J., Lange, G.: Ann. *571*, 167 (1951)
13. Roberts, J. D., Curtin, D. Y.: J. Am. Chem. Soc. *68*, 1658 (1946)
14 Wittig, G., Pickels and Droge: Chem. Ber. *71*, 1903 (1938)
15. (a) Narasimhan, N. S., Alurkar, R. H.: Indian J. Chem. *7*, 1280 (1969)
    (b) Narasimhan, N. S., Chandrachood, P. S., Shete, N. R.: Tetrahedron *37*, 825 (1981)
    (c) Narasimhan, N. S., Ranade, A. C.: Indian J. Chem. *7*, 538 (1969)
16. Barnes, R. A., Nehmsmann, L. J.: J. Org. Chem. *27*, 1939 (1962);
    Graybill, B. M., Shirley, D. A.: ibid. *31*, 1221 (1966)
17. Narasimhan, N. S., Ranade, A. C.: Tetrahedron Lett. 603 (1966); idem, Chemistry and Industry, 120 (1967)
18. (a) Narasimhan, N. S., Mali, R. S., Barve, M. V.: Synthesis, 906 (1979)
    (b) Klein, K. P., Hauser, C. R.: J. Org. Chem. *32*, 1479 (1967)
    (c) Narasimhan, N. S., Bhide, B. H.: Tetrahedron Letters, 4159 (1968)
19. Growther, G. P., Sundberg, R. J., Sarpeshkar, A. M.: J. Org. Chem. *49*, 4657 (1984)
20. Trost, B. M., Pearson, W. H.: J. Am. Chem. Soc. *103*, 2483 (1981)
21. Narasimhan, N. S., RadhaKrishnan, A.: Tetrahedron Lett. *24*, 4733 (1983)
22. Reed, J. N., Snieckus, V.: ibid. *24*, 3795 (1983)
23. Nishiyama, K., Tanaka, N.: J. Chem. Soc. Chem. Commun, 1322 (1983)
24. Seebach, D., Meyer, N.: Angew. Chem. Int. Ed. *17*, 521 (1978); Chem. Ber. *173*, 1304 (1980)
25. Fuhrer, W., Gschwend, H. W.: J. Org. Chem. *44*, 1133 (1979)
26. Eaborn, C., Golborn, P., Taylor, R.: J. Organometal Chem. *10*, 171 (1967)

27. Barve, M. V.: Ph. D. Thesis, University of Poona (1983)
28. Hillis, L. R., Gould, S. J.: J. Org. Chem. *50*, 718 (1985)
29. Narasimhan, N. S., Mali, R. S.: Synthesis, 797 (1975)
30. Mills, R. J., Snieckus, V.: J. Org. Chem. *48*, 1565 (1983)
31. Dean, F. M.: Naturally Occuring Oxygen Ring Compounds, Butterworths, London 1963; Turner, W. B.: Fungal Metabolites, Academic Press, London 1971
32. Puterbaugh, W. H., Hauser, C. R.: J. Org. Chem. *29*, 853 (1964)
33. Fitt, J. J., Gschwend, H. W.: ibid. *41*, 4029 (1976)
34. Beak, P., Brown, R. A.: ibid. *42*, 1823 (1977)
35. Comins, D. L.; Brown, J. D.: Tetrahedron Lett. *24*, 5465 (1983)
36. Meyers, A. I., Avila, W. B.: J. Org. Chem. *46*, 3881 (1981)
37. Watanabe, M., Snieckus, V.: J. Am. Chem. Soc. *102*, 1457 (1980)
38. Epsztajn, J., Berski, Z., Brzezinski, J. Z., Jozwiak, A.: Tetrahedron Lett. *21*, 4739 (1980)
39. Meyers, A. I., Gabel, R. A.: ibid. 227 (1978)
40. Vecchia, L. D., Vlattas, I.: J. Org. Chem. *42*, 2649 (1977)
41. Beak, P., Tse, A., Hawkins, J., Chen, C.-W., Mills, S.: Tetrahedron *39*, 1983 (1983)
42. Uemura, M., Tokuyama, S., Sakan, T.: Chemistry Letters, 1195 (1975)
43. Benkeser, R. A., Fitzgerald, W. P., Melzer, M. S.: J. Org. Chem. *26*, 2569 (1961)
44. Uemura, M., Nishikawa, N., Hayashi, Y.: Tetrahedron Lett. *21*, 2069 (1980)
45. Larock, R. C., Fellows, C. A.: J. Am. Chem. Soc. *104*, 1900 (1982)
46. Narasimhan, N. S., Mali, R. S., Kulkarni, B. K., Gupta, P. K.: Indian J. Chem. *22B*, 1257 (1983)
47. Cushman, M., Choong, T-C., Valko, J. T., Koleck, M. P.: J. Org. Chem. *45*, 5067 (1980)
48. Harris, T. D., Roth, G. P.: ibid. *44*, 2004 (1979)
49. Comins, D. L., Brown, J. D.: ibid. *49*, 1078 (1984) and references cited therein
50. Chakravarti, S. N., Perkin Jr., W. H.: J. Chem. Soc., 196 (1929)
51. Narasimhan, N. S., Bhide, B. H.: Tetrahedron *27*, 6171 (1971)
52. Slocum, D. W., Jennings, C. A.: J. Org. Chem. *41*, 3653 (1976)
53. Narasimhan, N. S., Mali, R. S.: Synthesis, 957 (1983)
54. Opitz, L., Hansel, R.: Arch. Pharmaz. *304*, 228 (1971)
55. Trost, B. M., Rivers, G. T., Gold, J. M.: J. Org. Chem. *45*, 1835 (1980)
56. Schmid, H., Karrer, P.: Helv. Chim. Acta *28*, 722 (1945)
57. de Silva, S. O., Reed, J. N., Snieckus, V.: Tetrahedron Lett., 5099 (1978)
58. Hung, T. V., Mooney, B. A., Prager, R. H., Trippett, J. M.: Aust. J. Chem. *34*, 383 (1981)
59. Kameda, K., Namiki, K.: Chemistry Letters, 1491 (1974)
60. Beak, P., Snieckus, V.: Acc. Chem. Res. *15*, 306 (1982)
61. Mali, R. S., Yeola, S. N.: Unpublished results
62. Knight, D. W., Portas, C. D.: Tetrahedron Lett., 4543 (1977)
63. Mali, R. S., Kulkarni, B. K.: Unpublished results
64. Kraus, G. A., Sugimoto, H.: Tetrahedron Lett. *20*, 2263 (1978)
65. Hauser, F. M., Combs, D. W.: J. Org. Chem. *45*, 4071 (1980)
66. (a) Narasimhan, N. S., Gokhale, S. M.: Unpublished results
(b) Iwao, M., Inoue, H., Kuraishi, T.: Chemistry Lett., 1263 (1984)
(c) Narasimhan, N. S., Patil, P. A.: Unpublished results
(d) Narasimhan, N. S., Joshi, R. R.: Unpublished results
67. (a) Snieckus, V.: Heterocycles *14*, 1649 (1980)
(b) Watanabe, W.: Yuki. Gosel, Kagaku Kyokaishi *41*, 728 (1983)
68. Dodsworth, D. J., Caliagno, M. P., Ehrmann, E. U., Devadas, B., Sammes, P. G.: J. Chem. Soc. Perkin Trans. I., 2120 (1981)
69. Baldwin, J. E., Bair, K. W.: Tetrahedron Lett., 2559 (1978)
70. deSilva, S. O., Snieckus, V.: ibid. 5103 (1978)
71. Broadhurst, M. J., Hassall, C. H.: J. Chem. Soc. Perkin Trans I, 2227 (1982)
72. Kende, A. S., Rizzi, J. P.: J. Am. Chem. Soc. *103*, 4247 (1981)
73. Kende, A. S., Rizzi, J. P.: Tetrahedron Lett. *22*, 1779 (1981)
74. Hauser, F. M., Prasanna, S.: J. Org. Chem. *47*, 383 (1982)
75. Russell, R. A., Warrener, R. N.: J. Chem. Soc. Chem. Comm., 108 (1981)
76. Kim, K. S., Spatz, M. W., Johnson, F.: Tetrahedron Lett., 331 (1979)

77. Kim, K. S., Vanoti, E., Suarato, A., Johnson, F.: J. Am. Chem. Soc. *101*, 2483 (1979)
78. Townsend, C. A., Davis, S. G., Christensen, S. B., Link, J. C., Lewis, C. P.: ibid. *103*, 6885 (1981)
79. Li, T., Wu, Y. L.: ibid. *103*, 7007 (1981)
80. Hauser, F. M., Rhee, R. P.: J. Org. Chem. *45*, 3061 (1980)
81. Uemura, M., Take, K., Hayashi, Y.: J. Chem. Soc., Chem. Commun., 858 (1983)
82. Mills, R. J., Snieckus, V.: Tetrahedron Lett. *25*, 479, 483 (1984)
83. Watanabe, M., Maenosono, H., Furukawa, S.: Chem. Pharm. Bull. *31*, 2662 (1983)
84. Dodsworth, D. J., Caliagno, M. P., Ehrmann, U. E., Sammes, P. G.: Tetrahedron Lett. *21*, 5075 (1980)
85. Narasimhan, N. S., Patil, P. A.: Unpublished results
86. Perkin Jr. W. H., Robinson, R.: Proc. Chem. Soc. London *26*, 46 (1910); J. Chem. Soc. *99*, 775 (1911)
87. Narasimhan, N. S., Joshi, R. R., Kusurkar, R. S.: J. Chem. Soc., Chem. Commun., 177 (1985)
88. Shono, T., Usui, Y., Hamaguchi, H.: Tetrahedron Lett. *21*, 3151 (1980)
89. Prager, R. H., Tippett, J. M., Ward, A. D.: Aust. J. Chem. *34*, 1085 (1981)
90. Iwao, M., Kuraishi, T.: Tetrahedron Lett. *24*, 2649 (1983)
91. Iwao, M., Watanabe, M., deSilva, S. O., Snieckus, V.: ibid. *22*, 2349 (1981)
92. Iwao, M., Mahalanabis, K. K., Watanabe, M., deSilva, S. O., Snieckus, V.: Tetrahedron *39*, 1955 (1983)
93. Govindachari, T. R., Rajadurai, S., Subramanian, M., Viswanathan, N.: J. Chem. Soc., 2943 (1957)
94. Marsden, R., MacLean, D. B.: Tetrahedron Letters *24*, 2063 (1983)
95. Smith III, A. B., Schow, S. R., Bloom, J. D., Thompson, A. S., Winzenberg, K. N.: J. Am. Chem. Soc. *104*, 4015 (1982)
96. Kusurkar, R. S.: Ph. D. Thesis, University of Poona (1983)
97. Freskos, J. N., Morrow, G. W., Swenton, J. S.: J. Org. Chem. *50*, 805 (1985)
98. Vaulx, R. L., Jones, F. N., Hauser, C. R.: ibid. *29*, 505 (1964)
99. Klein, K. P., Hauser, C. R.: ibid. *32*, 1479 (1967)
100. Gay, R. L., Hauser, C. R.: J. Am. Chem. Soc. *89*, 2297 (1967)
101. Slocum, D. W., Gierer, P. L.: J. Org. Chem. *41*, 3668 (1976)
102. Slocum, D. W., Rockett, B. W., Hauser, C. R.: J. Am. Chem. Soc. *87*, 1241 (1965)
103. Mali, R. S., Talele, M. I.: Unpublished results
104. Upton, C. J., Beak, P.: J. Org. Chem. *40*, 1094 (1975)
105. Keay, B. A., Rodrigo, R.: J. Am. Chem. Soc. *104*, 4725 (1982)
106. Corey, E. J., Das, J.: ibid. *104*, 5551 (1982)
107. Allahdad, A., Knight, D. W.: J. Chem. Soc. Perkin Trans. I, 1855 (1982)
108. Berlingozzi, S., Lupo, G.: Gazz. Chim. Ital. *57*, 255 (1927)
109. Gijbels, M. J. M., Scheffer, J. J. C., Svendsen, A. B.: Planta Medica Suppl., 41 (1980)
110. Natelson, S., Gottfried, S. P.: J. Am. Chem. Soc. *58*, 1432 (1936)
111. McAlees, A. J., McCrindle, R., Sneddon, D. W.: J. Chem. Soc. Perkin Trans I, 2030 (1977)
112. Braun, M.: Ann. Chem., 2247 (1981)
113. Duggal, J. K., Misra, K.: Indian J. Chem. *19B*, 1000 (1980)
114. Schwan, T. J., Gray, J. E.: J. Pharm. Sci. *67*, 863 (1978)
115. Narasimhan, N. S., Mali, R. S., Kulkarni, B. K.: Indian J. Chem. *22B*, 850 (1983)
116. Ludt, R. E., Hauser, C. R.: J. Org. Chem. *36*, 1607 (1971)
117. Rodriguez, H. R.: unpublished results, of Organic Reactions *26*, 1 (1979)
118. Houlihan, W. J., Nadelson, J.: U.S. Patent Appl. 126272 (1971); Chem. Abstr. *78*, 29485 (1973); *85*, 94225 (1976)
119. Watanabe, H., Mao, C. L., Barnish, I. T., Hauser, C. R.: J. Org. Chem. *34*, 919 (1969)
120. Katritzky, A. R., Rastgoo, S. R., Ponkshe, N. K.: Synthesis, 127 (1981)
121. Farbes, I., Pratt, R. A., Raphael, R. A.: Tetrahedron Lett., 3965 (1978)
122. Houlihan, W. H., Nadelson, J.: U.S. Patent 3, 879, 418 (1975); Chem. Abstr. *83*, 58651 (1975)
123. Ranade, A. C., Kurnawal, V. M.: unpublished results
124. Berry, R. D.: Chem. Rev. *64*, 229 (1964)
125. Bhide, B. H., Gupta, V. P., Shah, K. K.: Chemistry and Industry, 84 (1980)
126. Ellefson, C. R.: J. org. Chem. *44*, 1533 (1979)

127. Patel, H. A., MacLean, D. B.: Can. J. Chem. *61*, 7 (1983)
128. Ellefson, C. R., Prodan, K. A.: J. Med. Chem. *24*, 1107 (1981)
129. Bhide, B. H., Brahmbhatt, D. I., Shah, K. K.: Indian J. Chem. *20B*, 831 (1981)
130. Narasimhan, N. S., Bhide, B. H.: Chemistry and Industry, 75 (1974)
131. Harris, T. D., Neuschwander, Boekelheide, V.: J. Org. Chem. *43*, 727 (1978)
132. Bhide, B. H., Parekh, H. N.: Chemistry and Industry, 773 (1974)
133. Bhide, B. H., Gupta, V. P.: Indian J. Chem. *15B*, 30 (1977)
134. Vaulx, R. L., Puterbaugh, W. H., Hauser, C. R.: J. Org. Chem. *29*, 3514 (1964)
135. Kraus, G. A.: ibid. *46*, 201 (1981)
136. Watanabe, M., Sahara, M., Furukawa, S., Billedean, Snieckus, V.: Tet. Lett. *23*, 1647 (1982)
137. Watanabe, M., Sahara, M., Kubo, M., Furukawa, S., Billedean, R. J., Snieckus, V.: J. Org. Chem. *49*, 742 (1984)
138. Korte, D. E., Hegedus, L. S., Wirth, R. K.: ibid. *42*, 1329 (1977)
139. Beugelmans, R., Ginsburg, H., Bois-Choussy, M.: J. Chem. Soc. Perkin Trans I, 1149 (1982)
140. Nishikawa, H.: J. Agri. Chem. Soc., Japan *9*, 772 (1933)
141. Sibi, M. P., Miah, M. A. P., Snieckus, V.: J. Org. Chem. *49*, 737 (1984)
142. Blair, J., Newbold, G. T.: J. Chem. Soc., 2871 (1955)
143. Matsui, M., Mori, K., Arasaki, S.: Agril Biol. Chem. *28*, 890 (1964)
144. Yoshitsugu, A., Tadao, K., Takashi, K.: Bull. Chem. Soc. Japan *46*, 3311 (1973)
145. Harwood, L. M.: J. Chem. Soc. Chem. Commn., 1120 (1982)
146. Ballio, A., Barcellona, S., Santurbano, B.: Tetrahedron Lett., 3723 (1966)
147. Carpenter, R. C., Sotheeswaran, S., Sultanbawa, M. U. S., Balasubramanian, S.: Phytochemistry *19*, 445 (1980)
148. Chatterjea, J. N., Mukherji, J., Bhakta, C., Banerjee, B. K.: J. Ind. Chem. Soc. *49*, 797 (1972)
149. Bhide, B. H., Shah, K. K.: Indian J. Chem. *19B*, 9 (1980)
150. Sondheimer, F.: J. Am. Chem. Soc. *79*, 5036 (1957)
151. Logan, W. R., Newbold, G. T.: Chemistry and Industry, London, 1485 (1957); Slates, H. L., Webber, S., Wendlar, N. L.: Separatum of Chima. *21*, 468 (1967)
152. Bhide, B. H., Brahmbhatt, D. I.: Proc. Indian Acad. Aci. *89*, 525 (1980)
153. Govindachari, T. R., Patankar, S. J., Vishwanathan, N.: Phytochemistry *10*, 1603 (1971)
154. Bhide, B. H., Gupta, V. P., Narasimhan, N. S., Mali, R. S.: Chemistry and Industry, 519 (1975); Bhide, B. H., Gupta, V. P.: Indian J. Chem. *15B*, 512 (1977)
155. Govindachari, T. R., Patankar, S. J., Vishwanathan, N.: Indian J. Chem. *9*, 507 (1971)
156. Chatterjea, J. N., Bhakta, C., Sinha, N. D.: J. Indian Chem. Soc. *52*, 158 (1975)
157. Narasimhan, N. S., Bapat, C. P.: J. Chem. Soc. Perkin Trans I, 2099 (1982)
158. van der Merwe, K. J., Styne, P. S., Fourie, L.: J. Chem. Soc., 7083 (1965)
159. Steyn, P. S., Holzapfel, C. W.: Tetrahedron *23*, 4449 (1967)
160. Roberts, J. C., Woolven, P.: J. Chem. Soc. (C), 278 (1970)
161. Chandrachood, S. P.: Ph. D. Thesis, University of Poona (1973)
162. Asahina, Y., Asana, J.: Chem. Ber. *62*, 171 (1929); ibidem, *63*, 429 (1930); *64*, 1252 (1931)
163. Govindachari, T. R., Parthasarathy, P. C., Desai, H. K., Ramachandran, K. S.: J. Ind. Chem. Soc. *13*, 537 (1975)
164. Greger, H., Bohlmann, F., Zdero, C.: Phytochemistry, *16*, 795 (1977)
165. Mallabaev, A., Saitbaeva, I. M., Sidyakin, G. P.: Khim. Prir. Soeidin *7*, 1208 (1971); Chem. Abstr. *75*, 115871 (1971)
166. Chatterjea, J. N., Mukherjee, S. K., Bhakta, C.: Ind. J. Chem. *20B*, 359 (1981)
167. Chatterjea, J. N., Bhakta, C., Mukherjee, S. K.: ibid. *20B*, 992 (1981)
168. Batu, G., Stevenson, R.: J. Org. Chem. *45*, 1532 (1980)
169. Crawley, G. C.: J. Chem. Soc. Perkin Trans. I, 221 (1982)
170. Narasimhan, N. S., Mali, R. S., Kulkarni, B. K.: Tetrahedron Lett. *22*, 2797 (1981); idem, Tetrahedron *39*, 1975 (1983)
171. Mali, R. S., Patil, S. D., Patil, S. L.: Unpublished results
172. Nagata, W., Itazaki, H., Okada, K., Wakabayashi, T., Shibata, K., Tokutake, N.: Chem. Pharm. Bull. Japan *23*, 2867 (1975)
173. Braun, M., Ringer, E.: Tetrahedron Lett. *24*, 1233 (1983)
174. Pandey, G. D., Tiwari, K. P.: Polish J. Chem. *53*, 2159 (1979)
175. idem: Indian J. Chem. *18B*, 544 (1979)

176. idem: ibid. *19B*, 160 (1980)
177. Mali, R. S., Yeola, S. N.: ibid. *23B*, 79 (1984)
178. idem: ibid. *23B*, 818 (1984)
179. Patil, S. D., Mali, R. S.: ibid. *24B*, 360 (1985)
180. Mali, R. S., Patil, S. D.: Unpublished results
181. Meise, W., Pfisterer, H.: Arch. Pharm. *310*, 495 (1975); Chem. Abstr. *88*, 7156f (1978); ibidem *310*, 501 (1975); Chem. Abstr. *88*, 7157 g (1978)
182. Laitem, L., Christiaens, L., Welter, A.: Heterocycles *4*, 1107 (1976); Chem. Abstr. *85*, 123789 (1976)
183. Spangler, R. J., Beckmann, B. G., Kim, J. H.: J. Org. Chem. *42*, 2989 (1977); and references cited therein
184. Cushman, M., Dekow, F. W.: J. Org. Chem. *44*, 407 (1979)
185. Cushman, M., Choong, T. C., Valko, J. T., Koleck, M. P.: Tetrahedron Lett. *21*, 3845 (1980)
186. Pai, B. R., Natarajan, S., Suguna, H., Rajeswari, S., Chandrasekaran, S., Nagarajan, K.: Ind. J. Chem. *21B*, 607 (1982)
187. deSilva, S. O., Ahmed, I., Snieckus, V.: Can. J. Chem. *57*, 1598 (1979)
188. Nadkarni, D. R., Usgaonkar, R. N.: Indian J. Chem. *15B*, 185 (1977)
189. Nadkarni, D. R., Usgaonkar, R. N.: ibid. *16B*, 320 (1978)
190. Tirodkar, R. B., Usgaonkar, R. N.: ibid. *10*, 1060 (1972); Curr. Sci. *45*, 832 (1976)
191. Haimova, M. A., Ognyanov, V. I., Mollov, N. M.: Synthesis, 845 (1980)
192. Iida, H., Kawano, K., Kikuchi, T., Yoshimizu, F.: Yakugaku Zasshi *96*, 176 (1976)
193. Cushmann, M., Gentry, J., Dekow, F. W.: J. Org. Chem. *42*, 1111 (1977)
194. Iwasa, K., Gupta, Y., Cushman, M.: Tetrahedron Lett. *22*, 2333 (1981); Cushman, M., Dekow, F. W.: Tetrahedron *34*, 1435 (1978)
195. Cushman, M., Wong, W. C.: J. Org. Chem. *49*, 1278 (1984)
196. Murray, R. D. H., Mendez, J., Brown, S. A.: The Natural Coumarins, J. Wiley and Sons, Chichester, New York 1982
197. Johnson, J. R.: Org. React. *1*, (Ed. Adams, R.) John Wiley and Sons 1942
198. Kelkar, S. L., Phadke, C. P., Marina, S.: Indian J. Chem. *23B*, 458 (1984)
199. Guillaumel, J., Demerseman, P., Royer, R.: Tetrahedron *37*, 4215 (1981)
200. Gilman, H., Avakian, S.: J. Am. Chem. Soc. *67*, 349 (1945)
201. Kraus, G. A., Pezzanite, J. O.: J. Org. Chem. *44*, 2480 (1979)
202. Taylor, R. T., Cassell, R. A.: Synthesis, 672 (1982)
203. Bestmann, H. J., Schmid, G., Sandmeier, D.: Angew. Chem. *88*, 92 (1976)
204. Dreyer, D. L., Munderloh, K. P., Thiessen, W. E.: Tetrahedron *31*, 287 (1975)
205. Basyauni, S. E., Towers, G. H. N.: Can. J. Biochemistry *42*, 493 (1964); Chem. Abstr. *61*, 2183a (1964)
206. Murhurg, S., Tolman, R. L.: J. Het. Chem. *17*, 1333 (1980)
207. Narasimhan, N. S., Paradkar, M. V.: Chemistry and Industry, 1520 (1963); idem; Indian J. Chem. *7*, 536 (1969)
208. Cagniant, P., Cagniant, D.: Advances in Heterocyclic Chemistry, Vol. 18, (Ed. Katritzky, A. R., Boulton, A. J.) Academic Press, New York 1975. Friedrichsen, W.: Advances in Heterocyclic Chemistry, Vol. *26* (Ed. Katritzky, A. R., Boulton, A. J.) Academic Press, New York 1980
209. Donnelly, D. M. X., Hirotani, M., Reilly, J. O., Polonsky, J.: Tetrahedron Lett., 651 (1977)
210. Duffley, R. P., Stevenson, R.: J. Chem. Res. (s), 468 (1978)
211. Shima, K., Hisada, S., Inagaki, I.: Yakugaku Zasshi *92*, 1410 (1972)
212. Kanithong, B., Robertson, A.: J. Chem. Soc., 925 (1939)
213. Cooke, R. G., Rae, I. D.: Aust. J. Chem. *17*, 379 (1964)
214. Schreiber, F. G., Stevenson, R.: J. Chem. Soc. Perkin Trans. I, 90 (1977)
215. Duffley, R. P., Stevenson, R.: ibid. 802 (1977)
216. Bonner, W. A., DeGraw, J. I.: Tetrahedron *18*, 1295 (1962)
217. Hercouet, A., LeCorre, M.: ibid. *37*, 2867 (1981)
218. Bohlmann, F., Zdero, C., Grenz, M.: Chem. Ber. *107*, 2730 (1974)
219. Bohlmann, F., Fritz, G.: Tetrahedron Lett. *22*, 95 (1981)
220. Ronald, R. C.: ibid. 4413 (1976)
221. Ronald, R. C., Gewali, M. B., Ronald, B. P.: J. Org. Chem. *45*, 2224 (1980)
222. Marino, J. P., Dax, S. L.: ibid. *49*, 3671 (1984)

223. O'Brien, S., Smith, D. C. C.: J. Chem. Soc., 2907 (1963)
224. Birch, A. J., Hextall, P.: Aust. J. Chem. *8*, 253 (1955)
225. Birch, A. J., Salahud-Din, M., Smith, D. C. C.: J. Chem. Soc. (c), 523 (1966)
226. Narasimhan, N. S., Mali, R. S.: Synthesis, 796 (1975)
227. Whaley, W. M., Govindachari, T. R.: Org. React. *6*, (Ed. Adams, R.) John Wiley and Sons 1951
228. Gensler, W. J.: Org. React. *6*, (Ed. Adams, R.) John Wiley and Sons 1951
229. Narasimhan, N. S., Ranade, A. C., Bhide, B. H.: Indian J. Chem. *20B*, 439 (1981)
230. Kashadan, D. S., Schwartz, J. A., Rapoport, H.: J. Org. Chem. *47*, 2638 (1982)
231. Abrahum, T.: Monatsh. Chem. *113*, 371 (1982)
232. Sandberg, E.: Chem. Scripta 7, 223 (1975); Chem. Abstr. *83*, 114255f (1975)
233. Modi, P. R., Tirodkar, R. B., Usgaonkar, R. N.: Ind. J. Chem. *20B*, 813 (1981)
234. Dean, R. T., Rapoport, H.: J. Org. Chem. *43*, 2115 (1978)
235. Houlihan, W. J., Nadelson, J.: U.S. Patent 3, 870, 722 (1975); Chem. Abstr. *83*, 9819v (1975); U.S. Patent 3, 872, 125 (1975); Chem. Abstr. *83*, 28123e (1975)
236. Houlihan, W. J., Nadelson, J.: U.S. Patent 3, 892, 752 (1975); Chem. Abstr. *83*, 178856k (1975)
237. Shamma, M., Moniot, J. L.: Isoquinoline Alkaloid Research 1972—1977, Plenum Press, New York—London 1979;
Krane, B. D., Shamma, M.: J. Natural Products (Lloydia) *45*, 377 (1982)
238. Mali, R. S., Kulkarni, B. K., Shankaran, K.: Synthesis, 329 (1982)
239. Poindexter, G. S.: J. Org. Chem. *47*, 3787 (1982)
240. Houlihan, W. J., Nadelson, J.: U.S. Patent 53, 878, 215 (1975), C.A. *83*, 79099 (1975)
241. Venkov, A. P., Mollov, N. M.: Synthesis, 216 (1982)
242. Venkov, A. P., Lukanov, L. K., Mollov, N. M.: ibid. 486 (1982)
243. Narasimhan, N. S., Mali, R. S., Kulkarni, B. K.: Synthesis, 114 (1985)
244. Pandey, G. D., Tiwari, K. P., Tandon, S. P.: J. Indian. Chem. Soc. *59*, 532 (1982);
Pandey, G. D., Tiwari, K. P.: Indian. J. Chem. *18B*, 545 (1979)
245. Hunziker, F., Kunzle, F., Schmutz, J.: Helv. Chim. Acta *49*, 1433 (1966)
246. Nagarajan, K., Kulkarni, K., Venkateswarlu, C. L.: Ind. J. Chem. *6*, 225 (1968);
Jacques, R., Rossi, A., Urech, E., Bein, H. J., Hoffmann, K.: Helv. Chim. Acta *42*, 1265 (1959);
Hanze, A. R., Strube, R. E., Greig, M. E.: J. Med. Chem. *6*, 767 (1963):
Schmutz, J., Kunzle, F., Hunziker, F., Burki, A.: Helv. Chim. Acta *48*, 336 (1965)
247. Shete, N. R.: Ph. D. Thesis, University of Poona (1971)
248. Narasimhan, N. S., Chandrachood, P. S.: Synthesis 589 (1979)
249. Shirley, D. A., Gilmer, J. C.: J. Org. Chem. *27*, 4421 (1962)
250. Gilman, H., Spatz, S. M.: ibid. *17*, 860 (1952)
251. Iwao, M., Reed, J. N., Snieckus, V.: J. Am. Chem. Soc. *104*, 5531 (1982)
252. Johne, S., Groger, D.: Pharmazie *27*, 195 (1972)
253. Adams, J. H., Brown, P. M., Gupta, P., Khan, M. S., Lewis, J. R.: Tetrahedron *37*, 209 (1981)
254. Narasimhan, N. S., Paradkar, M. V., Alurkar, R. H.: ibid. *27*, 1351 (1971)
255. Narasimhan, N. S., Mali, R. S.: ibid. *30*, 4153 (1974)
256. Collins, J. F., Gray, G. A., Grundon, M. F., Harrison, D. M., Spyropoulos, C. G.: J. Chem. Soc. Perkin Trans I, 94 (1973)
257. Grundon, M. F.: The alkaloids, Vol. 17, (Ed. Manske, R. H. F., Rodrigo, R. G. A.) Academic Press 1979
258. Narasimhan, N. S., Mali, R. S., Gokhale, A. M.: Indian J. Chem. *18B*, 115 (1979)
259. Ranade, A. C., Mali, R. S., Kurnawal, V. M.: ibid. *21B*, 528 (1982)
260. Narasimhan, N. S., Bhagwat, S. P.: Synthesis, 903 (1979)
261. Joag, S. D.: Ph. D. Thesis, University of Poona (1980)
262. Houlihan, W. J., Pieroni, A. J.: J. Heterocycl. Chem. *10*, 405 (1973)
263. Muchowski, J. M., Venuti, M. C.: J. Org. Chem. *45*, 4798 (1980)
264. Sundberg, R. J., Russell, H. F.: ibid. *38*, 3324 (1973)
265. Ranade, A. C., Mali, R. S., Deshpande, H. R.: Experientia *35*, 574 (1979)
266. Ranade, A. C., Mali, R. S., Gidwani, H. R., Deshpande, H. R.: Chemistry and Industry, 310 (1977)
267. Ranade, A. C., Deshpande, H. R.: unpublished results

268. Blatcher, P., Middlemiss, D., Murray-Rust, P., Murray-Rust, J.: Tetrahedron Lett. *21*, 4193 (1980)
269. Houlihan, W. J., Parrino, V. A., Uike, Y.: J. Org. Chem. *46*, 4511 (1981);
Newkome, G. R., Pandler, W. W.: Contemporary Heterocyclic Chemistry, John Wiley and Sons, New York 1982
270. LeCorre, M., Hercouet, A., Le Baron, H.: J. Chem. Soc. Chem. Comm., 14 (1981)
271. Beugelmans, R., Roussi, G.: ibid. 950 (1979);
Bard, R. R., Bunnett, J. I.: J. Org. Chem. *45*, 1547 (1980)
272. Turner, J. A.: ibid. *48*, 3401 (1983)
273. Reed, J. N., Snieckus, V.: Tetrahedron Lett. *25*, 5505 (1984)
274. Rosenberg, S. H., Rapoport, H.: J. Org. Chem. *49*, 56 (1984)
275. Gribble, G. W., Saulnier, M. G., Sibi, M. P., Nutaitis, J. A. O.: J. Org. Chem. *49*, 4518 (1984)
276. Chandrachood, P. S.: Ph. D. Thesis, University of Poona (1980)

Electrochemistry of Solvated Electrons

Ninel M. Alpatova, Lev I. Krishtalik, and Yuri V. Pleskov

A. N. Frumkin Institute of Electrochemistry, Academy of Sciences of USSR,
117071 Moscow, Leninsky prospekt 31

Table of Contents

1 Introduction 151

2 Energetic Characteristics of Excess Electrons in Polar Media 153
2.1 Excess Electron Energy Levels and Their Methods of Determination 153
2.2 Determination of the Energy of a Delocalized Electron 159
2.3 The Difference in Delocalized and Solvated Electron Energies 162
2.4 Determination of the Energetic Characteristics of Solvated Electrons and Their Associates in Hexamethylphosphotriamide Solution 163

3 General Conditions for Cathodic Generation of Solvated Electrons 167

4 The Nature of Particles that Form on the Cathode in Liquid Ammonia and Hexamethylphosphotriamide. Properties of Solvated Electrons in These Systems 172

5 Equilibrium Electron Electrode 177

6 Anodic Reactions in Solvated Electron Solutions 180
6.1 Anodic Oxidation of Solvated Electrons and their Associates 180
6.2 Using Solvated Electrons in Batteries 186

7 Kinetics and Mechanism of Electrochemical Generation of Solvated Electrons 187
7.1 Specific Features of Cathodic Generation of Solvated Electrons in Different Solvents 187
7.2 Primary Nature of Cathodic Generation of Solvated Electrons 192
7.3 Mechanism of Electrochemical Generation of Solvated Electrons . . . 194
7.4 Effect of Electrode Passivation on the Generation of Electrons 196

8 Possible Role of Solvated Electrons in "Ordinary" Electrochemical Reactions 201

9 Application of Electrochemically Generated Solvated Electrons for the Reduction of Organic Compounds 205

10 Conclusion 212

11 References 213

Over the past 10–15 years a new trend has been developed in theoretical electrochemistry: the electrochemistry of solvated electrons. In this review theoretical concepts of the electrochemical properties of solvated electrons and the results of experimental studies are considered from a unified position. Also discussed are: energy levels of localized (solvated) and delocalized electrons in solutions and methods for their determination; conditions of electrochemical formation of solvated electrons and properties of these solutions; equilibrium on an "electron electrode". The kinetics and mechanisms of cathodic generation of solvated electrons and of their anodic "oxidation" are discussed in detail. In the last sections participation of solvated electrons in "ordinary" electrode reactions is discussed, and the possibilities of cathodic electrosyntheses utilizing solvated electrons are considered.

1 Introduction

The best known and most studied systems in which the formation of solvated electrons is observed are the alkali metal — liquid ammonia systems. Shortly after his discovery of alkali metals, Davy initiated studies into their reactions with dry gaseous ammonia. In November 1808 he noticed that potassium "assumed a beautiful metallic appearance and gradually became of a fine blue colour" when heated in an ammonia atmosphere. It is today difficult to interpret the processes that occured in his experiment. Most probably, he noticed a form of the electron localized in a condensed phase. Unfortunately, this and other analogous records of Davy remained unknown for long; only in 1980 were they discovered and later published [1].

Actual acquaintance with the systems in which solvated electrons play a significant role started in 1864 when Weyl discovered the metal/liquid ammonia solutions [2]. However, only after several decades, thanks to the work of Kraus [3,4], it became clear that in these solutions the alkali metal atoms dissociate into cations and "anions" that are solvated electrons. Although in several features solvated electrons differ distinctively from usual anions, it is nonetheless advantageous to draw an analogy between them.

Solvated electrons were considered to be exotic over several decades and that is why comparatively little time was devoted to their study. However, the situation drastically changed early in the 60's when the crucial role of solvated electrons in radiation chemistry was reliably ascertained. This triggered numerous theoretical and experimental studies into the structure and properties of solvated electrons and their solutions, the pathways of formation of solvated electrons, the kinetics and the mechanism of reactions involving their participation. The findings of these studies have been summarized in several monographs [5-7] and in the proceedings of international conferences dedicated to the physico-chemistry of solvated electrons (the last, 6th Weyl Symposium [8], was held in 1983); therefore, these will not be specifically considered in this overview. We shall juxtapose these with electrochemical data, of course. The same relates also to electron photoemission from metal into solution, to which monographs [9,10] are devoted.

In this overview we aim at considering the processes that occur at the metal/electrolyte solution interface, involving the participation of solvated electrons. A number of reviews [11-17] deal with this problem: part of these concerns the earlier stages of development of this field, much later works consider but special aspects of the problem.

Electrochemical generation of solvated electrons was first observed in 1897 by Cady [18], who found that when sodium solutions in liquid ammonia are electrolysed the blue coloration intensity increases at the cathode. All information on cathode generation of solvated electrons remained at this qualitative level for over half a century until Laitinen and Nyman [19] made the first attempt to quantitatively investigate the kinetics of this process. This work, however, remained isolated for a long time and only after 20 years, with the awakening of interest in the chemistry of solvated electrons, were systematic studies into the kinetics of electrode reactions of solvated electrons started, almost simultaneously by three groups of researchers: in Southampton [20], Tokyo [21], and Moscow [22]. In Moscow these studies were started at the initiative of Professor A. N. Frumkin. Later, other scientists also undertook such studies.

The studies into the electrochemical kinetics of solvated electrons were to some extent stimulated by the hypothesis put forward in the second half of 60's (see Sect. 8) for explaining the role of solvated electrons as intermediate products of electrode reactions, and also by the development made at that time in organic synthesis involving the participation of solvated electrons (see Sect. 9). Undoubtedly, knowledge of the mechanism of electrode generation of solvated electrons is of fundamental importance. "Electrochemistry is the chemistry of the electron", Professor A. N. Frumkin once said. In fact, electron reactions at the interface of electronic and ionic conductors are inevitably associated with the electron addition or detachment process. In a solvated electron reaction no heavy particle (atom or molecule) acts as electron acceptor, or donor. In this sense, the electrode reactions of solvated electrons are "the most simple" electrode processes. Therefore, an insight into the solvated electron reaction mechanism is necessary for electrochemical kinetics as a whole.

As will be seen later, along with the features common to all electrode reactions, the electrochemistry of solvated electrons also has its own distinctive characteristics. These are first of all related to the specificities of the state of excess electrons in polar liquids.[1] Unlike other particles, electrons can exist in a polar medium both in localized and delocalized states.

A proper solvated electron is a particle localized in the potential well of a polar medium, the well being created by the interaction of electron charge with the permanent and induced dipole moments of the nearest as well as remote neighbours. This notion of the nature of a solvated electron, based on the idea [25)] that the Landau-Pekar theory [23,24)] initially advanced for solid bodies can be applied also to liquid systems, was advanced in 1948; since then considerable efforts have been made to develop it and verify it experimentally. In most liquid systems, localization of an electron is followed by the formation of a cavity where most of the density of the solvated electrons is concentrated. The cavity is surrounded by the orientated dipoles of the solvent. Usually, the radius of this cavity equals about 3–3.5 Å which conforms to a solvated-electron molar volume of 70–100 cm^3. This is the reason why solutions with large concentrations of solvated electrons have a lower density.

Another state of an excess electron is the delocalized which is sometimes called "dry" electron. The electron in this state is not localized in a definite microscopic region, but freely travels inside the liquid. This state is analogous to that of an electron in a conduction band. As the electron is not localized for a sufficiently long time, there is no possibility of the appearance of a corresponding polarization of a nuclear subsystem, in particular of orientation of dipoles. A delocalized electron interacts

1 The term "excess electron" is an accepted term for identifying an electron in a condensed nonmetallic phase which is not bound to any definite molecule of the medium. The electron in this sense contrasts with the bound electrons occupying definite orbitals of molecules. By this term it should not be meant that the phase in which there is an excess electron has some negative charge. The electroneutrality of the system is not violated; for instance, in metal-liquid ammonia solutions the solvated eletron charge is compensated by the alkali metal cation charge. During cathodic generation of solvated electrons either cations are formed at the anode or anions are consumed in equivalent amounts, i.e. as in any process of electrolysis the electroneutrality of the system is preserved.

only with the electronic polarizability of the medium whose response is almost intertialess.

In some methods of obtaining excess electrons, for instance, during electron photoemission the excess electrons are injected into a solution having higher-than-thermal kinetic energy. In this state they can remain only as delocalized electrons. After their excess kinetic energy has been dissipated, relatively slow processes of trapping electrons and forming solvated electrons can proceed. However, it cannot be said that delocalized electrons completely disappear — thermal motion maintains an equilibrium between the localized and delocalized electron states. In other generation methods where deviations from thermal equilibrium are not significant, excess electrons can be immediately generated in one or the other state. For kinectic reasons, these can in general appear in nonequilibrium ratios, equilibrium being established in the sequel (see Sect. 7).

For discussing the electrode processes of solvated electrons it is necessary to know their energy characteristics, on the one hand. On the other hand, the electrochemical data permit the energy of the particles participating in the reactions to be determined. Therefore, at the beginning of this overview we have summarized the data on various energy characteristics of excess electrons. Further we shall consider some properties of the solvated electron solutions, electrochemical conditions of their formation, and equilibrium on an "electron electrode". Much attention has been given to the kinetics and mechanism of the cathodic and anodic reactions, respectively, of formation of excess electrons in a solution and their transition from the solution into a metal. This is the central part of this overview as it explains the main features of electrochemistry of solvated electrons; in particular, it has been shown that along with the reactions of localized electrons (solvated electrons proper) there occur processes involving the participation of delocalized electrons. In the last sections the participation of solvated electrons in other electrode processes is discussed, and the possibilities of cathodic electrosyntheses utilizing solvated electrons are considered.

2 Energetic Characteristics of Excess Electrons in Polar Media

2.1 Excess Electron Energy Levels and Their Methods of Determination

Two trends are observed in the investigations of the energy characteristics of excess electrons. Starting with Jortner's work, a number of authors, by using physical models, have tried to calculate these characteristics theoretically (see e.g. [26]). In other work, these have been found on the strength of the results of experimental investigations into the kinetics of electron transitions of the emission kind [10,27] at the solution/electrode and solution/vapour interfaces, and also of electron equilibria in solutions. Within the framework of the second approach, this problem has been systematically considered for the first time in [28].

We shall briefly consider the main relationship that describe the energy characteristics of excess electrons in electrolyte solutions, and the experimental techniques for measuring these characteristics.

As reference point of energy U (level $U^V = 0$, Fig. 1) the electron energy in vacuum

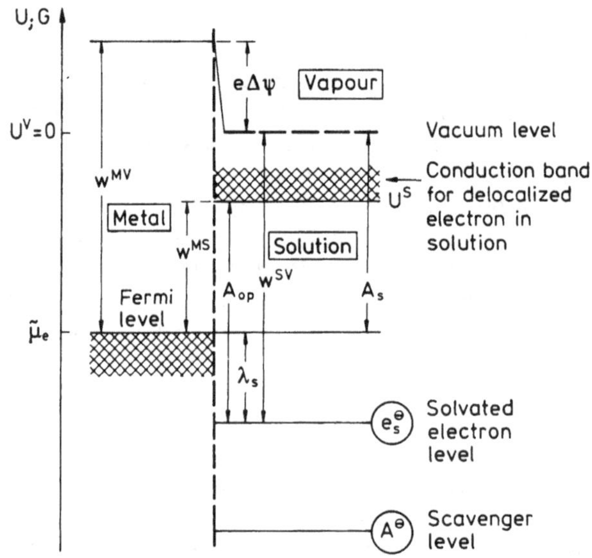

Fig. 1. Real energy levels for excess electrons in the metal-solution-vapour system

is taken.[2] In the case of a charged phase (when in contact the solution as well as the electrode are charged), the U^V level shows the energy of an electron which is not at infinity, but is in the immediate neighbourhood of the solution's surface (but beyond the range of action of purely surface forces, in particular of electrical image forces) [29,30]. This energy scale is related to the scale of electrode potentials E by

$$U = -eE + \text{const} \tag{1}$$

where e is the absolute value of the electron's charge and const $= \tilde{\mu}(\text{RE})$ is the electrochemical potential (per electron) of electrons in the reference electrode ("Fermi level of the reference electrode"). This constant depends on the nature of the solvent and the choice of the reference electrode. It can be determined by the relation [31]

$$-\tilde{\mu}(\text{RE}) = w^{MV} + e\Delta\Psi \tag{2}$$

in which w^{MV} denotes the (uncharged) metal-to-vacuum electronic work function and $\Delta\Psi$ is the Volta potential difference between the metal (at the reference electrode potential, E = 0) and the solution. (In particular, for a normal hydrogen electrode in aqueous solution $\tilde{\mu}(\text{NHE}) = -4.4$ eV) [31].

2 Here we shall not distinguish between energies of electron in vacuum and vapour phase (which can solely be realized in contact with the solution under the conditions of the emission experiment). For simplicity's sake it is assumed that the energy characteristics of electrons in solution are entirely determined by the solvent and are independent of the solute added to make the liquid phase conductive. In Figs. I and 3 both inner energies U (work functions) and free energies G (electrochemical potentials) are given; their interrelations are discussed below (see p. 7).

As mentioned in Section 1, a delocalized electron in solution interacts only with electronic polarization of the medium, but not with orientational polarization. Let us denote by U^S the lower limit to a delocalized electron energy which with certain stipulations can be called the bottom of a conduction band in solution.[3] It should be stressed that U^S is the real energy (as other energy levels in Fig. 1) of a delocalized electron, and not the chemical (ideal) energy, as it includes, besides the energy of interaction between the solution bulk proper and the electron, the work done in transferring the electron through the solvent surface potential χ^S.

The U^S level relative to the Fermi level for the metal (and hence to the U^V level related to it by Eqs. (1) and (2)) can be determined from w^{MS} — the metal-to-solution electronic work function. It should be noted that the work functions comprise variations in internal energy. For going from these to free energies, a correction has to be made for the entropy of delocalized electrons (in a gas or liquid)[4], determined by the formula for the entropy of ideal gas

$$S^\circ = k \ln [(2\pi mkT)^{3/2}/h^3] - k \ln n_0 \tag{3}$$

Here m is the electron mass (for delocalized electrons in solution — effective mass; in further calculations it has been taken equal to the mass of a free electron; this, however, should not introduce a large error); n_0 is the concentration (number of particles in 1 cm^3) in the chosen standard state; k and h are respectively the Boltzmann's and Planck's constants.

The (Gibbs) free energy level of a solvated electron in a solution in equilibrium with the electrode is equal to the level of electrochemical potential of electron in metal $\tilde{\mu}_e$. We shall call this equilibrium electrode the electron electrode. Suppose that we are concerned with a standard solution (1 mol/l) and hence the standard potential E_0°. In this case, w^{MS} determined at E_0° and corrected for the entropy of delocalized electrons (at n_0 conforming to 1 mol/l) is the difference between standard chemical potentials of localized and delocalized electrons.

The level of a localized (solvated) electron is generally below the bottom of the band of delocalized states because the solvated electron is in the energy well formed by appropriate spacing and orientation of the solvent's dipole molecules. The energy λ_s spent in forming a solvent of pertinent configuration is known as the medium's reorganization energy. Within the framework of a simple continuum model of a homogeneous dielectric, λ_s ist the sum of work done in creating nonequilibrium polarization, in particular for dipole orientation, and work done to form a cavity of radius R. The former is expressed as $e^2(1/\varepsilon_0 - 1/\varepsilon_s)/2R$; here, ε_0 and ε_s are

3 Without going into the details of energy band structure of a liquid solution, we shall note only that this structure can probably be described by the theory of disordered semiconductors (see, e. g. [32]). In a disordered semiconductor the conduction band does not have a sharp edge and the density of states extends to the forbidden band in the form of the so-called "tail". It would be more correct to talk not about the bottom of a conduction band, but about the "mobility band boundary" (that separates the delocalized and localized states). In view of this U^S is a certain amount of effective energy; its exact value generally depends on the method employed for its determination (see Sec. 2.2.).

4 With an accuracy sufficient for our studies the entropy of a degenerate electronic gas in metal can be assumed to be equal to zero.

respectively the optical and static dielectric permittivities. This expression represents the difference between total energy spent in charging a sphere of radius R in the given dielectric and the energy of a similar charging process associated only with electronic, i.e. inertialess, polarizability. It is this difference that conforms to the interaction of charge with the inertial part of polarization (orientational and atomic polarizabilities). The cavity formation energy equals $4\pi R^2\sigma$, where σ is the solvent's surface tension.[5]

Although energy is spent in forming a cavity, on the whole a gain in energy takes place due to interaction of an electron with orientated dipoles. Thus, localization of an electron proves to be energetically advantageous.

The difference in the energy levels of a solvated electron in solution and of an electron in vacuum A_s may be considered as solution-to-vacuum equilibrium electronic work function. The quantity $-A_s$ may be considered as free energy of electron solvation (i.e. with the formation of solvated electrons); it relates to an equilibrium process. That is, in the initial state the electron is in the equilibrium solvate cavity while in the final state it is in vacuum, close to the solution surface, and the solvent is in its equilibrium disordered state (i.e. there is no solvate cavity in it).

A different situation arises during electron photoemission from a solvated electron solution. After absorbing a high-energy light quantum, a localized electron can go from the solution to vacuum. This transition takes place without affecting the shape of the potential well proper, because the electron leaves the well very quickly, i.e., in a time much less than the relaxation time of the solvent molecule. Accordingly the solvent dipoles remain in their initial position. Therefore, the photoemission work function w^{SV} is not equal to A_s, i.e. to the difference in equilibrium energies of the electron in vacuum and of that localized in solution, but exceeds this difference by λ_s. Actually, the non-equilibrium energy level, at which the system is found as a result of photoemission, lies above U^V by value λ_s. To bring these to one datum level, we shall lay-off w^{SV} below the U^V level and obtain a conditional level [33)] lying below the equilibrium level of a solvated electron by λ_s (see Fig. 1). It is precisely from this conditional level the work of other fast processes for which nonequilibrium orientation of the solvent is retained should be counted off. Among these processes is photoionization, i.e. quick transition of an electron from a localized to a delocalized state within the solution. As is seen from Fig. 1, the photoionization work A_{op} exceeds the equilibrium work w^{MS} by λ_s; therein are also shown the relationships between A_{op}, w^{SV}, U^S, and other quantities.

For the determination of enumerated energy characteristics, use may be made (in a particular combination) of the following quantities measured by experiment: threshold of photo-induced and thermal electron emission from metal to solution and from solution to vapour phase; solvated electron photoionization threshold;

5 In the charge-transfer reaction theory by reorganization energy is implied a somewhat different quantity. It includes, along with λ_s for each reacting ion (for ionic reactions the contribution related to the formation of a cavity could generally be disregarded), the energy of interaction of reagents with a transferable charge. This interaction energy equals $-e^2(1/\varepsilon_0 - 1/\varepsilon_s)/R_{12}$, where R_{12} is the distance between the centers of charges; for electrode reactions, R_{12} is replaced by 4 R′ where R′ — the distance between the ion center and the electrode surface. When there are several localized states in the system (as, for instance, in hexamethylphosphotriamide), each state has its own reorganization energy (see Sect. 2.4.).

and equilibrium potential of the electron electrode. In so doing allowance must be made, as mentioned earlier, of that from the studies of equilibria can be determined the (Gibbs) free energies (ΔG) of corresponding processes and from the studies of non-equilibrium electron transitions the work functions or internal energies (ΔU). To compare these quantities, it is necessary to know the entropy (ΔS) of the corresponding transitions.

Calculation of entropy of a localized electron is a difficult and still unsolved task because of the absence of reliable information on microscopic structure of this particles. A rough estimation made for water in Ref. [34] reveals that ΔS° for a hydrated electron is small and $T\Delta S^\circ$ does not exceed 0.1 eV. For other solvents, ΔS° can be more (particularly, the estimation for liquid ammonia [35] yields $T\Delta S^\circ \simeq 0.38$ eV). Because of the non-availability of reliable values of solvated electron entropies we, as a rough approximation, have not ditinguished between the free and internal energies for the system considered (Figs. 1 and 3). The possible error that appears as a result of this is discussed in Section 2.4 where the reorganization energy in hexamethylphosphotriamide is determined.

Now we shall briefly dwell on the emission methods of studying excess electrons in polar (conducting) media [10,27]. Transfer of electrons (under the action of light energy or heat) through an interface is known as emission (electron photoemission or thermal electron emission). The final state of an electron during emission proper is a delocalized state. In particular, during electron photoemission from metal into electrolyte solution [36] the metal electron, after it has absorbed a light quantum having energy more than the metal-to-solution work function w^{MS}, leaves the metal and goes into the solution (i.e. into the conduction band). After thermalization it is at the U^S level. The time an electron remains delocalized in the solution is small, for example in water it is 10^{-11}–10^{-12} s [37]. The electron becomes solvated upon interacting with the orientational polarization of the medium. The duration of stay of the electron in this state depends on the nature of the solvent; sometimes it is very long.

Experimental measurements of photoemission currents are generally taken at far more positive potentials compared to the equilibrium potential of the electron electrode. Therefore, even when the solvated electrons are stable in the bulk of the solution, the electrode-emitter surface traps them effectively. For the electrode-to-solution transition of electrons to be irreversible (this is a necessary condition for measuring a stationary photocurrent), readily reducible substances — solvated electron acceptors (so-called scavengers) — are added to the solution. The electron level in a reduced acceptor (A^-) is quite low, and this makes this state very stable; trapping of electrons by a scavenger is the final transformation an emitted electron undergoes. H^+, N_2O, and NO_3^- are the extensively used scavengers for aqueous solutions.

The photoemission current I dependence on electrode potential E and quantum energy $h\nu$ is expressed by the so-called five-halves law [38]

$$I \sim (h\nu - w_0^{MS} - eE)^{5/2} \qquad (4)$$

Here, w_0^{MS} is the work function at $E = 0$ relative to the chosen reference electrode. Extrapolating the measured I–E dependence (for the given $h\nu$) in $I^{2/5}$–E coordinates to $I \to 0$ gives a photoemission threshold potential E_t. At $E = E_t$ the electrode-to-solution work function w_0^{MS} equals $h\nu$ — the energy of the light quantum that causes photoemission.[6]

Although the validity of physical prerequisites, on which is based the photoemission theory [38] that lead to the discovery of the five-halves law, was disputed in [40], nonetheless the existence of this law confirmed experimentally can be considered to be a well-founded fact at least for mercury and mercury-like metal electrodes (used in our experiments) [10, 27].

Electron photoemission from solvated electron solution (in solvents such as hexamethylphosphotriamide and liquid ammonia the solvated electrons are fairly stable) to vapour phase has been studied by Delahay and co-workers [41] (whose works are reviewed in Ref. [10]). According to them, this process proceeds in three stages: solvated electron photoionization; diffusion of generated delocalized electrons to the solution's surface; and emission proper, i.e. transition of electrons to the vapour phase where they are transferred from the "cathode" surface (i.e., from the solution) to the anode by the external electric field.

Finally, in the solvated electron solution/vapour system electronic emission can occur in the equilibrium manner, i.e., as thermoionic emission [42].

Table 1 summarizes the basic relationships that link energy characteristics of excess electrons with the values measured by the aforementioned methods (see also Fig. 1). In the equations given therein, i.e. in Eqs. (5) and (6) w^{MV}, w^{MS}, and w^{SV} denote respectively metal-to-vacuum, metal-to-solution, and solution-to-vacuum photoemission work functions; $\Delta\Psi$ is the Volta potential difference for a metal-solution system; E_0 is the equilibrium potential of the electrode in solvated electron solution; and $\tilde{\mu}(RE)$ is the Fermi level of the reference electrode. Equation (6) is approximate (see above) because the solvated electron entropy has not been taken into consideration. The main error in equating the heat of electron solvation and the activation energy of the thermoemission current for the solvated electron solution is caused by the variation in the solution's surface potential with temperature; apparently, here specific adsorption of solvated electrons (or of an alkali metal) on the solution/vapour interface makes major contribution to the surface potential [28, 41]. This error can be probably neglected if measurements are taken in very dilute solutions ($< 10^{-3}$ mol/l, see [42]) of the alkali metal. This follows from the dependence measured in [41] between thermoemission current and the concentration of sodium in hexamethylphosphotriamide.

In conclusion it may be said that the real energies of a delocalized and/or solvated electron in solution can be computed only if the Volta potential difference in the electrode-solution system is known. The methods based on the estimation of surface potentials enable the chemical, or ideal energies to be evaluated.

---

6 In Ref.[39] it has been asserted that electrons are first very rapidly localized in shallow traps placed slightly below the bottom of the conduction band in the solvent and then solvated. It should be noted that in determining the electrode-to-solution work function and hence the U^S, i. e. the energy of the "bottom of the conduction band", the validity of this statement does not matter. Actually, the work function is determined by extrapolating the photocurrent from the region of large (1–2 eV) energies of an emitted electron; at these energies the electron is undoubtedly in the delocalized state and therefore the result should not be very sensitive to the details of the density-of-states distribution in the immediate neighbourhood of the bottom of the conduction band.

Table 1. Methods for determining characteristics of excess electrons

Method	Quantity determined	Equation	Ref.
Electron photoemission from metal to solution + measuring the Volta potential difference for metal/solution system	Real energy of a delocalized electron	$U^S = w^{MV} - w^{MS}(E) - e\Delta\Psi(E)$ (5)	[28]
Photoemission from solvated electron solution to vapour phase + measuring the equilibrium potential of the electron electrode	Solvent's reorganization energy for solvated electrons	$\lambda_s \simeq w^{SV} - eE_0 + \tilde{\mu}(RE)$ (6)	[43, 44] (see also [31])
Electron photoemission from metal to solution + measuring the equilibrium potential of the electron electrode	Difference in standard free energies of a delocalized and a solvated electron	$w^{MS}(E_0) - TS^\circ$	[16]
Electron thermoemission from solvated electron solution to vapour phase	Activation energy of thermoemission current $\simeq$ heat of electron solvation	A_s	[28, 41]
Photoionization of solvated electrons	Photoionization threshold energy	A_{op}	[28]

2.2 Determination of the Energy of a Delocalized Electron

The investigations on electron photoemission from metals to electrolyte solutions offer a unique possibility of determining the energy of delocalized electrons in polar media. As shown earlier, the photoemission current measurements make possible the determination of the electrode-to-solution electronic work function. Knowing this work function and the Volta potential difference for the metal-solution system one can compute the real energy of interaction between a solvent and an electron U^S by Eq. (5). According to the method used, U^S is the internal energy; it practically equals the standard free energy if the same standard concentrations are taken for an electron in a vapour phase and in the solvent conduction band. The difference may be related only to dissimilar electron effective masses (see Eq. (3)).

The discussed calculation procedure is not based on any extrathermodynamic assumptions and therefore the inaccuracy of the result obtained is determined only by the experimental errors of measuring a work function and the Volta potential difference. Furthermore, from the solvent surface potential χ^S determined by any estimation method we can find the ideal solvent-electron interaction energy $V_0 = U^S - e\chi^S$. Unlike U^S, V_0 is not a strictly thermodynamic quantity and the inaccuracy in determining it, besides experimental errors, is caused by the inaccuracy of model assumptions made for estimating χ^S.

In [45–48] the metal-to-various-solvent work function has been determined from the photoemission current measurements. Photocurrent was measrued on a hanging-drop

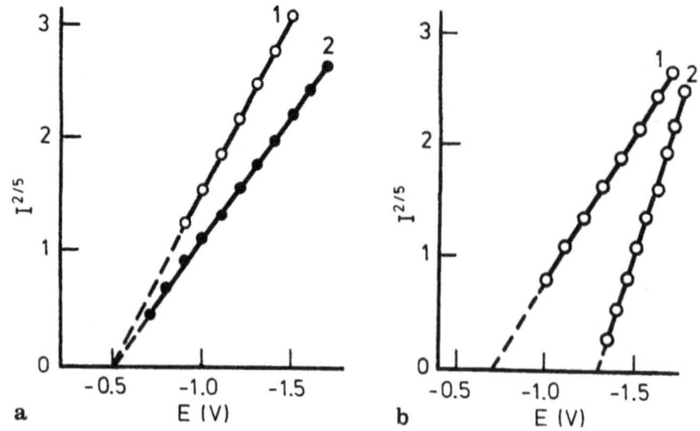

Fig. 2. Photoemission current (in arbitrary units) vs. mercury electrode potential (46)
(a) acetonitrile; scavenger: n-C_3H_7Br (1), N_2O (2); quantum energy: 3.4 eV; (b) methyl alcohol; scavenger: NO_3^-; quantum energy: 3.4 eV (1), 2.8 eV (2)
Reference electrode: aqueous SCE

mercury electrode; the measurement technique is discussed in Refs. [10] and [27]. The photoemission threshold potential E_t was found by extrapolating, in accord with Eq. (4), the I–E dependence in $I^{2/5}$–E coordinates.

It should be remembered that the linearity of $I^{2/5}$–E curves alone does not satisfactorily prove the emission nature of photocurrent. The arguments given in Refs. [10] and [27] suggest that use can be made also of the following two experimentally proved facts: the threshold potential is independent of the nature of solvated electron scavengers; the potential E and the quantum energy hv are additive in affecting the magnitude of photocurrent (cf. Eq. (4)). As is seen from Fig. 2a, the straight lines are extrapolated to one and the same value of E_t, irrespective to the nature of scavengers used. This value differs by 0.6 V when light sources with quantum energies 3.4 and 2.8 eV are used (which conforms to the difference in quantum energies) (see Fig. 2b).

The threshold potential can be measured to ± 0.1 V.[7]

The computed values of real (U^S) and ideal (V_0) energies of interaction between solvents and a delocalized electron, and also the surface potentials χ^S[8] used in calculating V_0 are listed in Table 2. The mercury-solvent Volta potential differences needed in the calculations of U^S were computed from the Volta potential differences for the solvent-water system; these have been measured for methanol, ethanol, dimethylsulfoxide, ethylene glycol, and propylene glycol-1,2 by Damaskin and co-

7 Processing of photocurrent-potential curves found experimentally for a number of solvents [46] directly on the strength of the five-halves law and by a more general technique [49] (a variant of the method suggested in Ref. [50]), ignoring the five-halves law, gives threshold potentials differing no greater than the indicated value.

8 Recall that the surface dipole with its positive pole directed toward the solution phase gives a positive χ^S.

Table 2. Energy of delocalized electron in different solvents [46]

Solvent	χ^s, V	U^s, eV	V_0, eV
Water	+ 0.13	−1.25	−1.12
Methanol	−0.2	−0.19	−0.39
Ethanol	−0.25	−0.15	−0.4
Isopropanol	−0.26	−0.21	−0.47
Formamide	+0.05	−1.1	−1.05
N-methylformamide	−0.11	−0.33	−0.44
Hexamethylphosphotriamide	−0.6	+0.22	−0.38
Acetonitrile	−0.1	−0.16	−0.26
Formic acid	−0.46	−1.73	−2.2
Dimethylsulfoxide	—	−0.05	—
n-Butanol	—	−0.1	—
Ethylene glycol	—	−0.22	—
Propylene glycol-1,2	—	−0.33	—

workers [51,52], and for isopropanol, n-butanol, formamide, N-methylformamide, acetonitrile, formic acid, and hexamethylphosphotriamide by Parsons and colleagus [53−55]. Use was made of the value of Volta potential difference measured by Randles [56] for the mercury-water system.[9] The surface potentials have been reproduced from Refs. [53−55], and [60−62]. A total error of 0.2 eV is assumed in determining U^S.

It should be noted first of all that, along with the "chemical" interaction of an electron with its environments, surface potential makes a significant contribution to the real energy of an electron. Thanks to this, U^S varies over a wide range; in the case of hexamethylphosphotriamide it is a positive quantity and this suggests that the bottom of the conduction band lies above the "vacuum" level. Nonetheless, conduction electrons do not "flow out" of the liquid phase; they are retained by a high ($|V_0| = 0.4$ eV) potential wall at the solution surface.

As regards the ideal energy V_0 describing the electron-phase bulk interaction, all solvents, as is seen from Table 2, effectively "suck in" a delocalized electron ($V_0 < 0$). By the value of $|V_0|$, all solvents come under two groups. Water, formamide, formic acid (and liquid ammonia [10]) form Group One ($|V_0| > 1$ eV). Other solvents fall in Group Two ($|V_0| \lesssim 0.4$ eV).

Concerning the nature of interaction between a delocalized electron and polar solvents and the dependence of this interaction on the structure and properties of the solvent only a tentative conclusion can be made at present. V_0 does not show correlation either with the dimensions of the solvent molecule or with its characteristics such as acceptor and donor numbers, optical and static permittivity, and the molecule's dipole moment. It is obvious that the electron-medium electrostatic interaction alone cannot explain the results obtained.

9 The results of Randles were disputed [57], but a recent detailed study [58] proves their validity (see also [59]).

10 For liquid ammonia the chemical energy of electron solvation is estimated at 1.0–1.7 eV [35,63] and the delocalized electron energy level is by 0.31 eV higher than the solvated electron level (see Sect. 2.3), so that $|V_0|$ is most probably more than, or equal to, 1 eV.

At the same time it should be pointed out that the solvents (water, formamide, formic acid, and liquid ammonia) exhibiting the greatest interaction with an electron can create a three-dimensional branched structure due to H bonds between the molecules. Replacing an H atom in the molecule by a hydrocarbon residue (in going, for instance, from water to alcohols, from formamide to N-methylformamide, etc.) causes the solvent molecules to form only short linear chains. One might think that the introduction of a hydrocarbon residue into the solvent molecule initiates repulsion which is superimposed on the electron — polar medium electrostatic interaction. The repulsive energy, as follows from Table 2, amounts to about 0.6 eV or more. This is in agreement with that observed in the experiment [64], i.e., the metal-to-nonpolar hydrocarbon electronic work function is higher than the metal-to-vacuum work function.

Now we shall draw the reader's attention to an interesting fact. As shown in Section 2.3, the difference between the delocalized and solvated electron levels for hexamethylphosphotriamide is by almost the same value, i.e. about 0.4 eV higher than for water or liquid ammonia, i.e. for the solvents having a branched structure of H bonds. It follows that the introduction of a hydrocarbon residue into the solvent's molecule "forces out" only the delocalized electrons from the polar medium; the solvated electron energy level in all the enumerated solvents has almost the same value. (An independent confirmation to this is the closeness of equilibrium potentials of the electron in water, hexamethylphosphotriamide, and liquid ammonia — see Section 5 — vs. the reference electrode whose potential is independent of the solvent.)

2.3 The Difference in Delocalized and Solvated Electron Energies

The difference in the standard free energies of a localized (solvated) and delocalized electron in various solvents equals:

Water [28,65,66]	0.25 eV
Hexamethylphosphotriamide [16]	0.7 eV
Liquid ammonia (computed from the experimental data of Ref. [48])	0.31 eV

These have been obtained as photoemission work function w^{MS} (cf. Fig. 1) at the electron electrode equilibrium potential (which for hexamethylphosphotriamide and liquid ammonia was measured by experiment, and, for water, calculated from the thermochemical data [65]) by making a correction for the ideal gas entropy according to Eq. (3)[16][11]. It should be noted that the aforementioned value, computed in this manner, is independent of the solvated electron concentration (for the same standard concentration of localized and delocalized electrons).

The difference in standard free energies is possibly close to the energy of interaction between an electron and the orientational polarization of the solvent surrounding it. Its low value in water and ammonia may have the consequence that if the solvated and delocalized electrons are in equilibrium, then the solvated electron

11 This correction equals 0.09 eV.

solution contains an appreciable amount of delocalized electrons. This, in principle, can affect the properties of such a solution, in particular its electrical conductivity, reactivity, etc.

On the contrary, the ratio of delocalized and localized electron equilibrium concentrations in hexamethylphosphotriamide equals about 10^{-11}. Note that the value given earlier, i.e. 0.7 eV [even with the allowance made for possible inaccuracy which is estimated [16)] at ± 0.1 eV] is by an order of magnitude greater than the activation energy of diffusion of solvated electrons in this solvent (about 0.07 eV); this eliminates the mechanism proposed for diffusion of solvated electrons via their excitation into the conduction band.

2.4 Determination of the Energetic Characteristics of Solvated Electrons and Their Associates in Hexamethylphosphotriamide Solution

In Section 4 we have discussed in detail that in hexamethylphosphotriamide (as in some other solvents) an electron exists as a solvated electron proper and as an associate formed by two solvated electrons and an alkali metal cation (say Na^+), which is most probably the alkali metal anion (Na^-). A comparison of their energy characteristics is of interest [43,44)] [12].

The standard free energy of solvation A_s (for the same standard concentrations of electrons in the gas and the solution) is determined from the delocalized-electron energy U^S (Sect. 2.2) and the difference in standard free energies of delocalized and solvated electrons (Sect. 2.3): $-A_s = 0.22 - 0.7 \simeq -0.5$ eV[13], see the left-hand part of Fig. 3 (cf. Fig. 1). (This again is a real energy; the chemical energy for electron solvation is appreciably higher; it equals $-A_s + e\chi^S = -1.1$ eV.)

The reorganization energy of a solvent can be estimated [see Fig. 1 and Eq. (6)] if we know the photoelectric work function for electron emission from a solvated electron solution to a vapour phase w^{SV}. This work function can be determined by extrapolating to zero current the dependence of photoemission current on the light quantum energy by making use of one or the other theory of electron photoemission from solution to vapour. The photocurrent vs. quantum energy curve has two peaks, i.e. at 1.7 and 2.8 eV, which obviously conform to two different types of emitters. Of these, the former is undoubtedly associated with solvated electrons; the nature of the latter will be discussed elsewhere. Delahay [41)], using the model theory of photoioniza-

12 The numerical values given in this section are somewhat different from those obtained earlier [43,44)], because use has been made of the improved Volta potential difference for the hexamethylphosphotriamide/electrode system.

13 In [41)] it has been pointed out that for a certain approximation the activation energy of the electron thermoemission from a solvated electron solution to a vapour phase can be taken as the (real) heat of solvation. (The sources of possible inaccuracy that appears with this assumption are discussed above, see p. 158). This activation energy equals 0.74 eV [43)]. The value 0.5 eV seems to be more substantiated; nonetheless, the difference ($\simeq 0.2$ eV) between these two values is the maximum error that occurs in such calculations, however inaccurate the estimation of effective mass of a delocalized electron may be.

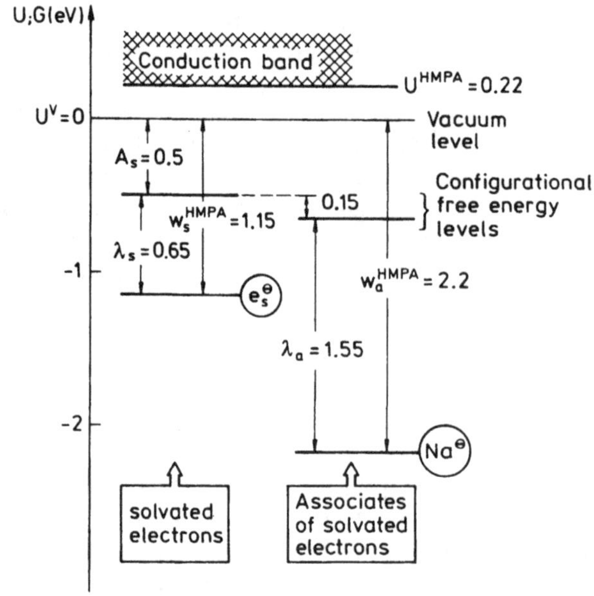

Fig. 3. Energy characteristics of solvated electrons and of their associates in hexamethylphosphotriamide (the values of energies are in electron volts)

tion of a hydrogen-like atom, has calculated w_s^{sv}[14] to be equal to 1.34 eV. In Ref. [67] a non-model photoemission theory based essentially on the use of a threshold approximation has been developed and the work function has been found to be equal to 0.9 eV. Both these approaches have weak points (see Refs. [10] and [44]). Due to the non-availability of an authentic value of w_s^{sv} we have assumed $w_s^{sv} = 1.15 \pm 0.2$ eV by considering the aforementioned values as being the upper and lower limits; this is in good agreement with the qualitative behaviour of the plot of photoionization cross-section for solvated electrons against quantum energy [68]. It follows that the solvent's reorganization energy for a solvated electron in hexamethylphosphotriamide $\lambda_s = 1.15 - 0.5 = 0.65$ eV.

Now we shall see to what extent the experimental estimation of λ_s agrees with the value that can be obtained theoretically. Using the literature data on physical constants for hexamethylphosphotriamide [14] and taking the cavity radius R = 3.2 Å [this conforms to the molar volume of a solvated electron, which is estimated at 80 cm^3 [69]], we find $\lambda_s = 1.31$ eV[15].

---

14 In this section the quantities relating to two localized states, i. e. to the solvated electron and the associate are denoted by subscripts s and a, respectively.

15 Note that for reasonable values of the cavity radius λ_s is rather less sensitive to variations in R; thus, at R = 3,0 Å λ_s = 1.33 eV and at R = 4.0 Å λ_s = 1.29 eV. The point is that the opposite effects of R on the polarization component of λ_s (1.05 eV and 0.79 eV, respectively) and the cavity formation work (0.28 eV and 0.50 eV) are compensated to a large extent. Thanks to these opposite effects, the plot of λ_s against R has a minimum at R = 3.8 Å (λ_s = 1.28 eV).

The computed value of λ_s is much more than that estimated by experiment; it exceeds even the upper limit of this estimate (0.85 eV). This discrepancy cannot be removed by varying the parameters in reasonable limits. It is much easier, one would think, to attribute this difference to insufficient accuracy of the continuum model. But it is known that for reactions of complex ions whose dimensions are close to those of solvated electron this model usually provides a better agreement between the values discussed [70].

The cause of this discrepancy may be that due regard is not given to the entropy factor. In fact, w_s^{sv} ist the variation in internal energy, while A_s (as well as the computed value of λ_s) is the free energy. If we assign all discrepancies to this factor, then ΔS for the electron transition from a gas to an already available cavity of the equilibrium configuration will be equal to +150 kJ/(mol · deg) at a minimum (for an average value of w_s^{sv} it will be still more by 67 kJ/(mol · deg)). It is hard to imagine the disordering process that would result in such a large increase in entropy. (Usually ΔS of ion solvation are less (in absolute value) and negative; this is due to the ordering of the medium caused by the electric field of the ions; some disordering effects are observed in water, but in solvents like hexamethylphosphotriamide, in which there are no H bonds, these are not observed (see Ref. [71]). And the process under consideration is not accompanied by any change in the medium's structure.)

Another significant discrepancy between the theory and experiment is that even the computed values of the reorganization energy, to say nothing of its experimental estimation, are too small to be used for explaining the rate of the cathodic generation of solvated electrons (see Sect. 7). All this indicates that some factors affecting the electrochemical behaviour of solvated electrons are still unknown.

Now we turn to the energy characteristics of Na^--type associates. The free energy of its formation from two solvated electrons and a Na^+ cation can be computed if we know the equilibrium constant K for the reaction $2e_s^- + Na^+ \rightleftarrows Na^-$. According to the data of Ref. [72], $K = 2.3 \times 10^4\ mol^{-2}l^2$. Since the reaction proceeds with a change in the total number of particles, this free energy depends on the chosen standard concentrations. It is therefore convenient to make use of the so-called configurational free energy (i.e. of the free energy after substracting the contribution made by transpositional entropy) which is independent of the choice of standard concentrations. This correction is made as per formula $2kT \ln X_0$, where X_0 is the molar fraction in the standard state (1 mol/l). The so-calculated configurational free energy levels for the two localized states (per electron) differ by about 0.15 eV. On the other hand, K depends but only to a small degree on temperature and hence the thermal effect of the reaction (which is virtually equal to the change in internal energy), in fact, is close to zero. Thus, the difference between the solvated electron and its associate levels ranges between 0 and 0.15 eV, i.e. relative to the vacuum level the free energy of the associate equals about −0.65 eV.

Now we return to the dependence of the current of photoemission, from sodium solutions in hexamethylphosphotriamide into vapour phase, on quantum energy. According to Ref. [73], the second peak in the photocurrent vs. energy curve has its origin in some complex containing Na^+. In Ref. [44] it has been proposed that it is the associate (of type Na^-, for instance) that acts as emitter. (In no case, significant concentration of some other complexes containing Na^+ and solvated electrons was detected in hexamethylphosphotriamide solutions [71]). The threshold energy of electron photoemission from hexamethylphosphotriamide solution of the associates into vapour phase, estimated by the technique suggested in Ref. [41] and using the experimental photocurrent spectra [73], yielded a value of 2.2 eV. This, with consideration of the earlier mentioned stipulations, can be taken as the upper limit of the work function w_a^{sv}. Hence, the solvent's reorganization energy, conforming to the formation of a potential well for Na^-, equals $\lambda_a \lesssim 1.6$ eV. The energy levels calculated for the associate are shown on the right in Fig. 3.

Large difference (about 1 eV) in reorganization energies for solvated electrons, λ_s, and their associates, λ_a, is remarkable. Irrespective of the assumptions made in calculating the reorganization energies, this difference exists because the difference ($\simeq$ 1 eV) in the photoemission current peak energies far exceeds that in the equilibrium levels of solvated electrons and their associates ($\simeq$ 0.1 eV). Apparently, this is caused by a stronger effect produced by the charge of the associate as compared to the solvated "monoelectron" on the surrounding solvent. This can be considered as an argument in favour of the fact that (see Sect. 4) the associate is indeed a particle significantly different, in its structure, from a solvated electron, namely Na^- anion (but not "bielectron"). The observed difference in the reorganization energies correlates to a lower reactivity of the associate compared to a solvated electron proper.

To conclude this section, we shall consider in brief the absorption spectra for a solvated electron solution in hexamethylphosphotriamide. In the general case the spectrum has 3 peaks. Of these, the one with the longest wavelength (peak energy 0.45 eV, oscillator strength 0.7 $\pm$ 0.2) [74] is caused by the solvated electron proper. The second (peak energy 1.6 eV) is caused by the Na^- type associates. (The nature of the third (peak energy 3 eV) is not quite clear; it, with a greater or lesser probability, is associated with the solvent's decomposition products upon its interaction with the solvated electrons) [75].

The nature of optical transitions constituing the adsorption spectra is a subject of long discussion. According to the widely accepted model (see [76]), the light absorption band conforms to phototransition of an electron from the ground bound (1s) state to the excited bound (2p) state. An alternative to this is the statement [41,67] that here photoionization, i.e. transition of an electron from the 1s-level to the delocalized state (i.e., into the conduction band), takes place.

In the case considered, by referring to Fig. 3, we see that the peak energy (0.45 eV) conforming to a solvated electron is much less not only than the photoionization energy, but even the difference in the equilibrium energies of localized and delocalized states. Hence, here only transition to the bound excited state can take place [16]. The same, though with less confidence (because of lesser reliability of the numerical values used), can be said also about the peak conforming to Na^--type associates.[16]

Transitions complying with photoionization of the localized states apparently do not show up in the absorption spectrum of solvated electron solutions in hexamethylphosphotriamide. Most likely this is explained by a relatively small photoioni-

16 Photodissocation of these associates onto "monoelectrons" proceeds apparently without any intermediate formation of a delocalized electron (say, via the excited bound state) [77, 78].

zation probability or by the fact that the photoionization bands are masked by absorption bands of different origin in the ranges of higher quantum energies.

3 General Conditions for Cathodic Generation of Solvated Electrons

Several methods of obtaining solvated electrons in a liquid phase are available. Of these, the most universal is evidently the radiation-chemical method in which electrons detach from molecules, ions or atoms under ionizing radiation to form solvated electrons. Similar to this is the photochemical method: solvated electrons are obtained under action of light on electron donors. But in the case of radiation-chemical or photochemical transformations, along with solvated electrons, there appear particles capable of reacting with them quickly. Both methods offer relatively small concentrations of solvated electrons and are therefore less suitable for obtaining stable concentrated solutions thereof.

Solvated electrons can also be obtained via atomic hydrogen [79–81]. The potential of the electrode which is in equilibrium with atomic hydrogen in an aqueous alkaline solution (pH 12) equals —2.8 V (NHE)[82]. It is close to the standard equilibrium potential of an electron in water (see p. 179). Therefore, hydrated electrons

$$H + OH^- \rightarrow e^-_{aq}$$

are produced on passing a mixture of atomic and molecular hydrogen through aqueous solutions of alkalies (pH over 12).

This method is known to be employed to obtain solvated electrons only for aqueous systems; in non-aqueous organic solvents, reduction or hydrogenation of a solvent with atomic hydrogen usually proceeds more readily.

Of interest is the Jolly reaction

$$1/2H_2 + NH_2^- \rightleftarrows e_s^-$$

which occurs at large H_2 pressures and high concentrations of KNH_2 [81, 83–85]. The reaction with molecular hydrogen is known only for liquid ammonia. The attempts to carry out this reaction in water even at high hydrogen pressures have ended in failure [85]. The advantage of liquid ammonia lies in that the difference between standard potentials of the hydrogen and electron electrodes in this solvent is by one volt less than the corresponding difference for water. At a molecular hydrogen pressure of about 100 kg/cm^2 and a NH_2^--concentration of 1 mol/l the hydrogen electrode will be in equilibrium with the 10^{-5} M solvated electron solution [81, 86, 87].

In the case of photoelectrochemical generation of solvated electrons, discussed in Section 2, the electrode/electrolyte interface is illuminated with light of quantum energy exceeding the electronic work function for this system and the electrons from the electrode go into the electrolyte — first in the delocalized and then in the solvated state. Since photoelectrochemical generation of electrons takes place at potentials far more positive than the equilibrium "electron" potential the electrode surface effectively traps (i.e. oxidizes) solvated electrons. This is the reason why in this method the solvated electrons exist near the illuminated electrode only for a limited period ($\lesssim 10^{-8}$ s) sufficient for the solvated electrons to travel by diffusion a distance of ca. 2 nm between the generation region in the solution and the electrode surface.

Obtaining solvated electrons by dissolving alkali metals and by electrochemical generation is of special interest. Back in the last century, the dissolution of alkali metals in liquid ammonia gave the very first evidence of obtaining solvated electrons. Electrochemical (cathodic) generation of solvated electrons is a process in which electrons are transferred from the electrode into solution under the action of high cathode potentials.

These methods enable stable systems with high concentration of solvated electrons to be obtained in a number of solvents. Electrolysis makes it possible to vary the concentration of solvated electrons in a definite manner, renew the system over and again, and make repeated measurements. The main limitation of the cathodic generation method (as of dissolution of an alkali metal) is that the experiment has to be carried out in a medium stable towards strong reducing agents. Thus, electrochemical generation of electrons proceeds at very negative potentials and this, of course, can be conveniently accomplished in systems where neither the solvent nor the solute undergo cathodic reduction. Another limitation is associated with the need of having a conducting medium, i.e. using indifferent electrolytes and solvents with a sufficiently large ionizing power.

These methods are linked in that the solvents which dissolve alkali metals yielding solvated electrons and solvated cations are most commonly used for electrochemical generation of solvated electrons. One of the first experiments on such a generation was performed in 1897 by Cady who noticed an increase in the intensity of the blue coloration at the Pt cathode in a dilute sodium solution in liquid ammonia at —34 °C [88].

Cathodic generation of solvated electrons in general competes with other cathodic reactions, say, in alkali metal salt solutions, with electrolytic deposition of the alkali metal on the electrode.[17] Therefore, it is necessary to find a criterion which could be used to estimate how effective the cathodic generation will be, if it is possible at all, for the given system.

A direct correlation exists between the ability of a solvent to dissolve alkali metals and the possibility of electrochemical generation of solvated electrons in the solutions of the metal salts. Quantitatively, it is expressed by the Makishima's equation [89]. A solid alkali metal and its saturated solution are in electrochemical equilibrium both for cations $M^+ + e^-(M) \rightleftarrows M^\circ$ and the solvated electron $e^-(M) \rightleftarrows e_s^-$. From the condition of equal potentials, caused by both these equilibria, we get

$$E_0^\circ(M/M^+) - E_0^\circ(e_s^-) = -\frac{RT}{F} \ln a(M^+)\, a(e_s^-) \tag{7}$$

where $E_0^\circ(M/M^+)$ is the standard potential of the alkali metal electrode reversible relative to its ions; $E_0^\circ(e_s^-)$ is the standard potential of electron electrode; and a stands

17 We mean namely the primary generation, unlike the formation of solvated electrons in the secondary process such as chemical dissolution of the deposited alkali metal.

for corresponding activities in the alkali-metal saturated solution. It follows from this equation that the more the solubility of the alkali metal the larger the difference in standard potentials and the more thermodynamically effective is the cathodic generation of solvated electrons compared to electrolytic deposition of the alkali metal. Usually, total solubilities of alkali metals in various solvents are known. But it is difficult to determine from these values the activity of a cation or solvated electron separately, the reason being the presence of complex equilibria. This hampers the quantitative application of the Makishima's equation; it can however be used to qualitively explain the observed relationships.

It may be deduced from [17,90–106] that alkali metals best dissolve in liquid ammonia and hexamethylphosphotriamide. In these solvents, cathodic generation of solvated electrons can be accomplished for all alkali metal salts and also for tetra-substituted ammonium salts. A correlation exists between the alkali metal solubility and the possibility of cathodic generation of solvated electrons also for glyme (1,2-dimethoxyethane) and methylamine. Methylamine is of special interest because all alkali metals dissolve in it, though with very different solubilities. According to [107,108] the solubility of alkali metals in methylamine decreases as follows: lithium > cesium > potassium > sodium [107,108]. According to the Makishima's equation, generation takes place at higher solubilities (i.e., for lithium or cesium); at a lower solubility (i.e. for sodium) the metal deposits. Potassium forms the intermediate case (mixed process).

Table 3. Physical properties of solvents (complied in Ref. [17])

Solvent	b.p., °C	Relative dielectric const.	Viscosity, cP	DN[a]	AN[a]
Liquid ammonia	−33.4	20	−34 °C 0.26	59	
Hexamethylphosphotriamide	232	28.7	25 °C 3.1	38.8	10.6
Thio-hexamethylphosphotriamide	94 (at 130 N/m²)	39.5	30 °C 5.6		
Methylamine	−6.7	12.7	−22.8 °C 0.35	55.5[b]	
Ethylenediamine	116.5	12.9	25 °C 1.54	55	
Glyme	82–85	7.2	25 °C 0.46	24	10.2

[a] Donor and acceptor numbers, after Gutmann [109–111]
[b] For ethylamine

Table 3 contains the physical properties of solvents that are used for dissolving alkali metals. Besides boiling point (b.p.) and viscosity (the data enable one to judge on the experimental potentialities inherent in a solvent) the Table contains values of dielectric constant and of donor and acceptor numbers (DN, AN). It is hard to notice any correlation between the macroscopic properties of the solvents, on the one hand, and their ability to dissolve alkali metals and the possibility of electrochemical generation of solvated electrons, on the other hand.

Thus, on replacing O by S in hexamethylphosphotriamide the dielectric constant and the dipole moment increase; the solvent does not dissolve alkali metals yet and therefore cannot serve as the medium for generating electrons. The authors who studied the properties of thiohexamethylphosphotriamide [112] have rightly related this to the decrease in the solvating ability of a thioderivative with respect to the sodium cation.

In the majority of cases it is precisely the solvation of a cation that is cruial for the medium's ability to dissolve alkali metals and for the solvated electron generation possibility. Enhancement of solvation shifts the equilibrium potential of the alkali metal in its ion solution to more negative values; this, according to the Makishima's equation, raises the thermodynamic probability of electrochemical generation of solvated electrons. Metallic lithium whose cation is always more strongly solvated than the cations of other alkali metals exhibits maximum solubility, as a rule, and it is in lithium salt solutions that generation usually takes place. In some cases generation of solvated electrons in lithium salt solutions is not observed; this however has not a thermodynamic but a kinetic reason — i.e. the formation of an hydroxide film inhibits the generation of solvated electrons, in organic solvents this film being less soluble in the case of lithium (see elsewhere).

The superiority of liquid ammonia and hexamethylphosphotriamide over all other solvents in dissolving alkali metals and serving as a medium for the generation of solvated electrons is linked to their high donor activity and strong solvation of cations as well as to their sufficiently high dielectric constants.

Strong complexing agents for alkali metal cations such as crown ethers and cryptands, have large effect on the considered processes of generation of solvated electrons and metal electrodeposition [100, 113–116]. Binding the alkali metal cations by cryptands and crowns stabilizes the unusual solid compounds containing electrons and alkali metal anions at anionic sites of the crystal lattice [116–118].

Metallic potassium dissolves in diethyl, diisopropyl, di-n-propyl ethers and also in diethylamine only in the presence of a crown or a cryptand [100, 119]. In the case of sodium which does not dissolve in ethylamine, a solution with a total sodium content of 0.4 mol/l can be obtained on adding cryptand-222 [113].

Binding of cations by cryptands and/or crowns hinders the cathodic reduction of cations. Thus, in propylene carbonate the presence of excess cryptand-222 markedly shifts the half-wave potential for the formation of alkali metal amalgams [120]. A maximum shift of about 1 V was observed for sodium ions. On adding 18-crown-6 to glyme and ethylenediamine, deposition of sodium is replaced by electrochemical generation of solvated electrons.

As mentioned earlier, for generation of solvated electrons it is necessary that the potentials of the cathode should be high enough. This can be realized for a number of systems (see the Scheme). Alkali metal cations serve as counterions in the systems where the solvated electrons are chemically stable; at low temperatures in liquid ammonia use can be made also of other cations.

Systems for cathodic generation of solvated electrons

A: solvated electrons are stable		B: solvated electrons are unstable	
A-1	A-2	B-1	B-2
Liquid ammonia HMPA $N(Alk)_4^+$	Methylamine Glyme Ethylenediamine	HMPA Dimethyl-sulfoxide	HMPA-water HMPA-ethanol
Li^+, Na^+, K^+, Rb^+, Cs^+	Li^+, Rb^+, Cs^+	$N(Alk)_4^+$	

In the systems listed under A-1, thermodynamically it would be more advantageous on solid electrodes if the cathodic generation of solvated electrons proceeds

$$e^-(M) \rightarrow e_s^-$$

and not the deposition of the alkali metal

$$M^+ + e^-(M) \rightarrow M$$

This is the reason why the reaction in the solution bulk

$$M^+ + e_s^- \rightarrow M$$

does not take place.

In the systems of subgroup A-2, cathodic reduction of a cation is thermodynamically disadvantageous, as a rule, for the salts of lithium, rubidium, and cesium, in whose solutions solvated electrons can be generated. In sodium salt solutions electrodeposition of the metal takes place. Potassium salt systems occupy an intermediate position.

All solvents suitable for generation of solvated electrons are thermodynamically unstable at high cathode potentials. However, cathodic reduction of a solvent kinetically slows down. In Group A solvents the chemical reaction of solvent and solvated electrons is also hindered; the reaction rate markedly increases as the temperature is raised and/or use is made of catalysts.

Among Group B systems are tetraalkylammonium (mainly, tetrabutylammonium) salt solutions in aprotic solvents (B-1) and in mixtures of an aprotic and protic solvent (B-2). In these systems the cathodic reduction of all the solution's components is appreciably retarded, while the electrons generated at the cathode disappear in the course of a homogeneous reaction with the cations of the background electrolyte or the solvent's molecules.

Of interest is also a qualitatively different behaviour of solvated electrons in the systems containing tetra-substituted ammonium salts: they are found to be chemically stable only in liquid ammonia. This happens because at low temperatures the reactivity of tetraalkylammonium cations towards the solvated electrons drastically decreases. Apparently, cathodic generation of solvated electrons is also possible in the melts. Thus, it has been shown [121] that cathodic polarization in molten nitrates containing water leads to the formation of hydrated electrons. The blue coloration appearing on the molten lead cathode in the NaCl—KCl (1:1) + 1% $PbCl_2$ melt at cathodic polarization exceeding 0.6 V was also ascribed to solvated electrons [122].

From what has been said it is clear that cathodic generation of solvated electrons is possible in a number of systems; hence, account should be taken of the electrochemical formation of solvated electrons in all cases where the cathode potentials are negative enough for this process to proceed.

4 The Nature of Particles that Form on the Cathode in Liquid Ammonia and Hexamethylphosphotriamide. Properties of Solvated Electrons in these Systems

Vast material on the properties of solvated electrons in liquid ammonia has been accumulated by now [78,123-127]. Ammonia was the first solvent for which it was shown that the properties of solvated electrons obtained by different methods (by dissolving alkali metals, by pulse radiolysis, and by cathodic generation) were identical (see Fig. 4).

In the past decade extensive studies into the properties of the particles that form in hexamethylphosphotriamide during cathodic polarization were carried out. On passing a current through solutions of lithium chloride, sodium bromide, sodium perchlorate, and potassium iodide in this solvent a blue (in the case of sodium and/or potassium salts) or dark blue (in the case of lithium salt) colour appears at the electrode. Optical spectra taken in these solutions are presented in Fig. 5. For comparison, the figure shows the spectra obtained during pulse radiolysis of the pure solvent and also upon dissolving metallic sodium in it. In all cases a longwave (1900–2300 nm) band is observed. It is the appearance of a longwave absorption

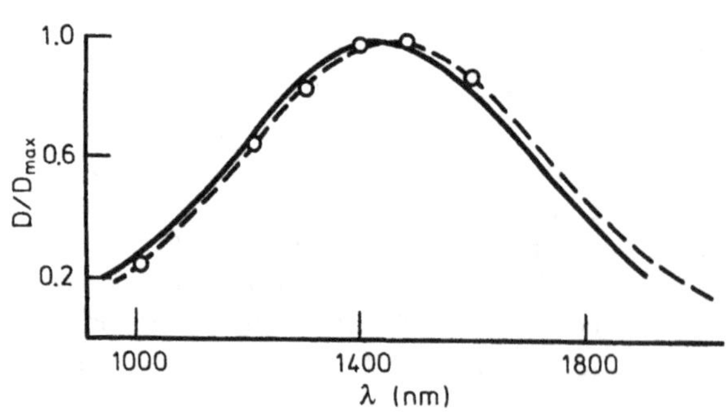

Fig. 4. Spectra of optical absorption of solvated electrons in liquid ammonia at —65–—70 °C: points: pulse radiolysis (128); solid line: electrochemical generation in solutions of alkali metal and tetrasubstituted ammonia halides (129, 130); dashed line: dissolution of alkali metal (131)

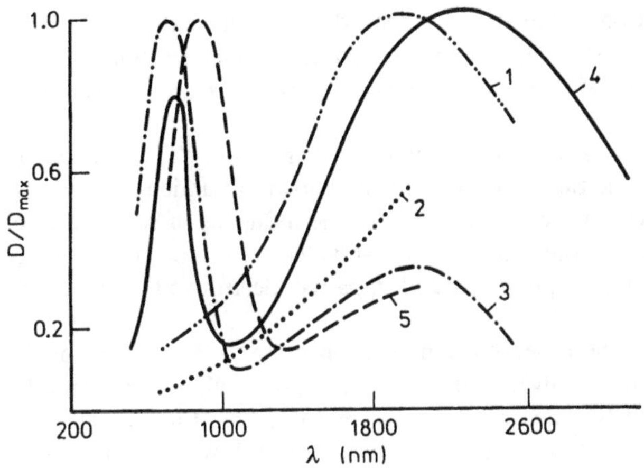

Fig. 5. Spectra of optical absorption of solvated electrons and of their associates in hexamethylphosphotriamide at room temperature:
1, 3, 5: electrochemical generation in 0.2 M solutions of LiCl, NaBr, and KI, respectively (72, 74); 2: pulse radiolysis of pure solvent (74, 132); 4 — dissolution of metallic sodium (133)

band during radiolysis of pure hexamethylphosphotriamide (curve 2) which serves as the most direct proof of the fact that this band belongs to solvated electrons and not to their associates containing alkali metal cations. Another proof of this is that, simultaneously with the passage of a current, a narrow single line appears in the ESR spectrum with a g-factor which coincides with the g-factor of a quasi-free electron [134-137]. Such a signal appears in all solutions, including lithium salt solutions whose light absorption spectra contain only the longwave band.

Comparison of the optical absorption curves shown in Fig. 5 reveals that solvated electrons are generated in alkali metal salt solutions during cathodic polarization. In fact, the particles that form during electrolysis are identical to those appearing under conditions for which the formation of solvated electrons is well proved, i.e., on dissolution of alkali metals and on pulse radiolysis of pure solvent and/or of the salt solutions.

Of interest are the results obtained in studies not of the excess electrons themselves, but of solvent (e.g. hexamethylphosphotriamide) molecules on introducing the solvated electrons. In the Raman spectrum, obtained by the coherent ellipsometry method, with the introduction of solvated electrons a positive shift in the C—H bond vibrational frequency is observed [138]. This has been attributed to the appearance of increased electron density at the C—H bond when a hexamethylphosphotriamide molecule enters into the solvate shell of an electron.

Reviewing the data on variation in the state of hexamethylphosphotriamide molecules under the influence of solvated electrons reveals [139] that in the molecule there is no center of preferential localization for the electrons, rather they are delocalized within the limits of the molecule (not to confuse with the above delocalization over the entire solution volume). The charge of a solvated electron may be assumed to be somewhat less localized compared to that of normal ions since the addition of bromide and perchlorate anions, in amounts exceeding by 2–3 orders the concentration of solvated electrons, does not cause the C—H bond vibration frequency to vary. Also, it has been ascertained that electrons interact only with the C—H fragment and do not affect the carbon-nitrogen and phosphorus-oxygen bonds [138].

The data of Ref. 138 agree well with the much earlier results [140] obtained by the method of dynamic polarization of hexamethylphosphotriamide protons in studying how the solvent's characteristics vary under the influence of solvated electrons. It was shown that in the presence of solvated electrons there exists a small (less than 10^{-4}) but finite spin density of unpaired electrons at hexamethylphosphotriamide protons.

These results coupled with the observed significant increase in the volume of the solution due to the formation of solvated electrons in hexamethylphosphotriamide, as reported by Gremmo and Randles [42,69], enable us to assume that the electron is in the cavity formed by the solvent's molecules oriented with their methyl groups toward the cavity. These data fit the accepted structure of hexamethylphosphotriamide molecule [141], where the positive charge sited mainly on phosphorus is well shielded. Indeed, at the positive pole of the molecule the methyl substituents at the amino group face outward; these methyls interact directly with the electron.

Cathodic generation of solvated electrons is a convenient method for studying their association, since it enables the concentration of solvated electrons to be varied independently over a wide range while preserving the cations concentration well in excess.

In hexamethylphosphotriamide systems the interaction of the dissolved salt cations with solvated electrons in general causes the solvated electron absorption peak to slightly shift towards the shortwave region. This can be put down to electrostatic interaction of cations and solvated electrons, which should cause a cationic atmosphere around the solvated electrons, and, at quite high concentrations of the electrolyte should form noncontacting (i. e. solvent-separated) ion pairs and triple ions [42,69]. Thus, it may be considered that a longwave absorption band in hexamethylphosphotriamide results both from unbound solvated electrons and the solvated electrons that weakly interact electrostatically with cations. And the data on ESR spectra [134–136] are indicative of the absence of contact interaction between them.

An analogous picture is observed in liquid ammonia in which, irrespective of the nature of cation, only one absorption band is noticed; this band conforms to solvated electrons and those which weakly interact with the cations. Lagovski [142], who has summarized data on optical electron spectra for liquid ammonia solutions, reported by various authors, has shown that in the range of concentrations about 10^{-5}–10^{-3} mol/l the absorption peak wavelength (1470 nm) does not vary; between 10^{-3} and 10^{-2} mol/l the peak abruptly shifts towards 1540 nm; further increase in concentration (up to 10^{-1} mol/l) has no effect on the absorption peak location. At that concentration of solvated electrons, at which the location of the absorption peak changes, an abrupt decrease is observed in the fraction of paramagnetic particles (i.e. unbound solvated electrons) in the total number of electrons present in liquid ammonia [143]. The most likely explanation thereof, given by Harris and Lagovski [144], is the formation of the triple ion $e_s^- M^+ e_s^-$ in which, despite the presence of some interaction that causes the spin moments to couple, both solvated electrons remain separate particles.

When electrochemical generation is accomplished in hexamethylphosphotriamide solutions of sodium and potassium salts there appears, along with the absorption band for solvated electrons, a peak at 750 or 880 nm (Fig. 5). A similar picture was seen earlier upon dissolving the alkali metals in hexamethylphosphotriamide [133] and also on carrying out pulse radiolysis of sodium bromide solution in it [145,146]. In lithium salt solutions the second absorption band never appears during electrochemical generation of electrons (Fig. 5). Note that the ESR spectrum for sodium bromide solution, recorded during electrochemical generation, is identical to that for lithium chloride solution [134–136]; this reveals that the particles responsible for the appearance of the band with absorption peak at 750 nm are not paramagnetic in nature. By different methods it has been proved that such a particle consists of two electrons and an alkali metal cation [73,147,148]: it is obtained according to the reaction

$$M^+ + 2\,e_s^- \leftrightarrows [\text{associate}] \tag{8}$$

where e_s^- is a solvated electron which can be free or weakly bound electrostatically with sodium cations.

A complex of similar composition was suggested by Gremmo and Randles for solutions of sodium in hexamethylphosphotriamide [42,69]. With the formation of an associate an abrupt change in optical characteristics and disappearance of paramagnetism occur. This enables one to affirm that the particle is a certain integral formation and not a mere combination of two solvated electrons which mostly retain their individual properties.

The absence of associates for lithium cations may be attributed to the stronger solvation of the latter compared to the cations of other alkali metals, and this should shift the equilibrium of Eq. (8) to the left.

The absorption spectra recorded during pulse radiolysis made it possible to determine the extinction coefficients for the particles formed and to compute the equilibrium constant $K = 2.3 \times 10^4\ \text{mol}^{-2}\text{l}^2$ for the reaction Eq. (8) [72,149].

For the complex in which the electrons completely lose their individual optical and paramagnetic properties two most-substantiated hypothetical structures have been proposed: a bielectron stabilized by the interaction with a cation, $M^+ \ldots e_2^{2-}$ [147], and an alkali metal anion M^- [116,150].

By a bielectron e_2^{2-} is implied a complex of two electrons having a common solvate shell. The gain in energy when such an associate is formed is due to the coupling of electron spins and the interaction of the double-charged "anion" (bielectron) with the alkali metal cation. In this variant of the structure, the cation and the bielectron seem partially to retain their individual solvate shells.

Within the framework of the hypothesis of the alkali metal anion it is assumed that both electrons are at the outer s-orbital of the alkali metal and the solvent interacts with the associate as with a single negatively charged particle [116].

Table 4 compares the location of absorption peaks of the associate in different solvents. If we restrict the analysis to well-studied metals, i.e. sodium and potassium, then it follows from the table that the absorption peaks are practically independent

Table 4. Absorption peak wavelength (nm) for the solvated electron and the associate, consisting of an alkali metal cation and two electrons, in hexamethylphosphotriamide [72], also complied in Ref. [151]

Solvent	e_s^-	Associate			
		Sodium	Potassium	Rubidium	Cesium
HMPA	2300	750	880	973	
Tetrahydrofuran	2049	740	900	940	1042
Glyme	1736	725	890		
			820 (−80 °C)	873 (−80 °C)	933 (−80 °C)
Diglyme	1610 (−55 °C)	666 (−60 °C)	885		

of the nature of the solvent (particularly when due allowance is made for its short-wave shift at low temperatures), while the peak location strongly depends on the nature of cation. A similar situation exists for other, though less-studied cations. This can be construed in favour of the structure M^-.

Comparing the donor and acceptor numbers (Table 3) for the solvents listed in Table 4 reveals that they should solvate the anions almost equally and the cations differently. This also enables one to assume that the associate does not contain a cation as such, but generally has the anion structure. Comparing the location of absorption peaks for solvated electron e_s^- and the associate shows that in all media the associate does not contain in its composition a solvated electron as such and the properties of the solvated electron e_s^- are more sensitive to the nature of the solvent than those of the associate.

Unlike the solvents enumerated in Table 4, in liquid ammonia only one absorption band caused precisely by solvated electrons is observed, irrespective of the nature of cation. The shift of the peak and the decrease in the fraction of paramagnetic particles as the alkali metal concentration is increased are attributed by a number of authors [144,152] to the formation of associates, the associate being a combination of two solvated electrons. The contribution of alkali metal anions to the formation of the single absorption band in liquid ammonia seems to be scarcely probable. Usually the location of the M^- absorption band is significantly different from that for solvated electrons. Rough estimates made by different authors in the last few years [153,154] reveal that the formation of alkali metal anions in liquid ammonia is not possible at all [153] or is possible only for sodium [154]. Taking the equilibrium constant for the reaction $Na^+ + 2\,e_s^- \leftrightarrows Na^-$ equal to 2.5×10^2 mol^{-2} l^2 [154] and assuming the ion activity coefficients for liquid ammonia to be small, an appreciable concentration of sodium anions could be expected only when the sodium cations are well in excess and for sufficiently large concentrations of solvated electrons.

Such a marked difference in the properties of solvated electron solutions in liquid ammonia and other solvents should be attributed to the extremely high donor number of ammonia (Table 3) and this shifts the equilibrium of Eq. (8) towards the cation.

Thus, analysis of the physico-chemical properties of systems containing solvated electrons suggests that the particles exist in the following two main forms: e_s^- and

M^-. In the former the solvated electrons are either free or weakly bound with an alkali metal cations; in the latter they completely lose their individual properties. The existence of such particles must be reckoned with in explaining the electrochemical properties of the systems containing solvated electrons.

5 Equilibrium Electron Electrode

On placing a metallic electrode in a solution containing solvated electrons, thermodynamic equilibrium is usually established between the metal electrons and the solvated electrons in the solution. Such an electron electrode acts as a reversible electrode of the first kind.

A knowledge of the electron-electrode equilibrium potential is necessary to form a judgement about the primary or secondary nature of the generation process (see Sect. 7) and the place of this process among other electrode reactions.

A reversible electron electrode for liquid ammonia was known long ago [19,155–157]. Recently, the electron electrode equilibrium in liquid ammonia has been studied anew by a number of authors in a series of works on the electrochemistry of solvated electrons.

Figure 6 shows the dependence of the equilibrium potential of an electron electrode on the logarithm of the total electron concentration in liquid ammonia against a background of sodium and potassium salts. Here, equilibrium is seen to be established for a single-charged particle (solvated electron) in the region of low concentrations. At concentrations over 10^{-3} mol/l double-charged associates contribute to the dependence. These associates may be assumed to have the structure of the $e_s^- M^+ e_s^-$ triple ions (see Sect. 4) because the variation in the slope of the $E_0 - \lg c_0$ curve

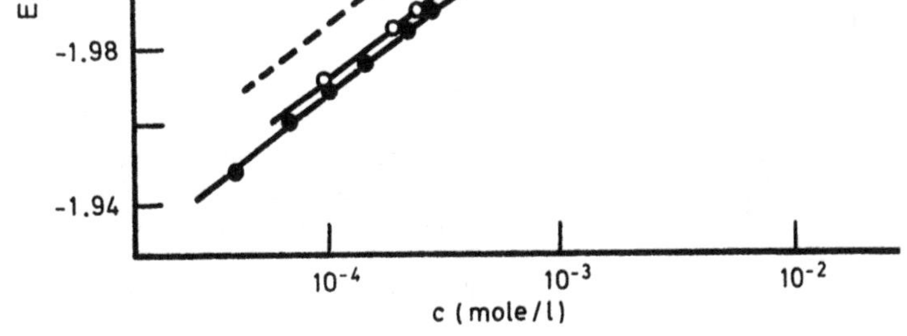

Fig. 6. Equilibrium potential of electron electrode versus logarithm of total concentration of electrons in liquid ammonia against a background of (○) 0.15 M NaI and (●) 0.5 M KI; dotted lines: theoretical slopes at n = 1 and n = 2; temperature: 40 °C; reference electrode: Pb/Pb^{2+} (0.05 mol/l) (158)

Table 5. Standard potentials E_0°, V in hexamethylphosphotriamide [17] and liquid ammonia [87]

System	$e_s^- \rightleftarrows e^-(M)$	$Li \rightleftarrows Li^+ + e^-(M)$	$Na \rightleftarrows Na^+ + e^-(M)$
E_0° in HMPA	−0.04 ± 0.02	−0.20	+0.08
E_0° in liquid NH_3	+0.11	−0.28	+0.17

System	$K \rightleftarrows K^+ + e^-(M)$	$Cs \rightleftarrows Cs^+ + e^-(M)$
E_0° in HMPA	−0.04	+0.04
E_0° in liquid NH_3	+0.02	−0.02

(c_0 = total concentration of solvated electrons) and the variation in location of the absorption peak, which is due to the formation of a triple ion with partially coupled spins, are observed in the same concentration ranges of the solvated electrons [142,158].

The standard potential of the electron electrode in liquid ammonia was measured by a number of authors [19,89,158–160]. In case it is necessary to recalculate the reference electrode scales of these papers, use should be made of the relations between standard potentials for different electrodes in the solvent [87,161,162].

Table 5 presents a comparison of potentials for the electron electrode and other types of electrode (for the discussion, see below).

The most reliable value was determined by Harima and Aoyagui [158] who obtained the same standard potentials against a backround of lithium, sodium, potassium, and cesium salts. Together with Kurihara [93], they have also determined the standard potential of the electron electrode in methylamine.

During the past few years it has been reported that the reversible electron electrode can be realized in solutions of solvated electrons in hexamethylphospotriamide against a background of lithium [163–165] and sodium [21,165,166] salts. That the system is reversible in these solutions is evidenced by the fact that the polarization curve in linear coordinates passes trough the origin of coordinates without any kink [163,167]. When the potential is more positive than the equilibrium potential,

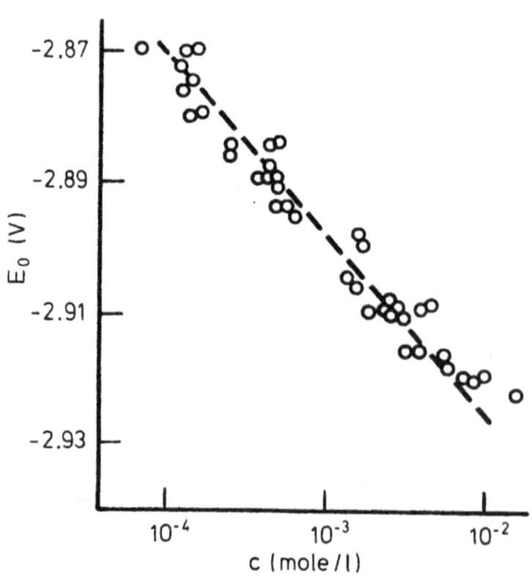

Fig. 7. Equilibrium potential of the electron electrode versus concentration of solvated electrons in hexamethylphosphotriamide against a background of 0.3–0.6 M LiCl points: solvated electron concentration calculated from Warburg impedance; dashed line: average concentration of solvated electrons, determined from galvanostatic measurements (164)

anodic oxidation of the solvated electrons starts; generation of solvated electrons starts at more negative potentials. These processes are discussed in detail in Sections 6 and 7.

The dependence of the electron-electrode potential on the concentration of solvated electrons against the background of a lithium salt in hexamethylphosphotriamide is shown in Fig. 7. It follows from this figure that the electrode behaviour obeys the Nernst equation for a single-charged particle. This is the most strict proof of the fact that in this solution thermodynamic equilibrium is established at the electrode.

The solvated electron concentration was estimated chronopotentiometrically on a rotating disc electrode and by measuring the diffusion impedance [164,165]; both methods gave values close to each other.[18] Knowing the concentration one can express the electron-electrode equilibrium potential at 5 °C (V, vs. aqueous SCE) as $E_0 = (-3.05 \pm 0.02) - 0.055 \lg c_0$, where c_0 is the bulk concentration of solvated electrons, mol/l. A close value ($E_0^\circ = -3.08$ V vs. SCE) has been obtained by Kanzaki and Aoyagui [21].

Table 5 compares the standard potential of the electron electrode in hexamethylphosphotriamide (5 °C) with the standard potentials of alkali metals (25 °C). Data for liquid ammonia are also given. In both solvents the rubidium electrode potential serves as a reference point since it depends very little on the solvent. It is seen from the Table that in both solvents the standard equilibrium potential of the electron electrode is more positive than that of a lithium electrode and is close to the potentials of other alkali metals. In the course of experiment, cathodic production of dilute solutions ($10^{-4} - 10^{-2}$ mol/l) of solvated electrons takes place and this makes the electron electrode equilibrium potential more positive compared to the standard value. In case of hexamethylphosphotriamide the same happens when electrons are bound in strong non-paramagnetic associates by the cations of all alkali metals except lithium (see Sect. 4). This enables one to assume that under the conditions of the experiments the electron-electrode equilibrium potential in liquid ammonia and hexamethylphosphotriamide is more positive than the equilibrium potential of all alkali metals. This makes thermodynamically possible primary cathodic generation of solvated electrons in solutions of all alkali metal salts in the two solvents.

The standard potential of the electron electrode in water is also known. Unlike liquid ammonia and hexamethylphosphotriamide, here it is not an experimental but a computed value. The most reliable estimates yield a value between −2.85 and −2.87 Vvs. NHE [65,168] [19]

From what has been said in this section it is evident that sufficient information on the electron electrode is now available for a number of solvents. Standard potentials for such an electrode in different solvents are known and this enables one to predict the potential region where the possibility of electrochemical generation of solvated electrons must be reckoned with.

---

18 Coincidence of the results indicates that the diffusion coefficient for solvated electron (2.7×10^{-6} cm²/s [164]) is correctly chosen, because for the determination of concentration by different methods use is made of different dependences relating the concentration to the diffusion coefficient: $c_0 \sim D^{-1/2}$ for chronopotentiometry [164]; $c_0 \sim D^{-2/3}$ for a rotating disc electrode [165].

19 The equilibrium potential of the electron electrode calculated in [65] somewhat differs from the earlier data (cited, e.g., in [5,6]) since allowance was made for the energy of atomic hydrogen dissolution in water. Note, that the original paper [65] contains an erratum; the true value of the equilibrium potential given above was recalculated from the hydration energy of the electron (see also [168]).

6 Anodic Reactions in Solvated Electron Solutions

6.1 Anodic Oxidation of Solvated Electrons and their Associates

As mentioned earlier, anodic oxidation of solvated electrons starts when the electrode potential becomes more positive than the equilibrium "electron" potential [169-171].

The results of the studies of this process in different media are summarized in Table 6. When investigated by cyclic voltammetry, one usually starts with solutions that initially do not contain solvated electrons; solvated electrons are then obtained during the cathodic sweep of potential. In other methods, the necessary bulk concentration of solvated electrons was attained by dissolving the alkali metal or by preliminary cathodic generation at an auxiliary electrode.

Of the results obtained over the past few years, those for hexamethylphosphotriamide are most interesting. For lithium salt solutions the shape of anodic curves conforms to oxidation of particles of only one type. This is in accord with the results of the studies on the state of solvated electrons in these systems. Indeed, as shown in Section 4, in the presence of a lithium salt the electrons exist exclusively as monoelectrons e_s^-; part of these electrons, when the salt is in excess, can be bound into noncontact ion pairs with lithium cations. The electrons in these pairs differ only slightly in their properties from non-associated electrons. And this yields a single-wave anodic curve.

The anodic limiting current in lithium salt solutions is determined by the diffusion of the solvated electrons to the electrode. This was quantitatively established by the measurements taken on rotating disc electrodes [165] and also by galvanostatic measurements [164]. In fact, as seen from Fig. 8, the limiting current density is proportional to the square root of the disc electrode rotation rate. This, in accordance with the rotat-

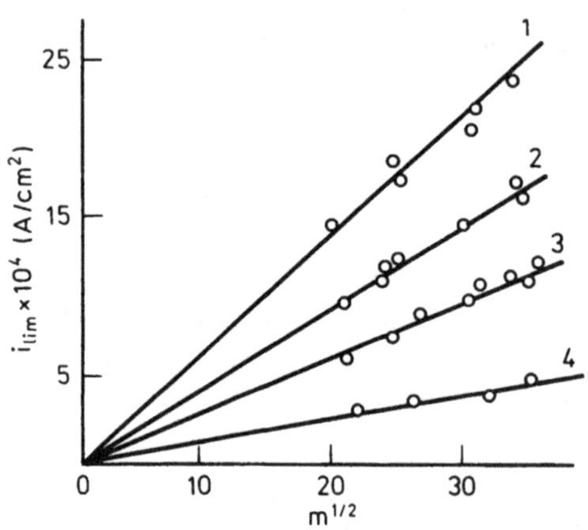

Fig. 8. Dependence of the limiting current density for anodic oxidation of solvated electrons on the square root of the rpm of a platinum disc electrode in hexamethylphosphotriamide solutions of solvated electrons against a background of 0.36 M LiCl at electrons concentrations corresponding to equilibrium potentials (vs. aqueous SCE): 1: —2.91; 2: —2.90; 3: —2.89; 4: —2.88 V. Temperature: 5.5 °C (165)

Table 6. Anodic oxidation of solvated electrons

System	Method	Shape of anodic curve	Year, Ref.
Liquid ammonia LiI, NaI, KI 0.5 mol/l; solvated electrons; Pt, Au, W, −40 °C	Potentiodynamic curves, 100–1000 mV/s	One peak ($n^a = 1$)	1977 [172]
	Stepwise deviation of potential from the equilibrium potential ($-10 < \eta^b < +100$ mV)	One plateau of limiting current	1979 [158]
Liquid ammonia KI, 0.1–0.4 mol/l; Pt, −55 °C	Cyclic voltammetry, 10–200 mV/s	One peak ($n = 1$)	1979 [159, 48] 1983 [173]
Liquid ammonia KI, 2 mol.%; Pt, −68 °C	Cyclic voltammetry, 500 mV/s	One peak	1979 [174]
Liquid ammonia KBr, 0.1 mol/l; Pt, −50 °C	Cyclic voltammetry, 80 mV/s	One peak	1982 [175]
Liquid methylamine KI, 0.5 mol/l; solvated electrons; W, −50 °C	Stepwise deviation of potential from the equilibrium potential up to $\eta = +100$ mV	At low concentrations of solvated electrons there is one plateau; at solvated electrons concentration more than 5×10^{-6} mol/l there are two plateaus in $it^{1/2}$-lgt coordinates[c]	1980 [93] 1981 [176]
HMPA LiCl, 0.5 mol/l; glassy carbon; room temperature	Cyclic voltammetry, 0.3–30 V/s	One peak	1973 [20]
HMPA LiCl, 0.3–0.4 mol/l; solvated electrons, 10^{-4}–10^{-2} mol/l; Pt, Cu, 5 °C	Steady state polarization curves on rotating disc electrode and on a stationary disc electrode in stirred solution	One wave	1976 [165]

System	Method		Shape of anodic curve	Year, Ref.
HMPA LiCl, 0.3–0.6 mol/l; glassy carbon	Cyclic voltammetry, 10–200 mV/s		One peak	1979 [177]
HMPA $NaClO_4$, 0.05–0.40 mol/l; solvated electrons, 10^{-3}–10^{-2} mol/l; Pt, 5 °C	Potentiodynamic curves	0.7–36 mV/s up to $\eta = +500$ mV	One peak	1972 [21]
		40 mV/s up to $\eta = +1000$ mV	One peak	
		67 and 134 mV/s up to $\eta = 1000$ mV	Two peaks	1973 [178, 679]
	Steady state polarization curves in a stirred solution up to $\eta = +1000$ mV		Two waves	
HMPA NaBr, 0.3–0.4 mol/l; solvated electrons, 10^{-3}–10^{-2} mol/l; Pt, Cu, 5 °C	Steady state polarization curves	on a stationary electrode in unstirred solution	One wave	1976 [165]
		on a stationary electrode in stirred solution	Two waves	
HMPA $NaClO_4$, 0.3 mol/l; solvated electrons, 10^{-4}–10^{-2} mol/l; Pt, Cu, 5 °C		on a rotating disc electrode, and on a stationary electrode in stirred solution	Two waves	1977 [166]
HMPA $NaClO_4$, 0.1 mol/l; Pt, 20 °C	Cyclic voltammetry, 150–200 mV/s		One peak	1978 [180]
HMPA $NaClO_4$, 0.1 mol/l; Au, Pt	Cyclic voltammetry, 10–200 mV/s		One peak	1979 [177]

[a] Charge of diffusing particle; [d] Deviation from the equilibrium potential; [c] t-time

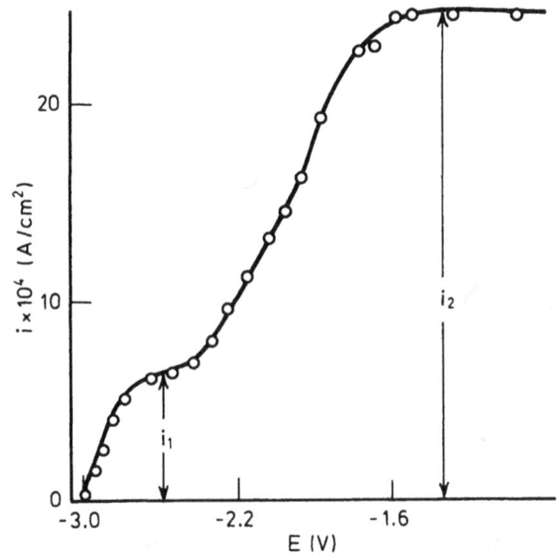

Fig. 9. Dependence of the anodic current density on the potential of a rotating platinum disc electrode (960 rpm) in hexamethylphosphotriamide solution of solvated electrons against a background of 0.3 M $NaClO_4$ at electrons concentration corresponding to an equilibrium potential of —3.00 V (vs. aqueous SCE). Temperature: 5.5 °C (166)

ing disc electrode theory [181], proves the diffusion nature of the limiting current. Anodic oxidation of solvated electrons under controlled hydrodynamic conditions has proved to be the most reliable method for determining their concentration.

It follows from Table 6 that the shape of anodic curves obtained by several authors for sodium salt solutions in hexamethylphosphotriamide is indicative of oxidation of two different kinds of localized electrons in the solution (cf. e.g., Fig. 9). The first wave conforms to the oxidation of solvated electrons proper and the second to the oxidation of non-paramagnetic associates containing one cation and two electrons. It is seen that the reactivity of monoelectrons far exceeds that of the complexes with respect to the anodic oxidation reaction. Potentials at which they oxidize differ by more than 0.5 V; this is consistent with the large difference in the energies of the optical absorption peaks for these particles.

The limiting current of the second wave (i_2) is a diffusion current [166]. Curve 2 of Fig. 10 shows the relationship between this current and the equilibrium potential. In the range of high electron concentrations, this relationship can be expressed as: $E_0 = \text{const} - (0.029 \pm 0.002) \lg i_2$. The value of the proportionality constant $0.029 = 2.3\,RT/nF$, where $n = 2$ (see Eq. (8)), indicates that the electrode is in thermodynamic equilibrium with the electrons in solution, which are practically completely bound to form non-paramagnetic complexes containing two electrons. On decreasing the concentration of electrons, the slope of curve 2 is observed to increase (which means a decrease in the effective value of n). This happens due to partial dissociation of the associate. In Ref. 178 an increase in the height of the first

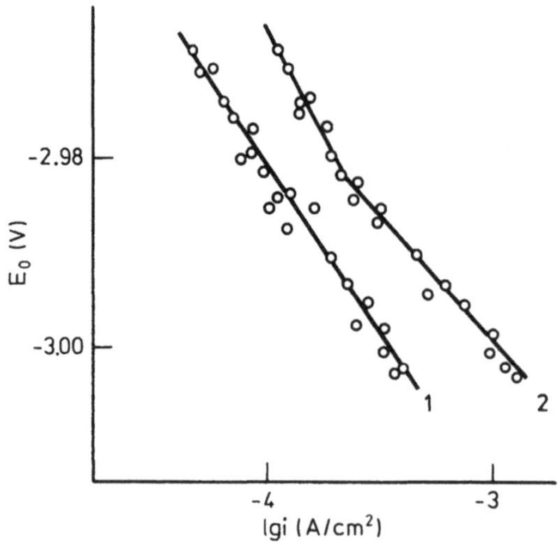

Fig. 10. Relationship between the logarithm of limiting anodic currents i_1 (curve 1) and i_2 (curve 2) and the equilibrium potential of a platinum rotating disc electrode (960 rpm) in hexamethylphosphotriamide solutions of solvated electrons against a background of 0.3 M $NaClO_4$. Temperature: 5.5 °C (166)

wave was observed with a decrease in the sodium ion concentration in the solution; this also points to dissociation of the complex.

The limiting current of the first wave (i_1) is independent of the electrode rotation rate [166]; this current is a kinetic one. It has been proposed [21,179] that the dissociation of the ion pair $Na^+ \dots e_s^- \rightarrow Na^+ + e_s^-$ is the rate determining process. However the experimental relationship of the equilibrium potential and current i_1, $E_0 = \text{const} - (0.037 \pm 0.004)\,\lg i_1$, (Fig. 10), conforms to the theoretical dependence for the current that appears due to the decomposition of an associate of different composition, viz. $Na^- \rightarrow Na^+ + 2\,e_s^-$ [166]. This is in accord with the results of a physico-chemical study of the composition of associates in sodium salt solutions (Sect. 4), and enables i_1 to be considered as kinetic current caused by slow dissociation of the complex, containing one sodium cation and two electrons, into a cation and two solvated electrons [166]. That oxidation potentials of the associate greatly differ from those of a solvated electron (Fig. 9) is indicative of large bond strenght in the complex and is an additional argument in favour of the alkali metal anion structure.

Slowness in the dissociation of the associate during anodic oxidation is qualitatively consistent with the small value of the dissociation rate constant $Na^- \xrightarrow{k_{dis}} Na^+ + 2e_s^-$ in hexamethylphosphotriamide, determined by laser photolysis [149] to be equal to $2.7 \times 10^2\ s^{-1}$. Knowing this value one can theoretically estimate the kinetic current $i_1 = 2F10^3 \sqrt{D_{s.e.} k_{dis} c_1 c_2}$, where c_1 and c_2 are the concentrations of solvated electrons and associates respecively [166]. c_2 was estimated from the diffusion current i_2 in the range of most negative equilibrium potentials when all electrons are bound to form associates. However different diffusion coefficients for the associate may be used (2.7×10^{-6} to 1.4×10^{-5} cm^2 s^{-1}, see [166]), in all cases the calculated value is always greater by 10–20 times than the experimental one. This is perhaps due to the electrode passivation that cuts off part of the electrode surface from the process. The effect of passivation on the electrode reactions of solvated electrons is discussed in detail in Section 7. Here we shall only mention that passivation does not alter the limiting diffusion current. The different effect of passivation on diffusion and kinetic currents is related to the fact that at the employed rotation rates the thickness of diffusion layer (that determines i_2) far exceeds that of the reaction layer (that determines i_1) [161].

Analyzing the data of Table 6 reveals that the results obtained by different authors in sodium salt solutions, at first sight, do not qualitatively agree with each other. Thus, in Refs. [177] and [180] only one oxidation peak (of solvated electrons or their associates) was observed on cyclic voltammetric curves during anodic sweep of potential. The "slow" potentiodynamic curves obtained initially by Kanzaki and Aoyagui [21] have also one peak. In the much later works of these authors [178,179] and in our works [165,166] it has been shown that a stationary anodic polarization curve for the stirred solution has two waves and the fast potentiodynamic curve has two peaks conforming to oxidation of two different particles. This disparity is explained differently in different cases. As mentioned above, current i_1 is independent of stirring and i_2 does depend on it. When taking a polarization curve in the unstirred solution, current i_2 is less than i_1 and the curve is transformed into a one wave curve [165]. Low sweep velocities correspond to large diffusion limitations and one peak or one wave is also observed on the curves. In Ref. [177] and [180] the anodic portion of cyclic voltammetric curves has one peak, though use was made of sufficiently high sweep velocities. These measurements were, however, taken under conditions when there were no solvated electrons in the bulk of solution. This leads to low near-cathode concentrations of electrons which, for this reason, did not practically form non-paramagnetic associates.

Listed in Table 6 the data on anodic oxidation of solvated electrons in different solvents point towards the different natures of non-paramagnetic associates, on the one hand, in liquid ammonia and, on the other hand, in hexamethylphosphotriamide and methylamine. In liquid ammonia, even the fast sweep (dynamic) voltammogram does not detect the second kind of particles at that bulk concentration of solvated electrons when non-paramagnetic associates exist [158,172,174]. Thus, the electrons that have gone into the associate do not markedly differ from an unbound solvated electron in their ability to undergo anodic oxidation. This is consistent with the hypothesis [144] concerning the appearance of triple ions $e_s^- M^+ e_s^-$. Unlike liquid ammonia, strong non-paramagnetic associates M^- are formed in methylamine as in hexamethyphosphotriamide. In these associates solvated electrons do not retain their individual properties. As shown in Refs. [93] and [176], at low concentrations of electrons in methylamine only monoelectrons are detected in anodic processes while monoelectrons and M^- associates are detected at high concentrations. It is of interest that the electrochemically determined [93] dissociation rate constant, when the potassium-containing associate dissociates into a cation and two electrons, equals 10^2 s^{-1}; this agrees in magnitude with the corresponding value 2.7×10^2 s^{-1} for the sodium associate in hexamethylphosphotriamide, determined by the laser photolysis method [149]. Closeness of these values can be considered as an argument in favour of the assumption that non-paramagnetic associates have a similar structure in these two solvents.

As two kinds of particles capable of undergoing oxidation are observed [182], both solvated electrons and alkali metal anions may be assumed to be formed also during cathodic generation of electrons in anhydrous solutions of alkali metal iodides in ethylenediamine.

That anodic oxidation does not reveal the existence of M^- associates for alkali metals in liquid ammonia is consistent with the decrease in the strenght of associates in this solvent. The experimentally determined association constant for Na^- in hexamethylphosphotriamide equals 2.3×10^4 [149] and the computed value for ammonia comes to 2.5×10^2 mol^{-2} l^2 [154]. To decide whether alkali metal anions exist in liquid ammonia, the oxidation of electrons must be studied at high concentrations of electrons in an intensively stirred concentrated background electrolyte solution.

At the same time, the M^- particles in liquid ammonia can be well detected in the presence of several other metals. It was shown [183–185] that gold could be dissolved in ammonia containing solvated electrons, with the formation of aurate-anion Au^-. This anion oxidizes by losing one electron. Anodic deposition of gold, in accordance with the reaction $Au^- - e^-(M) \rightarrow Au$, yields a bright coating with high current efficiency. The coating can be stripped by reverse scanning of potential: $Au + e^-(M) \rightarrow Au^-$. Unlike alkali metals, the large electropositivity of gold causes the stable form — metallic gold — to exist in liquid ammonia.

From what has been said in this section it follows that the anodic behaviour of excess electrons is determined by the form in which they exist in solution. Binding of solvated electrons with cations to form strong non-paramagnetic associates M^-, whose absorption spectra greatly differ from those of solvated electrons, involves an appreciable decrease in their reactivity with respect to anodic oxidation. Thus, the study of this reaction, coupled with an investigation of the optical and magnetic properties of electrons in liquid phase, will help towards obtaining valuable information on the properties of both the solvated electrons themselves and their associates.

6.2 Using Solvated Electrons in Batteries

The study of anodic oxidation of solvated electrons serves as theoretical basis for using solvated electrons in secondary chemical current sources. Recently it was suggested [186,187] that an inert metallic (tungsten or gold) electrode immersed in a concentrated solution of sodium [186] or lithium [186,187] in liquid ammonia can be used as a negative electrode in storage batteries. Specific energy of a $Li—NH_3/S—NH_3$ storage battery amounts to about 100 W h/kg. Significant polarization is observed on discharging; it is mainly linked with the work of the positive (sulphur) but not of the electron electrode [187].

It has been proposed that one could use a cation-exchange polyethylene membrane [186,187] or beta-alumina [186] for separating a solution containing solvated electrons from that present in the positive electrode compartment (solution of sulphur or polysulfide in liquid ammonia). Teflon-based cation-exchange membranes are chemically unstable with respect to solvated electrons.

As the proposed variants of membranes do not fully meet the work requirements of actual current sources, a liquid-ammonia battery was developed using the ternary system K—KI—NH_3 for the negative electrode. This system separates into layers, and a heavy layer rich in KI may safeguard the membrane against the K—NH_3 solution having high reductive power. With the ternary system K—KI—NH_3, intercalates, for example $NiPS_3$, can be used as the positive electrode [187,188].

7 Kinetics and Mechanism of Electrochemical Generation of Solvated Electrons

7.1 Specific Features of Cathodic Generation of Solvated Electrons in Different Solvents

By the beginning of 70's only one work on the kinetics of electrochemical generation of solvated electrons was known [19]. Later, however, a number of papers reflecting the increased interest in this problem appeared. A brief annotation of the works pertaining to the cathodic generation of solvated electrons in liquid ammonia and methylamine and also in hexamethylphosphotriamide is given in Table 7.

In some cases the generation of solvated electrons proceeds under diffusion control. A study of the cathodic process under these conditions yielded information on equilibrium standard potentials of the electron electrode (Sect. 5), and for methylamine — on the competition of electron generation and alkali metal deposition processes. Also, information has been obtained on the stoichiometry of the associates formed by electrons and on the tendency of various systems to association.

Comparing the data of Sections 4 and 6 with those of Table 7 reveals that in the presence of lithium cations, associates containing two electrons are not formed in any of the studied solvents (hexamethylphosphotriamide, liquid ammonia, and methylamine). For other alkali metal cations, such associates have been detected. By the example of methylamine it has been shown that the tendency to formation of associates with two electrons increases while proceeding from cesium to potassium. The higher the tendency the less the concentration of electrons, i. e. the smaller the density of the cathode current at which the formation of associates (as revealed by the change in the slope of the voltammetric curve) can be observed. Note that lithium cations never form associates with two electrons.

According to their tendency to form associates containing two electrons, the solvents may be arranged as follows:

liquid ammonia < hexamethylphosphotriamide < liquid methylamine

However, it should be remembered that at equal stoichiometry the nature of the associates in different solvents may differ. In liquid ammonia an associate represents, apparently, a triple ion $e_s^- M^+ e_s^-$ in which the electrons lose their paramagnetic properties, but retain other individual characteristics (optical spectrum, excess volume), whereas in other solvents (hexamethylphosphotriamide, liquid methylamine) new compounds — metal anions M^- are most likely to be formed.

Some authors have reported the results concerning the kinetics of the stage of

Table 7. Kinetics of cathodic generation of solvated electrons

System	Method of study	Main results	Mechanism proposed by the authors	Ref.
1	2	3	4	5
Liquid ammonia KI, 0.1–0.4 mol/l; Pt, —55 °C	Cyclic voltammetry, 10–200 mV/s	Tafel line with a slope of 43 mV	Diffusion control (by removal of a single-charged particle)	1979 [159]
Same solution + solvated electrons	Single galvanostatic pulse	$i_0 = (6 \pm 2) \times 10^{-3}$ A/cm² at solvated electron concentration 1.2×10^{-4} mol/l		
Liquid ammonia	Potentiodynamic curves	Tafel lines satisfying the equation:	Diffusion control;	1979 [158]
LiI, NaI, KI, CsI, KBr, 0.5 mol/l; Pt, Au, W, —40 °C	10–100 mV/s	$E = -\frac{2.3\,RT}{nF}\lg i + \frac{2.3\,RT}{2nF}\lg v + A$ where $A = E_0^0 + \frac{2.3\,RT}{nF} \times \left(\lg nF + \frac{1}{2}\lg\frac{nFD}{RT}\right)$, v is the scanning rate, D is diffusion coefficient	for lithium salt n = 1 at all conditions; for sodium, potassium, and cesium salts n = 1 at low currents; then $n \to 2$ (for potassium at $\lg i_0$ (mA/cm²) < —0.5)	
Liquid ammonia, NaI, 0.4 mol/l; solvated electrons; W, —35 — —75 °C	Single galvanostatic pulse Dependence of i_0 on solvated electrons concentration yields α	$\alpha = 0.5 \pm 0.1$ $i_0 = 10^{-2} - 10^{-1}$ A/cm² for solvated electrons concentration $2 \times 10^{-4} - 1.2 \times 10^{-3}$ mol/l	Electrochemical dissolution of electrons (fluctuation of solvent is a source of traps)	1980 [169]
		Diffusion coefficients were found to be underestimated by about 10 times	Effective concentration of traps has been introduced	
Liquid ammonia KI, 0,5 mol/l; Pt, —40 °C	Double galvanostatic pulse Dependence of i_0 on solvated electrons concentration yields α	$\alpha = 0.75$	Activation control	1982 [170]

Liquid methylamine W, —50 °C LiCl, 0.5 mol/l	Potentiodynamic curves, 50 and 100 mV/s	Tafel lines with a single slope, which satisfy the equation of Ref. 158 (see above)	Diffusion control (for a single-charged particle). Double-charged associates are absent up to lg i (mA/cm^2) = 0.5 (limit of measurements)	1980 [176] × 1981 [93]
Liquid methylamine KI, 0,5 mol/l	Potentiodynamic curves, 12.5 and 25 mV/s	Tafel lines having two portions with different slopes, which satisfy the equation of Ref. 158 (see above)	Diffusion control (for a single-charged particle) up to lg i (mA/cm^2) = —2.5; at higher current densities n → 2, then potassium is deposited	1980 [176] 1981 [93]
CsI, 0.1 mol/l	Potentiodynamic curves, 25 mV/s	Tafel line having two portions with different slopes	Diffusion control (for a single-charged particle) up to lg i (mA/cm^2) = —0.9; at higher current densities n → 2	
HMPA LiCl, 0.2 mol/l; Cu, 25 °C	Steady-state polarization curves on a rotating disc electrode	Tafel line with a slope of 60 mV; the stirring effect is absent	Electron thermoemission	1971 [22] 1972 [189]
HMPA $NaClO_4$, 0.2 mol/l; Pt, 5 °C; solvated electrons, 10^{-3} — 10^{-2} mol/l	Potentiodynamic curves, 0.7–36 mV/s	Tafel line with a slope of 60–70 mV	Diffusion control by removal of single-charged particles, i. e., solvated electrons	1972 [21]
	Stepwise variation of potential up to ±10 mV; i_0 as function of concentration of solvated electrons	α = 0.15–0.20	Analogy to cathodic process in ordinary redox systems	
$NaClO_4$, 0.05–0.4 mol/l	Single and double galvanostatic pulse; i_0 as a function of concentration of solvated electrons	α = 0.25–0.30		1974 [167]
HMPA LiCl, 0.2 mol/l; solvated electrons; Pt, Cu, 5 °C	Dependence of E_0 of electron electrode on i_0, determined from the initial portions of polarization curves	α = 1	Electron thermoemission	1973 [163]

Table 7. (continued)

HMPA LiCl, 0.5 mol/l; glassy carbon	Cyclic voltammetry, 100–1000 mV/s	Cathodic current at the initial stage of "anode" half-cycle of scanning exceeds the "cathode" scanning current corresponding to the same potential. The current decreases with increase in scanning rate	Direct transfer of electron from metal to a cavity in solvent. Activation control. Strong effect of ohmic losses which abruptly decreases as the concentration of solvated electrons increases	1973 [20]
	Potentiodynamic curves at slow scanning and small current densities (till the hysteresis loop arises)	Tafel line with a slope of 83–93 mV (α = 0.64–0.72)		
HMPA LiCl, 0.2 mol/l; LiCl, 0.2 mol/l + HCl, 0.16 mol/l; NaBr, 0.2 mol/l solvated electrons; Cu, Pt, Cd; 25 °C	Steady-state polarization curves in stirred and unstirred solution	Single Tafel line with a slope of 60 mV	Electron thermoemission	1972 [190] 1973 [191]
HMPA LiCl, 0.3–0.4 mol/l; solvated electrons; Cu, Pt; 5 °C	Steady-state polarization curves on a stationary electrode in the stirred solution (with allowance made for concentration polarization and back reaction)	Tafel line with a slope of 122 ± 6 mV (α = 0.45 ± 0.05)	Electrochemical dissolution of electrons	1975 [15] 1976 [192]
	Dependence of i_0, determined from initial portions of polarization curves, on E_0 of electron electrode	α = 0.55 ± 0.04		
HMPA $HClO_4$, HCl, 0.02–0.05 mol/l; Cu, Hg, Pt, 25 °C	Steady-state polarization curve on stationary and droping electrodes	After limiting diffusion-migration current plateau the current increases at potentials characteristic for cathodic generation	Cathodic generation of electrons in solutions which do not contain alkali metal cations	1975 [193]

HMPA, $(CH_3)_4NI$ / $(C_2H_5)NBF_4$ 0.04–0.20 mol/l; Cu, 25 °C; DMSO[a], $(C_4H_9)_4NBF_4$, 0.23 mol/l; Cu, Hg—Cu, 25 °C	Steady-state polarization curves in stirred and unstirred solution	Beyond the dependence on solution composition and electrode material the cathodic curve is located in the generation potential range of solvated electrons. It consists of two Tafel portions having different slopes: 120–140 mV (lower) and 60 mV (upper)	Cathodic generation of solvated electrons according to two parallel processes, i.e., electrochemical dissolution and thermoemission	1976 [194] 1978 [195]
HMPA, $NaClO_4$, 0.1 mol/l; Pt, 20 °C	Cyclic voltammetry, 150–200 mV/s	Tafel line with a slope of 60 mV	Diffusion control by removal of a single-charged particle	1978 [180]
HMPA, LiCl, 0.3–0.6 mol/l; solvated electrons; Cu, 5 °C	Determination of impedance components (calculated according to the Ershler-Randles equivalent circuit) at different concentrations of solvated electrons	$\alpha = 0.55$	Predominating process — electrochemical dissolution of electrons	1982 [164]

[a]DMSO = Dimethylsulfoxide

electron generation proper (Table 7). At first glance, the possibility of obtaining such information is at variance with the data of those who by using analogous methods have observed that the reaction is controlled purely by diffusion. But it should be borne in mind, as has been shown by a number of authors [158,170,192], that the rate and the kinetics of generation of solvated electrons strongly depend on the state of the electrode surface. It is most likely that the discrepancy between the data of various authors is only apparent and is caused by the use of electrodes having different degrees of passivation (see Sect. 7.4).

The data on the kinetics of the stage of solvated electron generation proper enable some conclusions to be drawn on the mechanism of this process.

7.2 Primary Nature of Cathodic Generation of Solvated Electrons

It follows from Section 4 that solvated electrons are generated during cathodic polarization of inert metal electrodes, e. g., in liquid ammonia and hexamethylphosphotriamide solutions of alkali metals salts.

In the early days of the electrochemistry of solvated electron it was taken for granted that cathodic generation of solvated electrons proceeds via deposition of alkali metal, $M^+ + e^-(M) \rightarrow M$ with its subsequent dissolution, $M \rightarrow M^+ + e_s^-$. However, besides this "secondary" process, a basically different way is possible, i. e., direct transition of electrons from electrode to solution, $e^-(M) \rightarrow e_s^-$ — a "primary" process [20,21,196].

The relation between standard potentials of the electron electrode and the alkali metals (Sect. 5) is such that generation in liquid ammonia and hexamethylphosphotriamide thermodynamically is more likely to proceed by the primary mechanism. However, this does not prove that this mechanism can be realized in practice yet.

An attempt to distinguish between the two variants of generation (primary and secondary) was initially made [189,190] on the basis of the absence of limiting currents in dilute solutions of alkali metal salts in hexamethylphosphotriamide. However, a detailed analysis dismissed it as a criterion. Indeed, at a fast rate of dissolution of the deposited alkali metal, regeneration of the cation takes place and the near-the-electrode layer of solution is not depleted in it.

If the electrode process is not complicated by adsorption or other phenomena, depending directly on the nature of the electrode, then the rate of the charge transfer stage proper should be independent of the electrode material. Indeed, this type of independence has been earlier ascertained experimentally for the electroreduction of anions [197] and the electron photoemission into solution [9]. To such processes should belong also the cathodic generation of solvated electrons by the primary mechanism.

The fact that the process rate under activation control conditions is independent of the nature of the background salt cation can serve as an additional criterion for the primary nature of cathodic generation of solvated electrons. Systematic studies into the effect of this factor were made in hexamethylphosphotriamide [14–17], and they have revealed that in this solvent the generation process is highly complicated by passivation caused by the shielding film coating the electrodes. The passivation phenomenon is discussed below in detail. We shall only mention here that comparison

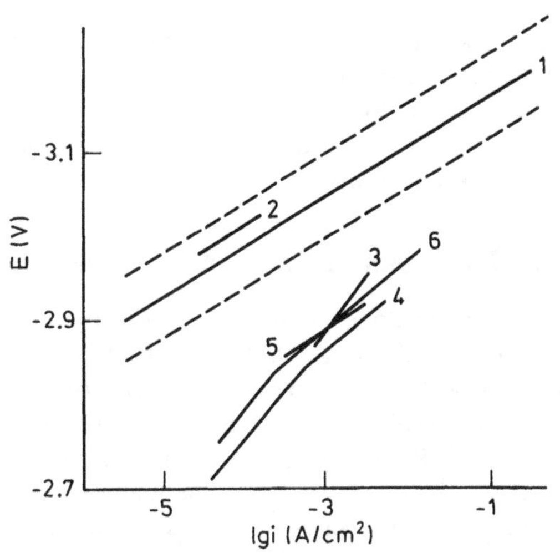

Fig. 11. Dependence of potential on current density in hexamethylphosphotriamide (curves 1–3, 6) and dimethylsulfoxide (curves 4 and 5):
1: copper, platinum, and cadmium passive electrodes in HCl containing solutions of LiCl, NaBr, and LiCl at 25 °C (scatter of experimental data is restricted by dashed lines); 2: passive copper electrode in LiCl solution containing solvated electrons at 5.5 °C; 3: active copper electrode in LiCl solution containing solvated electrons at 5.5 °C; 4: active copper and amalgamated copper electrodes in tetrabutylammonium salt solutions at 25 °C; 5: upper portion of curve 4 after making a correction in the value of the current (i.e., substracting of current values obtained by extrapolation of the lower portion of this particular curve); 6: active copper electrode in tetramethyl-, tetraethyl-, and tetrabutylammonium salt solutions at 25 °C (17)

should be made between the electrodes having the same passivation degree that is predetermined by the same electrode treatment.

Curves 1 and 2 of Fig. 11 have been taken on passive electrodes in hexamethylphosphotriamide solutions of different composition (sodium and lithium salts); and curve 6 — on a nonpassivated electrode (tetramethyl-, tetraethyl- and tetrabutylammonium salts). The rate of the process is equal for sodium and lithium, and is independent of the alkyl radical chain lenght. The generation rates in tetraalkylammonium salt solution were close to those in hexamethylphosphotriamide solutions of lithium salts on a copper electrode specially activated by anodic treatment [15,192] (curve 3 in Fig. 11). The practical equality of the currents in these systems opposes the idea that intermediate adatoms are participating.

Another direct experimental verification of the primary nature of cathodic generation of solvated electrons was done by carrying out this process in solutions that do not contain metal ions at all [193]. In dilute acids a rise in current was observed on different electrodes beyond the limiting diffusion-migration current plateau due to discharge of a proton donor and the potentials of the new process coincided with those of generation of solvated electrons on a nonpassivated surface in salt solutions. In Ref. [195] (curve 4, Fig. 11) it has been shown that the process rate is practically independent of the electrode material (copper or amalgamated copper); cathodic generation on mercury and gallium electrodes in dimethylsulfoxide solutions of tetrabuthylammonium tetrafluoroborate proceeds also in the same range of potentials. In these systems the electrode surface is free from passivating films.

Thus, it can be unambiguously affirmed that cathodic generation of solvated electrons in a number of cases proceeds by direct transition of electron from electrode into solution, i. e., cathodic generation is a primary process.

7.3 Mechanism of Electrochemical Generation of Solvated Electrons

In the last few years several authors have developed concepts for the possible mechanism of the primary cathodic generation of electrons. Of these electron thermoemission and "electrochemical dissolution of electrons" seem to be the most probable.

A mechanism of electron thermoemission into solution was considered by Brodsky and Frumkin [198] and then in Refs. 22, 195, and 196. This mechanism has no analogue among other electrode processes. Thermoemission proceeds via thermally excited metal electrons whose energy exceeds the electron energy level outside the metal (Fig. 12). They tunnel through the surface barrier and go into the solution. It is important that in the final state of this process the outside-metal electron remains delocalized and propagates in the solution as a plane wave. Thermoemission is a well-studies process for the metal-vacuum boundary where it usually proceeds at an appreciable rate only at sufficiently high temperatures. Considering the emission into solution account must be taken of the fact that the electron level in solution is lower than in vacuum owing to the electron interaction with solvent (more exactly, with its electronic polarization, see Sect. 2). The difference between the energies of the electron in metal and in solution (i.e., metal-to-solution electronic work function) decreases also when a negative potential is applied to the electrode. Precisely, the work function linearly depends on the electrode potential with a proportionality factor equal to 1. The effect of these two factors may be strong enough so that thermoemission in solution is observed even at room temperature. As indicated in Section 1, the energy level of the immediate product of the emission process, i.e., of the delocalized electron in solution is determined only by its interaction with the medium's electronic polarization. The orientational part of polarization plays a secondary role because "slow" dipoles do not manage to adjust themselves to the "fast"

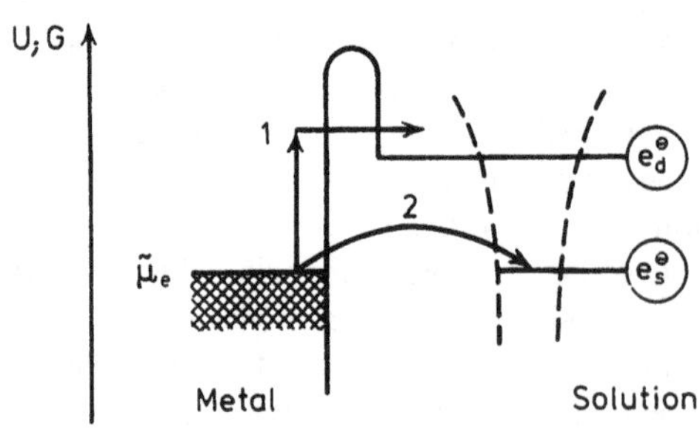

Fig. 12. Energy diagram for the process of electron transfer from electrode into solution during electrochemical generation of solvated electrons. 1: thermoemission; 2: dissolution of electrons; e_d^- : delocalized electron; e_s^- : solvated electron; $\tilde{\mu}_e$: Fermi level in metal. Dashed line shows the solvated electron potential well in solution

subsystem — electron. Nonetheless, when the electron is in motion a weak interaction with dipoles, particularly with suitable fluctuations of orientational polarization takes place. This finally causes the electron to localize. Thus, in cathodic generation by a thermoemission mechanism solvated electrons are formed via an intermediate stage of a delocalized electron. It is the thermoemission stage that determines the kinetics of the generation process, which is described by the Richardson-Sommerfeld equation

$$i = AT^2 e^{-\frac{w^{MS}+FE}{RT}} \tag{9}$$

Here, the constant A equals 120 A cm^{-2} deg^{-2} and w^{MS} is the metal-to-solution electronic work function at $E = 0$.

The second mechanism for the cathodic generation of solvated electrons has been suggested in a number of works [15, 20, 21, 192, 198]. The term "electrochemical dissolution of electrons" used to identify this mechanism points out an analogy of this process with usual electrode reactions. The direct cathodic reduction is the immediate transfer of metal electron to the acceptor present in the solution. On the acceptor, the electron is localized on a definite orbital. When the particle is reduced, a change in its charge is inevitably accompanied by a change in the orientation of dipoles surrounding the particle, i.e., by the reorganization of the solvent. The energy of the latter process determines mainly the activation energy of the electrode reaction. The orientation of solvent dipoles continuously changes due to thermal fluctuations. At the instant when the configuration of dipoles ensures equality of energy levels of the initial (electron is at the Fermi level in metal) and final (electron is on the acceptor) states, tunnelling from the initial into the final state becomes possible at a fixed position of dipoles.

Electrochemical dissolution of electrons proceeds in exactly the same manner. A cavity in the solvent acts as an acceptor, whose nucleus appears at a favourable orientation of dipoles owing to the thermal motion of the solvent's molecules. The electron tunnels when the electron energy level in the cavity which is not the equilibrium cavity equals the Fermi level in metal. After a solvated electron has been formed, the surrounding solvent relaxes to the equilibrium state.

Transfer of electron directly from metal into the localized state is described by a standard equation of electrochemical kinetics:

$$i = ke^{-\frac{v^{\neq}+\alpha FE}{RT}} \tag{10}$$

in which the activation energy is a linear function of the electrode potential and α is the transfer coefficient (usually it equals about 0.5).

The third mechanism for electron transfer into solution could, in principle, be electron autoemission, or emission under the action of electric field. In electron autoemission from metal into vacuum the electric field outside metal extends to macroscopic distances and appreciably changes the shape of the potential barrier, in particular its width, but for the metal-solution boundary the potential barrier width depends only slightly on the potential, as a rule, and is determined by the structure of the electric double layer. Therefore, in electrochemical systems a direct analogue of autoemission is less probable. Besides, in autoemission the current-potential dependence is a power dependence; this does not fit the experimental data on electrochemical generation of solvated electrons (Table 7 and Fig. 11).

Electrochemical dissolution of electrons and electron thermoemission may be regarded as two parallel and independent processes. It is seen from Fig. 12 that electrochemical dissolution is preferred from the thermodynamic viewpoint, because electrons are transferred to a lower level. The mechanism of the process is, however, dictated not only by the thermodynamic factor, but mostly by the activation energy for one or the other pathway of reaction. Electrochemical dissolution demands reorganization of the solvent, while thermoemission does not. From Table 7 and Fig. 11 it follows that electrochemical dissolution is observed in liquid ammonia and hexamethylphosphotriamide solutions (transfer coefficient $\alpha = 0.5$–0.75).

On applying a more negative potential, the activation energy for the electrochemical dissolution of electrons, as in any other electrode reaction, decreases by $\alpha F \Delta E$. For thermoemission, with increase in the negative potential, the work function decreases faster than the activation energy of dissolution (cf. Eqs. (9) and (10)). Therefore, at sufficiently high negative potentials the thermoemission rate exceeds the electrochemical dissolution rate and the thermoemission process dominates. This has been confirmed experimentally: curves 4 and 6 of Fig. 11 consist of subsequent portions with slopes of 120–140 and 60 mV.

This reasoning is valid only when the activation energy plays a decisive role; more precisely, when the tunnelling probability is high and the pre-exponential factors for the processes discussed are close to each other. This can be achieved on a clean electrode surface.

The presence of a passive film is equivalent to an increase in the barrier width. For a lower level — the solvated electron level — the barrier height is more than for a delocalized electron (Fig. 12). Therefore, an increase in barrier width will more strongly lower the tunnelling probability in direct formation of localized (solvated) electrons than in thermoemission. It may be expected that at a sufficient thickness of the barrier the tunnelling probability rather than the activation energy will play a decisive role. This will make thermoemission a predominant process at relatively low polarization (curves 1 and 2 of Fig. 11). Indeed, at passive electrodes in the whole range of potentials the kinetics of the generation process is described by straight lines having a slope of 60 mV.

7.4 Effect of Electrode Passivation on the Generation of Electrons

At present the formation of passive films on electrodes in the process of generating solvated electrons may be considered to be proved. More precisely, a different degree of electrode passivation explains why different authors obtained qualitatively and quantitatively different results (e.g., diffusion or activation regimes; different values of exchange current; different values of α — see Table 7).

Passivation of cathodes in hexamethylphosphotriamide has been studied in greater detail. Measurements of double-layer capacity have revealed [199] that cathodic polarization of electrodes in alkali metal salt solutions or the enrichment of the bulk of these solutions in solvated electrons leads to passivation which decreases the capacity and the rate of solvated electron cathodic generation. Anodic polarization, particularly in solutions containing solvated electrons, activates the surface. Polaromicrotribometrically it has been ascertained [200] that during solvated electron generation in a hexamethylphosphotriamide solution of lithium perchlorate a passivating film is formed on cathodes of different metals. Possibly there are several causes of passivation in this solution. By spectral methods it has been shown [177, 201] that films of different composition may be formed depending on the solution purity and the electrode potential.

First, cathodic reduction of traces of water causes the lithium hydroxide layer to precipitate:

$$Li^+ + H_2O + e^-(M) \rightarrow 1/2\, H_2 + LiOH\downarrow$$

According to Ref. [103], the solubility product of hydroxides in hexamethylphosphotriamide increases in the following order: lithium < sodium < tetraethylammonium < tetrabutylammonium. In the same order the tendency to passivation decreases.

The second cause for the formation of passivating layers may be the polymerization of the solvent itself under the action of solvated electrons in the near-electrode layer of solution. Polymerization is particularly typical of compounds with P—Cl bonds [202] which may be present as impurities in hexamethylphosphotriamide.

The third cause of passivation is the destruction of solvent immediately on the electrode surface. Such a mechanism is characteristic for a catalytically active metal — platinum. At cathodic polarization of a platinum electrode in well-purified hexamethylphosphotriamide solution of lithium perchlorate an organic film is formed; this film contains nitrogen, phosphorus, carbon, and oxygen [201], i.e., the elements forming the solvent molecule. In an ill-purified hexamethylphosphotriamide solution of sodium perchlorate the film formed contains sodium, chlorine, oxygen, and carbon, the film thickness may exceed 20–30 Å. The formation of the film from the products of hexamethylphosphotriamide polymerization and destruction is suspected to be the reason for the scattering of the data obtained by different authors and presented in Table 7.

The formation of a passive film and its thickening in systems where cathodic generation of solvated electrons is possible produces various effects:

1) increase in the barrier thickness, which governs the tunnelling, results in a decrease in the generation rate and causes the mechanism to change; this effect has been discussed above (transition from curves 5 and 6 to curve 1 in Fig. 11);

2) shielding of the surface cuts off some of its portions from the cathodic generation process;

3) complete suppression of the generation of solvated electrons changes the nature of the cathodic process; in this case, electrodeposition of alkali metal takes place instead of generation.

In a number of aprotic solvents an almost reversible lithium electrode can be realized (see, for example, [203, 204]). The potential of such an electrode is more negative than that at which solvated electrons are generated. Nonetheless, generation of solvated electrons does not take place at such electrodes. The fact is that in these systems the lithium electrode is covered with a passive film consisting of lithium hydroxide and basic salts, its conductivity is caused by Li^+ cations. This film is no bar to cathodic deposition and to anodic dissolution of lithium, but slows down chemical dissolution of lithium and the generation of solvated electrons. It is interesting that even in solvents like hexamethylphosphotriamide and liquid ammonia in which solvated electrons can be generated both cathodically and by dissolving any alkali metals, the cathodic generation can be suppressed by modifying the electrode surface in spite of fairly negative potentials. In fact, as is shown in Ref. [20], a lithium electrode in a saturated hexamethylphosphotriamide solution of lithium chloride is chemically stable and is, probably, reversible with respect to lithium ions. Generation of electrons at such an electrode does not take place though its potential is more negative than the generation potential. Obviously, the electrochemical generation of solvated electrons is suppressed by the formation of a film; the high concentration of lithium ions in the solution favours the formation of this film.

The potentials at which electrons are generated in a liquid ammonia solution of sodium salts are also more positive than the reversible potential of the sodium electrode [160], but in this system too the authors of Refs. [205] and [206] succeeded in realizing a reversible sodium electrode consisting of metallic sodium covered with a layer of beta-aluminate. The conductivity of the latter is due to sodium ions.

Thus, considering the ways by which a passive film affects the cathodic generation of electrons (thin films decrease the tunnelling probability; thick films suppress the generation completely) reveals that a change in the state of electrode surface can affect not only the rate and mechanism of the generation process but can also bring about complete suppression of generation and can thus change the very nature of the cathodic process.

It is evident from what has been said in Section 7 that the study of the cathodic generation of solvated electrons is in the making. The fundamental problem of the primary nature of generation has been solved unambiguously, but the same cannot be said about the quantitative study of particular mechanism of generation. Specific difficulties are faced in taking account of the possible effect of passivation of electrodes. Thus, in determining the transfer coefficient from the dependence of the equilibrium potential of the electron electrode on the exchange current it remains to be seen whether the state of the surface changes on passing from one potential to another. By impedance measurements in hexamethylphosphotriamide [164] on moderately passivated electrodes one succeeds in measuring the differential resistance characterizing the interfacial charge transfer; however, determining the bulk concentration of solvated electrons from the Warburg impedance components yielded too low values because of the surface nonuniformity. In liquid ammonia a too low value of the solvated electron diffusion coefficient was also obtained [169]. To explain this, the authors introduced the notion of effective concentration of traps as are the cavities in a solvent, which diffuse to the electrode. However, it would be more correct to assign these too low values of "diffusion resistance" to passivation of the electrode surface, which cuts off part of the electrode from the process.

On a more active and homogeneous surface, the authors of a number of papers observed only purely diffusional kinetics (Table 7).

In one of the recent works, carried out by Aoyagui et al. [170], particular emphasis was placed on the difficulties caused by passivation; a successful attempt has been made to overcome these difficulties. Prior to every series of experiments, the spherical platinum electrode was fused in the reducing flame of a coal gas-oxygen torch; and to avoid passivation, it was immersed in a liquid ammonia solution of potassium iodide after the removal of impurities with the aid of solvated electrons generated by dissolving metallic potassium. Using this procedure of preparing the electrode surface, the authors of Ref. 170 obtained self-consistent results by the galvanostatic double pulse method. From the slopes of the curves, $\eta - t^{1/2}$(η — overvoltage) they managed to obtain on a microsecond scale undistorted values of the product $cD^{1/2}$, coinciding with those obtained by the potentiodynamic method (time scale in seconds). This is indicative of high degree of surface homogeneity. Linear $\lg i_0 - \lg c$ plots have been obtained (Fig. 13). As is seen, at a solvated electron concentration of 10^{-3} mol/l the exchange current equals 200 mA/cm^2. A similar value of the exchange current has been obtained for methylamine [207]. At these very concentrations, in hexamethylphosphotriamide solutions on a highly activated copper electrode the i–E curve (corrected for concentration polarization) gives $i_0 \simeq 1$ mA/cm^2 (see curve 3 of Fig. 11). In the same potential range similar values of currents are observed also in tetraalkylammonium salt solutions where the electrode passivation is expected to be relatively small.

Now we compare the experimental data on the kinetics of the generation of solvated electrons with some theoretical estimations of the generation rate. The current density is computed by the equation taken from Ref. [208]:

$$i = \frac{ekT}{h} \varkappa N e^{-\frac{(\lambda_e + \Delta\bar{G}_c)^2}{4\lambda_e kT}} \tag{11}$$

Here $\varkappa$ is the transmission coefficient; N is the number of reaction centers on a unit surface; λ_e is the reorganization energy for the electrode reaction; $\Delta\bar{G}_c$ is the electrochemical free energy of the elementary act.

In Section 2 we estimated the reorganization energy for an isolated solvated electron (in fact, we talked therein about the reorganization energy for the process of removing an electron from its solvation shell to infinity). In the case of the

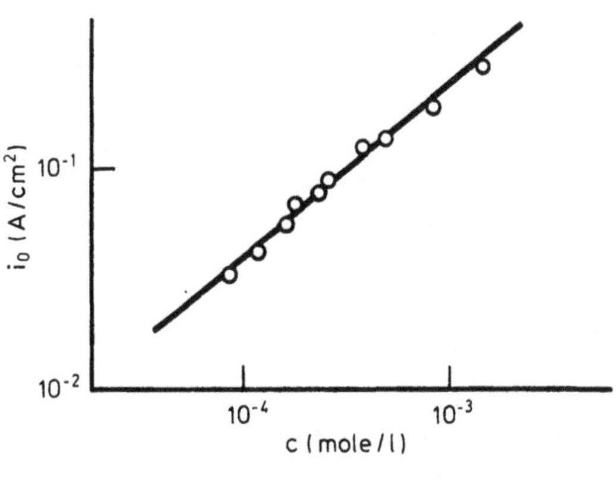

Fig. 13. Dependence of the exchange current at a platinum electrode on the concentration of solvated electrons (in bilogarithmic coordinates) in a liquid ammonia solution of 0.5 M KI at —40 °C (170)

electrode reaction it is diminished by the polarization energy of solvent due to the charge-in-electrode image field (see footnote at page 156). The distance up to the electrode R′ is usually equal to the ion radius, but in the case under consideration, as a solvated electron cannot be conceived without the surrounding solvent molecules, we take $R' = R + d$, where d is the diameter of the molecule. For hexamethylphosphotriamide $d \simeq 7$ Å; using these values, the contribution to the reorganization energy was estimated to be -0.15 eV. In all, we get $\lambda_e = 1.16$ eV.

The free energy of an elementary act is the configurational free energy, i.e., it takes account of all the entropy components, excepting the transpositional contribution which directly depends on concentration [208]. At the equilibrium potential, $\Delta G = 0$, the configurational free energy $\Delta \bar{G}_c$ differs from it by a value corresponding to the change in the free energy for formal transition from solution of known concentration to the hypothetical state of the pure solute (in the latter case there is no transpositional contribution to the free energy). For hexamethylphosphotriamide, the solvent concentration amounts to 5.8 mol/l; this value should be taken as limiting value for the dissolved substance concentration. Varying the solvated electron concentration from 10^{-3} to 5.8 mol/l increases ΔG by 0.21 eV. This will be the value of $\Delta \bar{G}_c$ for the electrode that is in equilibrium with a 10^{-3} M solution of solvated electrons.

We estimate the number of reaction centers on the electrode surface, assuming that the projection area of one solvated electron equals about 30 Å^2, i.e. $N \simeq 3 \times 10^{14}$ cm^{-2} $\simeq 5 \times 10^{-10}$ mol/cm^2. Substituting the values of the universal constants and taking $\varkappa = 1$, we find the exchange current density for a 10^{-3} M solution of solvated electrons to be equal to about 50 A/cm^2. This is by 4.5 orders higher than the experimentally determined value. An analogous estimation for solvated electrons in liquid ammonia yields an exchange current equal to 10 A/cm^2 (—40 °C); this also exceeds the experimental values (200 mA/cm^2) [170], though not so much as in hexamethylphosphotriamide.

It should be noted that if in calculating the exchange current we had used the experimental value of λ_s, even its upper limit (see Sect. 2), then the discrepancy between the computed and the experimentally determined values of i_0 would have increased still further by 2 orders. Thus, we deal not only with insufficient accuracy of theoretical calculation of λ_s — the impression arises that experimental data on energy characteristics of solvated electrons poorly correlate with the experimental data on the kinetics of cathodic generation of solvated electrons.

It may seem natural to ascribe the discrepancy between the theoretical and experimental values, first of all, to too low values of the experimental currents due to the effect of passivation. Thus, it may be thought that for liquid ammonia too low experimental currents are caused by the presence of a thin and very homogeneous film on the electrode surface. However, it is yet not possible to say with confidence that the passivation effect, which undoubtedly occurs, is completely responsible for the noticed discrepancy. It is quite probable that the theoretical model does not sufficiently well describe the process of electrochemical dissolution of electrons. Clearly, this problem calls for further experimental as well as theoretical studies.

For the thermoemission mechanism, the effect of a passivation film, as mentioned above, should not be so strong. Logically it may be assumed that for less passive surface which are still accessible for kinetic study, the hindering effect of thin films is very small. Finding the electronic work function in hexamethylphosphotriamide from photoemission measurements taken in the range of potentials, which is far from the passivation range of the electrode, we can recalculate it for any electrode potential, say for E = —2.88 V (this potential corresponds to the equilibrium concentration of solvated electrons equal to about 10^{-3} mol/l, i.e. the same potential at which the parameters of the electrochemical dissolution of electrons were compared earlier). This electronic work function equals 0.62 eV. To it, according to the Richardson-Sommerfeld formula (Eq. (9)), corresponds a current of $10^{-4.7}$ A/cm^2. The experimentally determined value (from curve 6, Fig. 11, corrected for the contribution of the parallel process) amounts to $10^{-3.3}$ A/cm^2. It was found to be even somewhat more than the computed one, but the difference between these will hardly significantly exceed the possible limits of experimental errors, particularly for the far extrapolation used. It is possible that a contribution in raising the experimental value compared to the computed one is made by some diffuseness of the bottom of a conduction band in liquid, which is not taken into account in extrapolating the photoemission currents to a formal threshold but has a definite role in thermoemission.

8 Possible Role of Solvated Electrons in "Ordinary" Electrochemical Reactions

After the discovery of hydrated electrons extensive experimental data in radiation chemistry have been accumulated, which show that the solvated electron acts as a universal primary reducing agent in the processes that take place in the liquid bulk under the action of ionizing radiation. This provoked interest among researchers in the role of solvated electrons in other physico-chemical processes also, in particular in electrochemical processes.

Thus, in Refs. [11,12], and [209–214] it has been suggested that electrochemical reactions of cathodic reduction and the processes of spontaneous dissolution of metals in aqueous media proceed via an intermediate stage of formation of hydrated electrons. This means that the electrons leave the electrode and go into the solution and then the hydrated electrons react in the solution bulk, reducing, for instance, the proton donors. The theoretical and experimental argumentation of these authors, as regards the possibility of the mechanism itself, was disputed by other researchers [13,198,215–219].

The strongest argument adduced in favour of a hydrated electron as an intermediate in cathodic hydrogen evolution is the statement of Walker [212] that in electrolysing an aqueous solution using a polished silver cathode he succeeded in observing optical absorption in the near-electrode layer, caused by a hydrated electron. However, later it was shown [215,219] that no reliable proofs of the formation of hydrated electrons during electrolysis of aqueous solutions under the conditions of the experiments [212] were available and the observed optical effect was due to the variation in the reflectivity of the electrode surface on cathodic polarization.

As a second argument in favour of the formation of a hydrated electron as an intermediate the data of Hills and Kinnibrugh [210] on the molar activation volume for the reaction of cathodic hydrogen evolution on mercury from a hydrochloric acid aqueous solution were considered. The value determined from the current density — pressure dependence was negative. This, according to Ref. [210], suggested the participation of hydrated electrons in the elementary act which determines the process rate (introduction of an additional charge into the solution should cause the system to shrink due to electrostriction; and the removal of hydroxonium ions would have caused the system to expand). However, Parsons [216] and Krishtalik [217] have proved that the activation volume obtained in Ref. 210 does not characterize the true change in the volume in the course of an elementary act, because this has been determined at a constant overvoltage but not at a constant potential. The true activation volume, estimated in Ref. [217], proved to be positive, and thus it was found unnecessary to use the hypothesis of intermediate formation of hydrated electrons in the reaction of cathodic hydrogen evoluton in the range of usual, not very negative, potentials.

Brodsky and Frumkin [198] estimated the possible thermoemission rates (from electrodes into aqueous solutions), using the Richardson-Sommerfeld equation and the electronic work function for the metal-water system, determined from photoemission measurements. They have shown that in the range of potentials typical of cathodic reactions in aqueous solutions the emission rates should be very small and need not ensure cathodic evolution of hydrogen via the thermoemission stage, i.e., via the intermediate formation of hydrated electrons.

Of importance is the thermodynamic aspect of the problem concerning the participation of solvated electrons in electrochemical processes. The standard potentials for the electron electrode equal (see Sect. 5) —2.85 V for water (vs. NHE

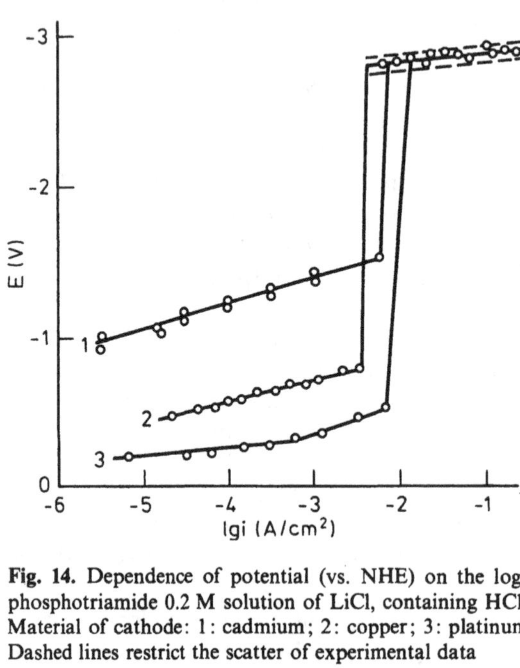

Fig. 14. Dependence of potential (vs. NHE) on the logarithm of current density in hexamethylphosphotriamide 0.2 M solution of LiCl, containing HCl additions: 1: 0.17; 2: 0.20; 3: 0.20 mol/l Material of cathode: 1: cadmium; 2: copper; 3: platinum; temperature: 25 °C (17, 191, 220, 221). Dashed lines restrict the scatter of experimental data

in water), —1.85—–1.95 V for liquid ammonia (vs. NHE in ammonia), —3.02–—3.05 V for hexamethylphosphotriamide (vs. aqueous SCE), and —2.90 V for liquid methylamine (vs. Ag/0.02 M Ag^+ in methylamine). Obviously, at potentials typical of ordinary cathodic reactions — far more positive compared to these values — it is unlikely that the processes occurring to a significant extent via the stage of formation of solvated electrons will take place in the above solvents. However, the aforementioned contrary viewpoint made it necessary to experimentally compare the two mechanisms: direct reduction of substance at the cathode and its reaction with the solvated electrons generated there. Figure 14 shows the cathodic polarization curves obtained for hexamethylphosphotriamide solutions of lithium chloride containing additions of hydrogen chloride. The shape of these curves suggests the occurrence of two different processes, one of which is replaced by the other as the cathodic polarization is increased. Namely, the Tafel plot sections at small current densities conform to the process of cathodic hydrogen evolution; at large current densities these sections conform to electrochemical generation of solvated electrons. At the generation potentials, a thin layer of the solution near the electrode surface acquires a blue coloration. The colour intensity of this layer in the presence of proton donors is much less than in their absence. The light blue colour qualitatively proves, on the one hand, the very fact of generation and points, on the other hand, towards the reaction of solvated electrons with proton donors present in the solution bulk.

The portions of the curves, which describe the hydrogen evolution have Tafel slopes equal to 60 and 140 mV for platinum, 120–140 mV for copper, and 150 mV for cadmium. The reaction rate depends on the nature of the cathode and the acid concentration. Overvoltage at which hydrogen is evolved on the copper electrode equals 0.5 V at pH 4 and a current density of 10^{-3} A/cm^2. This value is slightly less, but is rather close to that of the overvoltage in water (0.7 V [222]), particularly when allowance is made for the error that occurs due to recalculation of the potential scale. This error can be eliminated by comparing not the overvoltages but their differences (for different metals). The difference for copper and platinum electrodes (in the range of curves with a practically similar slope) equals 0.4 V; for copper and cadmium 0.6 V; these are close to the corresponding values for aqueous solutions: 0.5 ± 0.05 V [222].

As seen from Fig. 14, in hexamethylphosphotriamide the solvated electron generation potential is by 1.2–2.0 V more negative than the hydrogen evolution potential. Furthermore, the generation of electrons and the evolution of hydrogen are governed by essentially different laws: the relationship between the process rate and the nature of electrode and solution composition is different (unlike the hydrogen evolution rate, the generation rate is independent of the electrode material and the nature of the background electrolyte) and the slopes of the polarization curves are different.

Hence an unambiguous conclusion can be drawn that the cathodic hydrogen evolution on metals, including those with a high overvoltage, is a direct discharge of a proton donor at the electrode and does not involve the intermediate formation of solvated electrons and their subsequent chemical reaction with the proton donor in the bulk of the solution.

Although this conclusion has been drawn by considering an example of hexamethylphosphotriamide acid solutions, it holds for a wide range of systems. On going to

Table 8. Standard potentials of alkali metals, E_0°, and their amalgams, $E_0^\circ(am)$, in aqueous solutions [223, 224]

System	Li/Li^+	Na/Na^+	K/K^+	Rb/Rb^+	Cs/Cs^+
E_0°, V(NHE)	−3.045	−2.714	−2.924	−2.925	−2.923
$E_0^\circ(am)$, V(NHE)	−2.178	−1.956	−1.974	−1.968	−1.947

solutions with a much higher pH or to other solvents a variation is, of course, possible in the energy characteristics of electrons and proton donors. But these variations do not exceed 1 eV, as a rule, and hence do not overlap the above-indicated interval of potentials.

Table 8 contains standard potentials of alkali metals and their amalgams in water.

It is seen from the Table that a process involving the participation of solvated electrons is possible, in principle, for self-dissolution of alkali metals in water. (This process proceeds, of course, parallel to the processes occuring via other mechanisms). In fact, back in 1934, Wolthorn and Fernelius [225] had reported that a blue coloration temporarily appeared on the surface of metallic potassium when it reacted with water. This coloration would have been caused by hydrated electrons stabilized in the potassium hydroxide film covering the metal [6]. According to Ref. 226, a dark-blue product was formed on reacting alkali metal with ice at −196 °C, its ESR spectrum had a narrow single line typical of a trapped electron.

Also it follows from Table 8 that the assumption about the participation of solvated electrons in the reactions of alkali metal amalgams [209, 212] is not very likely.

It is interesting to compare the data on hydrogen evolution from aqueous and liquid ammonia solutions. As mentioned above, the distinction between liquid ammonia and water is that the difference in standard potentials for the electron and hydrogen electrodes in the former is less by one volt than in water (−1.90 and −2.85 V, respectively). With respect to the rubidium electrode potential, which weakly depends on the nature of the solvent [227], the standard potentials of the electron electrode do not practically differ (0.15 and 0.08 V, Sect. 5) as the accuracy of determining these values does not exceed ±0.05 V, in particular. Hence, the decrease in the difference of standard potentials for liquid ammonia is caused by the shift in the hydrogen electrode potential towards negative values; this is due to high solvation energy of the proton. In other words, this is due to the much higher energy of formation of the ammonium ion in liquid ammonia as compared to the energy of formation of the hydroxonium ion in water.

Although the evolution of hydrogen from liquid ammonia acid solutions even on a metal with high overvoltage such as lead obviously demands a higher overvoltage than in water (approximate values of overvoltage 1.1–1.2 V for a current density of 10^{-5}–10^{-4} A/cm^2 in a 0.1 M ammonium chloride solution [227, 228]), it nonetheless occurs at potentials much less than the reversible potential of the electron electrode in this solvent.

A different situation may be observed in liquid ammonia basic solutions (i.e., in a solution containing NH_2^-). As mentioned in Section 3, at a hydrogen pressure of 100 kg/cm^2 and amide ion concentration of 1 mol/l the hydrogen electrode will be in equilibrium with the solvated electrons, the concentration of the latter being 10^{-5} mol/l [81]. This indicates that the possiblity of solvated electrons participating in this process must be reckoned with when hydrogen is evolved from liquid ammonia basic solutions.

Thus, from what has been discussed in this section it is clear that in the range of ordinary, i.e., not very negative potentials, the cathodic reaction of hydrogen evolution proceeds as a direct discharge of proton donors without the participation of solvated electrons as intermediate particles. Such a mechanism is valid not only for hydrogen evolution, but also for the majority of cathodic processes in various solvents. There can be exceptions only in some particular cases when, thanks to the experimental conditions, it is possible to attain very negative potentials sufficient for the generation of solvated electrons. However, under these conditions also, electrochemical generation should be considered as a process which proceeds parallel to the direct cathodic reduction reactions.

9 Application of Electrochemically Generated Solvated Electrons for the Reduction of Organic Compounds

As is seen from the values of the standard potentials of the electron electrode in different media (Sect. 5), solvated electrons are strong reducing agent. They have practically the same reducing power as the alkali metals.

Solvated electrons obtained by dissolving alkali metals in liquid ammonia and similar solvents are now extensively used for the reduction of organic compounds (Birch reaction) [229]. To this end, the application of electrochemically obtained solvated electrons helps to avoid inconveniences associated with the use of alkali metals and to obtain readily a reducing agent in strictly predetermined amounts.

By considering an example of a proton donor, Fig. 14 illustrates the general situation which prevails when a reducing agent is added to the system where cathodic generation of solvated electrons can take place. In the range of small current densities and not very negative potentials direct reduction of the added reagent occurs at the electrode. After a certain current density has been attained the process rate does not increase any more and this abruptly shifts the potential towards the generation potential of the solvated electrons. The limiting current for direct electrode reduction of the reagent may be a diffusion (i.e., limited by the reagent supply) or a passivation current. Often passivation is caused by a change in the state of the surface due to the formation of hydroxides or other basic compounds of the background salt cations and also due to polymerization of organic compounds. At the solvated electron generation potentials the reagent is reduced by two parallel pathways: directly on the electrode at a rate conforming to the limiting current; and in the solution bulk (here a chemical reaction with solvated electrons takes place). As the generation current is increased the fraction of direct reduction decreases because its rate is independent of the potential. In experiments, direct electrode reduction of organic compounds seems to be hindered quite often for kinetic reasons, and that is why the major portion of the substance is reduced via the reaction with the solvated electrons.

Table 9. Diffusion coefficients for solvated electrons D in liquid ammonia, water, and hexamethylphosphotriamide

Solvent	$D \times 10^4$, cm^2/s	$D\eta^a \times 10^5$, cm^2cP/s	$\frac{D(s.e.)}{D(Na^+)}$	Ref.
Liquid ammonia (−36 °C)	1.74	4.45	6.7	[157, 230]
Water (at room temperature)	0.49	4.90	3.7	[6]
HMPA (at room temperature)	0.14	4.59	10	[132, 146]

^{a} η-viscosity

Reduction of organic compounds with the use of electrochemically generated solvated electrons has several distinctive features.

i) *Possibility of using larger current densities than in direct cathodic reduction.* The advantage of this method lies particularly in high diffusion coefficient of solvated electrons. Magnitudes of diffusion coefficients are listed in Table 9. As is seen from the table, solvated electrons are by several times more mobile than ordinary ions. This is, probably, due to the hopping mechanism of the motion of solvated electrons. The product of the diffusion coefficient and the solvent viscosity (the Walden product) is almost constant for three solvents. The Walden rule which, in the case of solvated electrons, is valid for a wide range of polar solvents [231], can be employed for a rough estimation of the diffusion coefficient of solvated electrons in the solvents where this coefficient has not been measured yet.

Another factor ensuring high current density is the rapid removal of solvated electrons from the electrode due to intense convection in the solution, which is caused by a decrease in its density at the cathode surface. This phenomenon is associated with an increase in solution volume caused by the introduction of electrons into it. Unlike electrostriction that accompanies the solvation of ordinary ions, the formation of solvated electrons increases the volume by 65–96 ml/mol for liquid ammonia[81] and by about 80 ml/mol for hexamethylphosphotriamide[42,69]. As a result, according to Avaca and Bewick [232], the current densities for the generation of solvated electrons can by 2500 or more times exceed the rate of mass transfer of organic compounds to the electrode.

ii) *Reduction of hard-to-reduce compounds and selectivity.* A typical characteristic of solvated electrons is their ability to react with such compounds which are not reduced at all at the cathode or are reduced with great difficulty.

For instance, benzene is one of the most hard-to-reduce compounds. It is not active polarographically [233] and is not subjected to direct electrode reduction in liquid ammonia [234]. Benzene does not react with solvated electrons obtained by dissolving alkali metals in liquid ammonia in the absence of proton donors [5, 229, 235]. However, in the presence of proton donors benzene is hydrogenated by solvated electrons (Table 10).

The data listed in Table 10 illustrate the reduction of benzene in media typical of the electrochemical generation of solvated electrons. High current efficiencies in

Table 10. Reduction of benzene with electrochemically generated solvated electrons (compiled in Ref. [17]; also [236–238])

System	Proton donor	Product	%	Current efficiency, %
Liquid ammonia NaCl, Pt, −33 °C	Ethyl alcohol	Cyclohexadi-1,4-ene	98	25–40
HMPA LiCl	Methyl alcohol	Cyclohexadi-1,4-ene	53	
		Cyclohexadi-1,3-ene	2	
		Hexene	21	
		Hexane	24	
HMPA LiCl, Al, 28 °C	Ethyl alcohol	Cyclohexadiene	20	95
		Cyclohexene	9	
		Cyclohexane	71	
HMPA $LiClO_4$, Pt	Water	Cyclohexadi-1,3-ene		
Methylamine LiCl, Pt, −70 °C	The solvent	Cyclohexene	100[a]	49
		Cyclohexadi-1,4-ene	95[b]	49
		Cyclohexene	4	
Ethylene diamine LiCl, Pt 25–27 °C	The solvent	Cyclohexadiene	17[a]	
		Cyclohexene	70	
		Cyclohexane	13	
Ethylene diamine 33 °C LiCl, carbon	The solvent	Cyclohexadi-1,4-ene	38[b] (calcul. from current)	52
		Cyclohexane	14 (calcul. from current)	

[a] With diaphragm
[b] Without diaphragm

some cases reveal that the fraction of side processes (in particular, reduction of the solvent) may be small. This is in agreement with the results of Ref. [239] where it is shown that solvated electrons obtained by laser photolysis in an ammonia-water mixture containing benzene do not react with the components of the solvent, but rather with benzene and reduce it. However, contradictions are found in the data of different authors on hydrogenation of benzene under the action of solvated electrons. As suggested by Birch and Smith [240] and corroborated by other authors [241], hydrogenation of benzene proceeds according to the reaction:

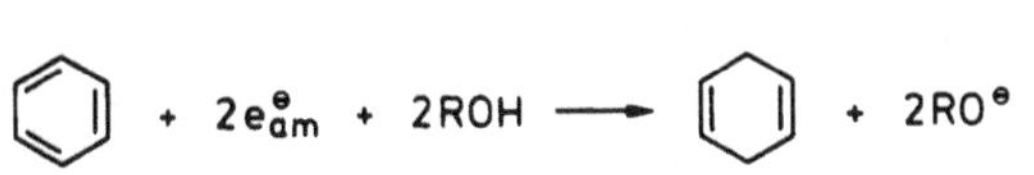

where R = H or C_2H_5 and e^-_{am} stands for solvated electron in ammonia. However, later it was shown [235, 242] that the process of hydrogen evolution could compete with this process. In Refs. [235] and [242] it has been proposed that in the case both of water and of alcohol hydrogen evolution takes place via the irreversible reaction of an electron with a proton donor:

$$2\,ROH + 2\,e^-_{am} \rightarrow 2\,RO^- + H_2$$

The ratio of the hydrogen and the 1,4-cyclohexadiene (dihydrobenzene) produced depends on the conditions under which the reaction is carried out. Accumulation of alkoxy ion increases the fraction of hydrogen; and the experiments conducted especially without maintaining a high vacuum, which is usually employed at the preparatory stages of the experiment with ammonia solutions, aids in increasing the yield of dihydrobenzene. It should, of course, be taken into consideration that when use is made of water, hydrogen may also appear as a result of the reaction of the solvated electrons with NH_4^+ ions [235].

The media represented in Table 10 contain, as a rule, a protic component but in small amounts. However, the content of proton donors can be increased appreciably; hydrogenation of benzene can be effected in mixtures of water and aprotic solvents (such as diglyme, ethylenediamine, sulfolan, Table 11). Furthermore, benzene can be reduced not only in a mixture of aprotic and protic solvents but often in protic media as well. Mono- and dibasic alcohols and also water (Table 11) proved to be suitable for the purpose. Although the presence of tetrabutylammonium cations enhances to some extent the dissolution of benzene in an aqueous phase,

Table 11. Reduction of benzene in protic media (compiled in Ref. 17; also [243, 244])

System	Product	%	Current efficiency, %	Conditions
Diglyme-water $(C_4H_9)_4NBr$; Hg; 24–30 °C	Cyclohexadi-1,4-ene Cyclohexene	96 4	71	Diaphragm, benzene in excess
Ethylene diamine-water NH_4Cl, $(C_6H_{13})_4NBr$; graphite	Cyclohexadi-1,4-ene Cyclohexane	50 2	52	
Sulfolan-water $(C_4H_9)_4NBr$; Hg; 40 °C	Cyclohexadi-1,4-ene	99	71	Diaphragm, conversion of benzene 33%
Diethyleneglycol $(C_4H_9)_4NBr$; Hg; 40 °C	Cyclohexadi-1,4-ene	91	68	Diaphragm, conversion of benzene 37%
Isobutyl alcohol $(C_4H_9)_4NBr$; Hg; 40 °C	Cyclohexadi-1,4-ene	91	49	Diaphragm, conversion of benzene 32%
Water $[(C_4H_9)_4N]_2SO_4$; $(C_4H_9)_4NOH$; Hg; 65 °C	Cyclohexadi-1,4-ene Cyclohexene	87 13	77	Cation-exchange membrane; benzene in excess
Water $(C_4H_9)_4NOH$ (pH > 11) Hg	Cyclohexadi-1,4-ene		>50	Without diaphragm

Table 12. Reduction of toluene with electrochemically generated solvated electrons [237, 245–247]

System	Proton donor	Product	%
HMPA LiClO$_4$ or NaClO$_4$, Pt	H_2O	1-Methylcyclohexadi-1,4-ene	
HMPA LiCl; Pt	C_2H_5OH	Methylcyclohexane	45
Methylamine LiCl, Pt, −70 °C	The solvent	Methylcyclohexadiene Methylcyclohexene	94[a] 5
		Methylcyclohexane	86[b]

[a] Without diaphragm
[b] With diaphragm

Table 13. Reduction of tetralin by electrolysis in ethylenediamine [236]

Background electrolyte	Temp., °C	Electrode	Products (current efficiency, %) Hexalin	Octalin
Lithium chloride	33	Platinum	41.6	7.5
		Graphite	55.4	11.6
		Carbon	67.3	9.5
Tetrabutylammonium iodide	28	Carbon	13.0	1.8
Ammonium chloride	20	Carbon	0	0

an emulsion is usually formed in water. The benzene phase contains a certain amount of proton donors; hence reduction can be thought to occur just in this phase.

Toluene behaves in much the same way as benzene. Polarographically it is also inert, and does not react with solvated electrons obtained by dissolving alkali metals in liquid ammonia [229], but it can be reduced by solvated electrons in the presence of proton donors (Table 12).

Toluene can be reduced also by using for the generation of solvated electrons a mercury cathode in a diglyme-water solutions of tetrabutylammonium bromide [248] as well as in diethyleneglycol [249], isobutanol [250] and water [244].

Like benzene and toluene, tetralin (tetrahydronaphthalene) is not reduced directly on the electrode, for example on a mercury electrode in ethylenediamine; it is however well hydrogenated when using solvated electrons in this solvent (Table 13) and also in hexamethylphosphotriamide [251]. In Ref. [251] it has been shown that in solutions of polarographically inert benzene and tetralin as well as of directly reducible naphthalene the potentials of the cathode processes coincide. This is also a distinctive indication of reduction of organic substances with the aid of electrochemically generated solvated electrons. The process rate is practically independent of the nature of organic compound.

As is evident from the examples given and also from the review on the preparative reduction of organic compounds using of solvated electrons [252], the processes in

Table 14. Rate constants for reactions of solvated electrons with water and alcohols

Medium	Electron acceptor	k, l/(mol · s)	Ref.
Water	H_2O	16	[5]
Ethyl alcohol	C_2H_5OH	4.7 10^3	[254]
Ethylenediamine	H_2O	24.7	[255]
Heavy water	D_2O	1.25	[5]
Water	CH_3OH	10^3	[6]
Water	C_2H_5OH	20	[6]

most cases represent hydrogenation. The possibility of carrying out such reactions is closely related to the selectivity typical of the action of solvated electrons. Thus, while hydrogenating the hard-to-reduce compounds, solvated electrons reduce only very slowly a number of substances containing active hydrogen. It is precisely the slowness of reduction of water and aliphatic alcohols by solvated electrons that enables these substances to be used as proton donors for the reduction of organic compounds. According to Ref. [237], water up to a concentration of 0.1 mol/l does not react with solvated electrons in hexamethylphosphotriamide solutions of sodium and lithium perchlorates. Aliphatic alcohols in liquid ammonia can be reduced by solvated electrons obtained upon dissolving alkali metals; the reduction takes place with hydrogen evolution and the formation of alcoholates. However, the reaction proceeds slowly and incompletely. Thus, ethyl alcohol does not completely react with a liquid ammonia solution of sodium but forms the complex $C_2H_5ONa \cdot C_2H_5OH$ [253].

Table 14 contains the rate constants for the reactions of solvated electrons with water and alcohols in different media. It is evident that reduction of proton donors in fact proceeds very slowly. Also, water reacts slowly with methylamine solutions of cesium and sodium [256].

iii) *Stereospecifity.* When organic compounds are reduced with solvated electrons, another distinctive feature of this process, i.e., its stereospecifity shows up. In greater detail this phenomenon was studied for reduction of alkylcyclohexanones and alkylcyclopentanones. Aliphatic ketones demand potentials more negative than —2.0 V (vs. SCE in water) for their reduction. The reduction products of alkylcyclohexanones and alkylcyclopentanones are alcohols which exist as two stereoisomers (cis- and trans-). The results of the preparative synthesis of alcohols are summarized in Table 15.

From this Table it follows that the material of the cathode and the reduction method do not affect the isomeric composition of the alcohols whereas in direct electrode reduction a mixture of trans- and cis-isomers is obtained, the ratio between these isomers strongly depends on electrolysis conditions [257]. This indicates that in all cases, represented in Table 15, ketones are reduced by solvated electrons. It is evident from this Table that such a reduction ensures the predominance of trans-forms.

iv) *Chemiluminescence.* In a number of cases the reaction of electrochemically generated solvated electrons with organic substances is accompanied by chemiluminescence. In taking cyclic voltammetric curves, luminescence appears concurrently with the attainment of a solvated electron generation potential.

Table 15. Reduction of alkylcyclopentanones and alkylcyclohexanones through the use of solvated electrons generated electrochemically in hexamethylphosphotriamide

Ketone	ketone-to-proton donor ratio[a]	Electrode	Current efficiency, %	Yield of trans isomer, %	Method[b]	Ref.
2-Methylcyclo-pentanone	1:2	Copper	84	90	2	257)
	1:2	Copper	88	91	1	
	1:4.5	Platinum	56	86	2	
	1:4.5	Copper	79	94	2	
2-Ethylcyclo-pentanone	1:2	Copper	72	93	2	
	1:2	Copper	85	89	1	
2-Methylcyclo-pentanone	1:1	Copper	Polymer		2	
4-tret-Butylcyclo-hexanone	4 mol. % of donor	Platinum	87	95	1	258)
3,3,5-Trimethyl-cyclohexanone	10 mol. % of donor	Platinum	50	91	1	

[a] In all cases the proton donor is ethyl alcohol.

[b] 1st Method: the entire amount of the proton donor and of the substance to be reduced was introduced into catholyte and electrolysis was carried out until a stable coloration due to solvated electrons appeared; 2nd Method: the substance to be reduced and the proton donor were added to the catholyte with pregenerated solvated electrons in such an amount that the solvated electrons remained in some excess.

The reactions which are accompanied by light emission can be placed in two groups. Those occurring between solvated electrons and cation radicals with the formation of neutral radicals or molecules constitute Group 1.

Aminoderivatives with stable cation radicals, for example N, N, N′, N′-tetramethyl-*p*-phenylenediamine, in the cyclic scanning of potential can be oxidized to the cation radical during an anodic halfcycle in a hexamethylphosphotriamide solution of sodium perchlorate; the latter reacts in the bulk of the solution with the solvated electrons generated during the cathodic halfcycle, forming thereby an excited molecule of the substance. This molecule changes to its ground state by emitting a light quantum [180, 259, 260].

Similar to this, luminescence appears in the galvanostatic pulse electrolysis of hexamethylphosphotriamide solutions of different salts without the separation of catholyte and anolyte. The authors of Ref. [261] believe that the cation $[(CH_3)_3N]_3P^+OY$, where Y stands for iodine or tosyl (Ts, $p-CH_3C_6H_4SO_2^-$) or mesyl ($CH_3SO_2^-$), appears at the anode. The iodine derivative was formed in a reaction between the solvent and the iodine liberated at the anode in potassium iodide solution. The other two compounds were obtained by adding tosyl and mesyl chlorides, respectively to the solution [261].

With Group 2 of the processes that cause light emission are classed the reduction reactions of uncharged molecules in hexamethylphosphotriamide, accompanied by splitting of the molecule [260]:

$$(C_6H_5)_3CCl + e_s^- \rightarrow (C_6H_5)_3C^{\cdot} + Cl^-$$

$$(C_6H_5)_3C^{\cdot} + e_s^- \rightarrow (C_6H_5)_3C^{*}_{-}$$

The excited particle deactivates with light emission. With this type are grouped the reduction reactions of N-tosylcarbazol, 9-chlorfluorene, benzylchloride, and diphenylmethylchloride in hexamethylphosphotriamide [180,209].

Chemiluminescence, under the action of electrochemically generated solvated electrons, was also observed in a mixture of ammonia and tetrahydrofurane [160,262]. At a volume ratio of 2:1 the system behaves essentially as would pure ammonia and the tetrahydrofurane ensures good solubility of the organic compounds. In such a mixture, α- and β-naptholtosylates and also N-tosylcarbazol (R—Ts) react with the solvated electrons by emitting light: $R{-}Ts + e_s^- \rightarrow R{-}Ts^- \rightarrow R\cdot + Ts^{\cdot}_{-}$; $R\cdot + e_s^- \rightarrow R^{*}_{-} \rightarrow R^- + h\nu$.

It is interesting to compare direct cathodic reduction with reduction of the same substance using solvated electrons. Thus, phosphine cations, which are formed in hexamethylphosphotriamide solutions of iodine, tosyl and mesyl chlorides (see above), and also N-tosylcarbazol during direct cathodic reduction (whose potential is by 2.5 and 0.8 V, respectively, more positive than the generation potential of solvated electrons) yield, unlike the reaction with solvated electrons, unexcited particles; chemiluminescence is absent in this case [180,261]. Thus, investigation of chemiluminescence is a useful method of studying the mechanism of reduction of organic compounds with the aid of solvated electrons.

The material of this section, though it is of illustrative nature, indicates that syntheses performed with the use of solvated electrons generated at the cathode of the electrochemical cell have, in a number of cases, important advantages over other types of reduction processes and are therefore finding ever-increasing application. For a more detailed acquaitance with the reduction of different classes of organic compounds through the use of solvated electrons the reader is referred to the review by Lund [252].

10 Conclusion

In the past two decades, thanks to the work of scientists in the U.K., the USSR, Japan, the USA, and other countries, solvated electrons have become an important subject of electrochemical investigations. These have been obtained in a number of solvents; in many of them, they are quite stable. Their typical chemical, physical, and optical properties have been well determined. Also, the fundamental relationships for electrochemical and photoelectrochemical generation of solvated electrons have been found. As is evident from this review, solvated electrons behave as individual chemical reagents, in particular they enter into electrochemical reactions at electrodes, and have their equilibrium potential, etc.

Thus, the appearance of a new branch of theoretical electrochemistry, that is the electrochemistry of solvated electrons, may be ascertained. It is significant that the reactions at the electron electrode, from the viewpoint of a researcher, have an important advantage over other electrochemical reactions. More precisely, they represent the simplest type of electrode processes because only the solvent molecules, besides the electron, participate in them. Therefore, they are of particular interest for elucidating fundamental problems of electrochemistry as a whole. On the other hand, practical applications of solutions of solvated electrons have become possible in the last few years.

At the same time, the picture of the electrochemical behaviour of solvated or, in the more general form, excess electrons (including also the delocalized states) can in no way be considered to be complete. The following are most important problems that need be solved: development of quantitative methods for determining the energy characteristics of solvated electrons; experimental study of the kinetics of the electrode reactions of solvated electrons on electrodes completely free of passivating films, and their quantitative comparison with the theory; further development of the theory of the electrode reactions of solvated electrons, which relates the kinetics of processes with the structure and energy characteristics of solvated electrons; continuation of the study of the nature of passivation of electrodes in the range of high cathode potentials and its effect on the process of cathodic generation of solvated electrons.

Undoubtedly, further advancement in solving these problems is inseparable from the task of providing a detailed physico-chemical description for the structure of solvated electrons and, in a more broad sense, inseparable from the development of our notions concerning the interaction of a polar liquid with the charged particles introduced into it.

11 References

1. Edwards, P. P.: Adv. Inorg. Chem. Radiochem. *25*, 135 (1982)
2. Weyl, W.: Poggendorffs Ann. *121*, 601 (1864)
3. Kraus, C. A.: J. Amer. Chem. Soc. *30*, 1197 (1908)
4. Kraus, C. A.: J. Amer. Chem. Soc. *43*, 749 (1921)
5. Pikaev, A. K.: Solvatirovannyi elektron v radiatsionnoi khimii, Moscow, Nauka 1969
6. Hart, E. J., Anbar, M.: The Hydrated Electron, New York e. a., Wiley-Interscience 1970
7. Thompson, J. C.: Electrons in Liquid Ammonia, Oxford, Clarendon Press 1976
8. The sixth international conference on excess electrons and metal — ammonia solutions. Colloque Weyl VI. June 26–July I, 1983. J. Phys. Chem. *88*, 3699 (1984)
9. Brodsky, A. M., Gurevich, Yu. Ya., Pleskov, Yu. V., Rotenberg, Z. A.: Sovremennaya fotoelektrokhimiya. Fotoemissionnye yavleniya, Moscow, Nauka 1974
10. Gurevich, Yu. Ya., Pleskov, Yu. V., Rotenberg, Z. A.: Photoelectrochemistry, New York—London, Consultants Bureau 1980
11. Antropov, L. I.: in: Itogi nauki, ser. Elektrokhimiya, Vol. 6, p. 5, Moscow, VINITI 1971
12. Kenney, I. A., Walker, D. C.: In: Electroanalytical Chemistry (Ed.) Bard, A. J., Vol. 5, p. 1, New York, M. Dekker 1971
13. Conway, B. E.: In: Modern Aspects of Electrochemistry (eds.) Conway, B. E., Bockris, J. O' M., Vol. 7, p. 83, New York, Plenum Press 1972
14. Alpatova, N. M., Kessler, Yu. M., Krishtalik, L. I., Ovsyannikova, E. V., Fomicheva, M. G.: in: Itogi nauki i tekhniki, ser. Elektrokhimiya, Vol. 10, p. 45, Moscow, VINITI 1975
15. Krishtalik, L. I., Alpatova, N. M.: J. Electroanal. Chem. *65*, 219 (1975)
16. Krishtalik, L. I., Alpatova, N. M.: Elektrokhimiya *12*, 163 (1976)

17. Alpatova, N. M., Krishtalik, L. I.: in: Itogi nauki i tekhniki, ser. Elektrokhimiya, Vol. 15, p. 132, Moscow, VINITI 1979
18. Cady, H. P.: J. Phys. Chem. *1*, 707 (1896—1897)
19. Laitinen, H. A., Nyman, C. J.: J. Amer. Chem. Soc. *70*, 3002 (1948)
20. Avaka, L. A., Bewick, A.: J. Electroanal. Chem. *41*, 395 (1973)
21. Kanzaki, Y., Aoyagui, S.: J. Electroanal. Chem. *36*, 297 (1972)
22. Krishtalik, L. I., Alpatova, N. M. Fomicheva, M. G.: Elektrokhimiya *7*, 1393 (1971)
23. Landau, L. D.: Phys. Z. Sov. Union *3*, 664 (1933)
24. Pekar, S. I.: Zh. eksp. i teor. fiz. *16*, 341 (1946)
25. Davydov, A. S.: Zh. eksp. i teor. fiz. *18*, 913 (1948)
26. Jortner J. In: Actions chimiques et biologiques des radiations. Paris, Masson, 1970, Ser. 14, p. 7–72.
27. Pleskov, Yu. V., Rotenberg, Z. A.: in: Advances in Electrochemistry and Electrochemical Engineering (eds.) Gerischer, H., Tobias, C., Vol. II, p. 3, New York e. a., Wiley-Interscience 1978
28. Dogonadze, R. R., Krishtalik, L. I., Pleskov, Yu. V.: Elektrokhimiya *10*, 507 (1974)
29. Parsons, R.: in: Modern Aspects of Electrochemistry (ed.) Bockris, J. O'M., Vol. I, p. 130, London, Butterworths 1954
30. Trasatti, S.: J. Electroanal. Chem. *139*, I (1982)
31. Gurevich, Yu. Ya., Pleskov Yu. V.: Elektrokhimiya *18*, 1477 (1982)
32. Bonch-Bruevich, V. L.: Uspekhi fiz. nauk *140*, 583 (1983)
33. Gerischer, H.: in: Physical Chemistry (ed.) Eyring, H., Vol. 9, p. 463, New York, Academic Press 1970
34. Jortner, J., Noyes, R. M.: J. Phys. Chem. *70*, 770 (1966)
35. Schindewolf, U.: J. Phys. Chem. *88*, 3820 (1984)
36. Barker, G. C., Gardner, A. W., Sammon, D. C.: J. Electrochem. Soc. *113*, 1182 (1966)
37. Aldrich, J. E., Bronskill, M. J., Wolff, R. K., Hunt, J. W.: J. Chem. Phys. *55*, 530 (1971)
38. Brodsky, A. M., Gurevich, Yu. Ya.: Teoriya elektronnoi emissii iz metallov, Moscow, Nauka 1973
39. Krohn, C. E., Antoniewicz, P. R., Thompson, J. C.: Surface Sci. *101*, 241 (1980)
40. Sass, J. K., Gerischer, H.: in: Photoemission and the Electronic Properties of Surfaces (eds.) Feuerbacher, B., Fitton, B., Willis, R. F., p. 469, Wiley 1978
41. Delahay, P.: Acc. Chem. Res. *15*, 40 (1982)
42. Gremmo, N., Randles, J. E. B.: J. Chem. Soc. Faraday Trans. *70*, Pt. I, 1478 (1974)
43. Pleskov, Yu. V.: Khim. vys. energii *12*, 296 (1978)
44. Pleskov, Yu. V.: J. Electroanal. Chem. *105*, 227 (1979)
45. Eletsky, V. V., Pleskov, Yu. V.: Elektrokhimiya *13*, 1376 (1977)
46. Kokilashvili, R. G., Eletsky, V. V., Pleskov, Yu. V.: Elektrokhimiya *20*, 1075 (1984)
47. Kokilashvili, R. G., Eletsky, V. V., Pleskov, Yu. V., Dzhaparidze, Dz. I.: Elektrokhimiya *17*, 262 (1981)
48. Itaya, K., Malpas, R. E., Bard, A. J.: Chem. Phys. Lett. *63*, 411 (1979)
49. Rotenberg, Z. A., Gromova, N. V.: Elektrokhimiya *22*, 152 (1986)
50. Lange, P., Sass, J. K., Unwin, R., Tench, D. M.: J. Electroanal. Chem. *122*, 387 (1981)
51. Kaganovich, R. I., Damaskin, B. B., Ganzhina, I. M.: Elektrokhimiya *4*, 867 (1968)
52. Damaskin, B. B., Kaganovich, R. I.: Elektrokhimiya *13*, 293 (1977)
53. Case, B., Parsons, R.: Trans. Faraday Soc. *63*, 1224 (1967)
54. Parsons, R., Rubin, B. T.: J. Chem. Soc. Faraday Trans. *70*, Pt. I, 1636 (1974)
55. Parsons, R., de Valera, E.: Nouv. J. Chim. *2*, III (1978)
56. Randles, J. E. B.: Trans. Faraday Soc. *52*, 1573 (1956)
57. Gomer, R., Tryson, G.: J. Chem. Phys. *66*, 4413 (1977)
58. Farrell, J. R., Mc Tigue, P.: J. Electroanal. Chem. *139*, 37 (1982)
59. Antropov, L. I., Gerasimenko, M. A., Khirkh-Yalan, I. F., Shkolnaya, E. F.: Elektrokhimiya *20*, 1357 (1984)
60. Frumkin, A. N., Iofa, Z. A., Gerovich, M. A.: Zh. fiz. khim. *30*, 1455 (1956)
61. Trasatti, S.: in: Modern Aspects of Electrochemistry and Electrochemical Engineering (eds.) Conway, B., Bockris, J. O'M., Vol. 13, p. 81, New York, Plenum Press, 1979
62. Case, B., Hush, N. S., Parsons, R., Peover, M. E.: J. Electroanal. Chem. *10*, 360 (1965)

63. Lepoutre, G., Jortner, J.: J. Phys. Chem. *76*, 683 (1973)
64. Holroyd, K. A., Tames, S., Kennedy, A.: J. Phys. Chem. *79*, 2857 (1975)
65. Rotenberg, Z. A.: Elektrokhimiya *8*, 1198 (1972)
66. Pleskov, Yu. V., Rotenberg, Z. A.: J. Electroanal. Chem. *20*, I (1969)
67. Brodsky, A. M., Tsarevsky, A. V.: Adv. Chem. Phys. *44*, 483 (1980)
68. Aulich, A., Nemec, L., Delahay, P.: J. Chem. Phys. *61*, 4235 (1974)
69. Gremmo, N., Randles, J. E. B.: J. Chem. Soc. Faraday Trans. *70*, Pt. I, 1488 (1974)
70. German, E. D.: Reviews in Inorganic Chemistry *5*, 123 (1984)
71. Barthel, J.: in: Ionen in nichtwäßrigen Lösungen. Fortschritte der physikalischen Chemie (ed.) Jost, W., Bd. 10 Darmstadt, Dr. D. Steinkoff Verlag 1976
72. Maltsev, E. I., Alpatova, N. M., Vannikov, A. V.: Elektrokhimiya *13*, 263 (1977)
73. Baron, B., Delahay, P., Lugo, R.: J. Chem. Phys. *55*, 4180 (1971)
74. Alpatova, N. M., Maltsev, E. I., Vannikov, A. V., Zabusova, S. E.: Elektrokhimiya *9*, 1034 (1973)
75. Shaede, E. A., Dorfman, L. M., Flynn, G. J., Walker, D. C.: Canad. J Chem. *51*, 3905 (1973)
76. Copeland, D. A., Kestner, N. B., Jortner, J.: J. Chem. Phys. *53*, 1189 (1970)
77. Maltsev, E. I., Golovanov, V. V., Zolotarevskii, V. I., Vannikov, A. V.: Khim. vys. energii *11*, 97 (1977)
78. Alpatova, N. M., Grishina, A. D.: Elektrokhimiya *9*, 1375 (1973)
79. Logan, S. H.: J. Chem. Educ. *44*, 344 (1967)
80. Rabani, J.: in: Solvated Electron, p. 242, Washington, Amer. Chem. Soc. 1965
81. Schindewolf, U.: Chemie in unserer Zeit *4*, 37 (1970)
82. Frumkin, A. N., Bagotsky, V. S., Iofa, Z. A., Kabanov, B. N.: Kinetika elektrodnykh protsessov, Moscow, Izd. MGU 1952
83. Belloni, J., Saito, E.: Nucl. Sci. Abstr. *28*, 150 (1973)
84. Saito, E.: in: Electrons in Fluids, The Nature of Metal-Ammonia Solutions (eds.) Jortner, J., Kestner, N. R., p. 139, Berlin-Heidelberg-New York, Springer-Verlag 1973
85. Belloni, J., Saito, E.: in: Electrons in Fluids. The Nature of Metal-Ammonia Solutions (eds.) Jortner, J., Kestner, N. R., p. 71, Berlin-Heidelberg-New York, Springer-Verlag 1973
86. Baxendale, J. H.: Zh. V. Kh. O. im. Mendeleeva *11*, 168 (1966)
87. Jolly, W.: J. Chem. Educ. *33*, 512 (1956)
88. Putnam, G. L., Kobe, K. A.: Trans. Electrochem. Soc. *74*, 609 (1938)
89. Makishima, S.: J. Faculty Eng. Tokyo Univ. *21*, 115 (1938); C. A. *32*, 8220 (1938)
90. Davidson, A. W., Kleinberg, J., Bennet, W. E., Mc Elroy, A. D.: J. Amer. Chem. Soc. *71*, 377 (1949)
91. Mc Elroy, A. D., Kleinberg, J., Davidson, A. W.: J. Amer. Chem. Soc. *72*, 5178 (1950)
92. Quinn, R. K., Lagowski, J. J.: Inorg. Chem. *9*, 414 (1970)
93. Harima, Y., Kurihara, H., Aoyagui, S.: J. Electroanal. Chem. *124*, 103 (1981)
94. Dainton, F. S., Wiles, D. M., Wright, A. N.: J. Chem. Soc. 4283 (1960)
95. Cafasso, F. A., Sundheim, B. R.: J. Chem. Phys. *31*, 809 (1959)
96. Glarum, S. H., Marshall, J. H.: J. Chem. Phys. *52*, 5555 (1970)
97. Morijiri, M., Itaya, K., Toshima, S.: Rev. Polarography *22*, 107 (1976)
98. Strong, J., Tuttle, T. R.: J. Phys. Chem. *77*, 533 (1973)
99. Down, J. L., Lewis, J., Moore, B., Wilkinson, G.: J. Chem. Soc. 3767 (1959)
100. Mei Tak Lok, Tehan, F. J., Dye, J. L.: J. Phys. Chem. *76*, 2975 (1972)
101. Seddon, W. A., Fletcher, J. W., Sopchyshyn, F. C., Selkirc, E. B.: Canad. J. Chem. *57*, 1792 (1979)
102. Dubois, J. E., Lacaze, P. C., de Fiquelmont, A. M.: C. r. Akad. Sci. Ser. C, *262*, 181 (1966)
103. Gal, J. Y., Yvernault, T.: Bull. Soc. chim. France 2770 (1971)
104. Gal, J. Y., Yvernault, T.: C. r. Acad. Sci. *272*, Ser. C, 42 (1971)
105. Izutsu, K., Sakura, S., Fujinaga, T.: Bull. Chem. Soc. Japan *45*, 445 (1972)
106. Dubois, J. E., Monvernay, A., Lacaze, P. C.: Electrochim. Acta *15*, 315 (1970)
107. Catterall, R.: in: Solutions metal-ammoniac. Proprietes physicochimiques (eds.) Lepoutre, G., Sienko, M., p. 41, New York, Benjamin 1964
108. Nakamura, Y., Yamamoto, M., Shimokava, S.: Bull. Chem. Soc. Japan *44*, 3212 (1971)

109. Gutmann, V.: Coordination Chemistry in Non-Aqueous Solutions, Wien—New York, Springer-Verlag 1968
110. Mayer, U., Gutmann, V., Gerger, W.: Monatsh. Chem. *106*, 1235 (1975)
111. Gutmann, V.: Electrochim. Acta *21*, 661 (1976)
112. Diggle, J. W., Bogsanyi, D.: J. Phys. Chem. *78*, 1018 (1974)
113. Tehan, F. J., Barnett, B. L., Dye, J. L.: J. Amer. Chem. Soc. *96*, 7203 (1974)
114. Dye, J. L., Andrews, C. W., Ceraso, J. M.: J. Phys. Chem. *79*, 3076 (1975)
115. Dye, J. L.: in: Electrons in Fluids. The Nature of Metal-Ammonia Solutions (eds.) Jortner, J., Kestner, N. R., p. 78, Berlin—Heidelberg—New York, Springer-Verlag 1973
116. Dye, J. L.: Scientific American *237*, 92 (1977)
117. Dinh Le, L., Issa, D., Van Eck, B., Dye, J. L.: J. Phys. Chem. *86*, 7 (1982)
118. Ceraso, J. M., Dye, J. L.: J. Chem. Phys. *61*, 1585 (1974)
119. Dye, J. L., De Baster, M., Nicely, V. A.: J. Amer. Chem. Soc. *92*, 5226 (1970)
120. Peter, F., Gross, M.: in: International Society of Electrochemistry. 27th Meeting, Extended Abstracts, No. 240, Zürich 1976
121. Espinola, A., Jordan, J.: in: Charact. Solutes Nonaqueous Solvants. Proc. Symp. Spectrosc. Electrochem. Charact. Solute Species Nonaqueous Solvents, 1976 (ed.) Mamantov, G., p. 311, New York, Plenum Press 1978
122. Isaeva, L. I., Polyakov, P. V., Mikhalev, Yu. G., Dubova, G. F.: in: Tezisy dokladov VIII Vsesoyuznoi konferentsii po fizicheskoi khimii i elektrokhimii ionnykh rasplavov i tverdykh elektrolitov, Vol. 2, p. 88, Leningrad (1983)
123. Solutions metal-ammoniac. Proprietes physico-chimiques. Colloque Weyl (I). Juin 1963 (eds.) Lepoutre, G., Sienko, M. J., New York, Benjamin 1964
124. Metal-Ammonia Solutions. Colloque Weyl II, June 1969 (eds.) Lagowski, J. J., Sienko, M., London, Butterworths 1970
125. Electrons in Fluids. The Nature of Metal-Ammonia Solutions. Colloque Weyl III. June 1972 (eds.) Jortner, J., Kestner, N. R., Berlin—Heidelberg—New York, Springer-Verlag 1973
126. Electrons in Fluids. The Nature of Metal-Ammonia Solutions. Colloque Weyl IV. June 30–July 3 1975, J. Phys. Chem. *79*, 2789 (1975)
127. The Fifth International Conference on Excess Electrons and Metal-Ammonia Solutions. Colloque Weyl V. June 1979, J. Phys. Chem. *84*, 1065 (1980)
128. Khaikin, G. I., Zhigunov, V. A., Dolin, P. I.: Khim. vys. energii *5*, 54 (1971)
129. Quinn, R. K., Lagowski, J. J.: J. Phys. Chem. *72*, 1374 (1968)
130. Quinn, R. K., Lagowski, J. J.: J. Phys. Chem. *73*, 2326 (1969)
131. Doughit, R. C., Dye, J. L.: J. Amer. Chem. Soc. *82*, 4472 (1960)
132. Mal'tsev, E. I., Vannikov, A. V.: Radiation Effects *20*, 197 (1973)
133. Broocks, J. M., Dewald, R. R.: J. Phys. Chem. *72*, 2655 (1968)
134. Alpatova, N. M., Grishina, A. D.: Elektrokhimiya, *7*, 853 (1971)
135. Alpatova, N. M., Grishina, A. D., Fomicheva, M. G.: Elektrokhimiya *8*, 253 (1972)
136. Grishina, A. D., Vannikov, A. V., Alpatova, N. M.: Elektrokhimiya *12*, 1507 (1976)
137. Grishina, A. D., Vannikov, A. V., Alpatova, N. M.: Radiat. Phys. and Chem. *11*, 289 (1976)
138. Akhmanov, S. A., Aslanyan, L. S., Bunkin, A. F., Zhuravleva, T. S., Vannikov, A. V.: Khim, vys. energii *14*, 417 (1980)
139. Zhuravleva, T. S., Kessenikh, A. V.: Khim, fizika 1464 (1982)
140. Pavlov, Yu. V., Fomicheva, M. G., Mishustin, A. I., Alpatova, N. M.: Elektrokhimiya *9*, 541 (1973)
141. Kessler, Yu. M., Emelin, V. P., Mishustin, A. I., Yastremskii, P. S., Verstakov, E. S., Alpatova, N. M., Fomicheva, M. G., Kireev, K. V., Gruba, V. D., Bratishko, R. Kh.: Zh. strukt. khimii *16*, 797 (1975)
142. Lagowski, J. J.: in: Electrons in Fluids. The Nature of Metal-Ammonia Solutions (eds.) Jortner, J., Kestner, N. R., p. 29, Berlin—Heidelberg—New York, Springer-Verlag 1973
143. Schindewolf, U., Werner, M., J. Phys. Chem. *84*, 1123 (1980)
144. Harris, R. L., Lagowski, J. J.: J. Phys. Chem. *84*, 1091 (1980)
145. Mal'tsev, E. I., Vannikov, A. V.: Dokl. Akad. Nauk SSSR *200*, 379 (1971)
146. Mal'tsev, E. I., Vannikov, A. V., Bach, N. A.: Radiation Effects *11*, 79 (1971)
147. Vannikov, A. V., Alpatova, N. M., Mal'tsev, E. I., Krishtalik, L. I.: Elektrokhimiya *10*, 830 (1974)

148. Vannikov, A. V., Alpatova, N. M.: Elektrokhimiya *11*, 996 (1975)
149. Mal'tsev, E. I., Vannikov, A. V.: Radiat. Phys. and Chem. *10*, 99 (1977)
150. Harima, Y., Kurihara, H., Nishiki, Y., Aoyagui, S., Tokuda, K., Matsuda, H.: Canad. J. Chem. *60*, 445 (1982)
151. Alpatova, N. M., Ovsyannikova, E. V., Zabusova, S. E.: in: Itogi nauki i tekhniki, ser. Elektrokhimiya, Vol. *23*, Moscow, VINITI 1986
152. Rubinstein, G., Tuttle, T. R., Golden, S.: J. Phys. Chem. *77*, 2872 (1973)
153. Evtushenko, N. E., Karasevskii, A. I.: Ukrain, fiz. zh. *27*, 1325 (1982)
154. Bratsch, S. G., Lagowski, J. J.: J. Phys. Chem. *88*, 1086 (1984)
155. Kraus, C. A.: J. Amer. Chem. Soc. *36*, 864 (1914)
156. Russel, J. B., Sienko, M. J.: J. Amer. Chem. Soc. *79*, 4051 (1957)
157. Dye, J. L.: in: Solutions metal-ammoniac. Proprietes physico-chimiques (eds.) Lepoutre, G., Sienko, M. J., p. 137, New York, Benjamin 1964
158. Harima, Y., Aoyagui, S.: Israel J. Chem. *18*, 81 (1979)
159. Teherani, T., Itaya, K., Bard, A. J.: Nouv. J. Chim. *2*, 481 (1979)
160. Bard, A. J., Itaya, K., Malpas, R. E., Teherani, T.: J. Phys. Chem. *84*, 1262 (1980)
161. Lagowski, J. J.: Pure and Appl. Chem. *25*, 429 (1971)
162. Ahrens, M., Heusler, K. E.: Electrochim. Acta *27*, 239 (1982)
163. Alpatova, N. M., Krishtalik, L. I., Ovsyannikova, E. V., Zabusova, S. E.: Elektrokhimiya *9*, 884 (1973)
164. Ovsyannikova, E. V., Krishtalik, L. I., Novitskii, S. P., Burenkov, I. I., Alpatova, N. M.: Elektrokhimiya *18*, 867 (1982)
165. Ovsyannikova, E. V., Krishtalik, L. I., Alpatova, N. M.: Elektrokhimiya, *12*, 1032 (1976)
166. Ovsyannikova, E. V., Alpatova, N. M., Krishtalik, L. I.: Elektrokhimiya *13*, 1094 (1977)
167. Kanzaki, Y., Aoyagui, S.: J. Electroanal. Chem. *51*, 19 (1974)
168. De Battisti, A., Trasatti, S.: J. Electroanal. Chem. *79*, 251 (1977)
169. Gross, W., Schindewolf, U.: J. Phys. Chem. *84*, 1266 (1980)
170. Harima, Y., Aoyagui, S.: J. Electroanal. Chem. *137*, 171 (1982)
171. Gordon, R. P., Sundheim, B. R.,: J. Phys. Chem. *68*, 3347 (1964)
172. Harima, Y., Aoyagui, S.: Rev. Polarography *36*, 47 (1977)
173. Uribe, F. A., Sharp, P. R., Bard, A. J.: J. Electroanal. Chem. *152*, 173 (1983)
174. Wilhelm, S. M.: J. Electrochem. Soc. *126*, 207 C (1979)
175. Van Amerongen, G., Guyomard, G., Heindl, R., Herlem, M., Sculfort, J.-L.: J. Electrochem. Soc. *129*, 1998 (1982)
176. Harima, Y., Kurihara, H., Aoyagui, S.: Canad. J. Chem. *58*, 1151 (1980)
177. Martin, G. W., Murray, R. W.: J. Electroanal. Chem. *98*, 149 (1979)
187. Kanzaki, Y., Aoyagui, S.: J. Electroanal. Chem *47*, 109 (1973)
179. Kanzaki, Y., Aoyagui, S.: in: Japan—USSR Seminar on Electrochemistry, Extended Abstracts, p. 155, Tokyo, 1974
180. Itaya, K., Kawai, M., Toshima, S.: J. Amer. Chem. Soc. *100*, 5996 (1978)
181. Pleskov, Yu. V., Filinovsky, V. Yu.: The Rotating Disk Electrode, New York—London, Consultants Bureau 1976
182. Worley, G., Lagowski, J. J.: in: 179th ACS Nat. Meet. Houston, Tex., Abstr. Pap. Washington, D. C. s. a., p. 571, 1980
183. Teherani, T. H., Peer, W. J., Lagowski, J. J., Bard, A. J.: J. Electrochem. Soc. *125*, 1717 (i978)
184. Peer, W. J., Lagowski, J. J.: J. Amer. Chem. Soc. *100*, 6260 (1978)
185. Teherani, T. H., Peer, W. J., Bard, A. J.: J. Amer. Chem. Soc. *100*, 7768 (1978)
186. Bennet, J., Harney, D., Mitchell, T.: in: 18th Intersoc. Energy. Convers. Eng. Conf., Energy Marketplace, Orlando, Fla, Aug. 21–26, 1983. Proc. Vol. 4, p. 166, New York, 1983
187. Bernard, L., Demortier, A., Lelieur, J. P., Lepoutre, G., t'Kint de Roodenbeke, F., Le Mehaute, A.: J. Phys. Chem. *88*, 3833 (1984)
188. Le Mehaute, A.: J. Power Sources *9*, 167 (1983)
189. Alpatova, N. M., Krishtalik, L. I., Fomicheva, M. G.: Elektrokhimiya *8*, 535 (1972)
190. Krishtalik, L. I., Alpatova, N. H., Fomicheva, M. G.: Croatica Chemica Acta *44*, 1 (1972)
191. Alpatova, N. M., Fomicheva, M. G., Ovsyannikova, E. V., Krishtalik, L. I.: Elektrokhimiya *9*, 1234 (1973)
192. Krishtalik, L. I., Alpatova, N. M., Ovsyannikova, E. V.: Elektrokhimiya *12*, 1493 (1976)

193. Zabusova, S. E., Fomicheva, M. G., Krishtalik, L. I., Alpatova, N. M.: Elektrokhimiya *11*, 1888 (1975)
194. Alpatova, N. M., Zabusova, S. E., Krishtalik, L. I.: Elektrokhimiya *12*, 625 (1976)
195. Zabusova, S. E., Fomicheva, M. G., Alpatova, N. M., Krishtalik, L. I.: Elektrokhimiya *14*, 1619 (1978)
196. Krishtalik, L. I.: Electrochim. Acta *21*, 693 (1976)
197. Frumkin, A. N., Petrii, O. A., Nikolaeva-Fedorovich, N. V.: Dokl. Akad. Nauk SSSR *147*, 878 (1962)
198. Brodsky, A. M., Frumkin, A. N.: Elektrokhimiya *6*, 658 (1970)
199. Fomicheva, M. G., Alpatova, N. M., Zabusova, S. E.: Elektrokhimiya *13*, 216 (1977)
200. Dubois, J. E., Monvernay, A., Lacaze, P. C.: J. Chim. Phys. *70*, 39 (1973)
201. Massignon, D., Le Gresus, C., Dubois, J. E.: in: International Society of Electrochemistry. 29th Meeting, Extended Abstracts, Pt. I, p. 337, Budapest 1978
202. Pianka, M., Owen, B. D.; J. Appl. Chem. *5*, 525 (1955)
203. Kuznetsova, T. V., Kedrinskii, I. A., Ivanov, E. G., Potapova, G. P.: Elektrokhimiya *12*, 1453 (1976)
204. Kedrinskii, I. A., Kuznetsova, T. V., Morozov, S. V., Ivanov, E. G., Grudyanov, I. I.: Elektrokhimiya *12*, 1458 (1976)
205. Ichikawa, K., Thompson, J. C.: In: Electrons in Fluids. Nature of Metal-Ammonia Solutions (eds.) Jortner, J., Kestner, N. R., p. 231, Berlin—Heidelberg—New York, Springer-Verlag 1973
206. Ichikawa, K., Thompson, J. C.: J. Chem. Phys. *59*, 1680 (1973)
207. Sato, H., Harima, Y., Aoyagui, S.: J. Electroanal. Chem. *144*, 449 (1983)
208. Krishtalik, L. I., Kuznetsov, A. M.: Elektrokhimiya *22*, 246 (1986)
209. Huges, G., Roach, R. J.: Chem. Communs. 600 (1965)
210. Hills, G. J., Kinnibrugh, D. K.: J. Electrochem. Soc. *113*, 1111 (1966)
211. Walker, D. C.: Canad. J. Chem. *44*, 2226 (1966)
212. Walker, D. C.: Canad. J. Chem. *45*, 807 (1967)
213. Pyle, T., Roberts, C.: J. Electrochem. Soc. *115*, 247 (1968)
214. Jansta, J., Dousek, F. P., Riga, J.: J. Electroanal. Chem. *44*, 263 (1973)
215. Postl, D., Schindewolf, U.: Ber. Bunsenges. phys. Chem. *75*, 662 (1971)
216. Parsons, R.: J. Electrochem. Soc. *113*, 1118 (1966)
217. Krishtalik, L. I.: J. Electrochem. Soc. *113*, 1117 (1966)
218. Conway, B. E., Mc Kinnon, D. J.: J. Phys. Chem. *74*, 3663 (1970)
219. Bewick, A., Conway, B. E., Tuxford, A. M.: J. Electroanal. Chem. *42*, Appl. II (1973)
220. Krishtalik, L. I., Fomicheva, M. G., Alpatova, N. M.: Elektrokhimiya *8*, 629 (1972)
221. Krishtalik, L. I., Alpatova, N. M., Fomicheva, M. G.: in: International Society of Electrochemistry, 22nd Meeting, Extended Abstracts, p. 113, Dubrovnik 1971
222. Krishtalik, L. I.: in: Advances in Electrochemistry and Electrochemical Ingineering (ed.) Delahay, P., Vol. 7, p. 283, New York, Interscience 1970
223. Korshunov, N. V.: in: Sovremennye problemy polarografii s nakopleniem, p. 27, Tomsk 1975
224. Rabinovich, V. A., Khavin, Z. A.: Kratkii khimicheskii spravochnik, Moscow, Khimiya 1977
225. Wolthorn, H. J., Fernelius, W. C.: J. Amer. Chem. Soc. *56*, 1551 (1934)
2261 Bennet, J. E., Mile, B., Thomas, A.: Nature *201* 919 (1964)
227. Pleskov, V. A.: Uspekhi khimii *16*, 254 (1947)
228. Pleskov, V. A.: Zh. fiz, khim. *13*, 1449 (1939)
229. Smith, H.: in: Chemistry in Nonaqueous Ionizing Solvents Vol. I, Pt. 2. New York—London, Interscience Publishers 1963
230. Dye, J. L., Sankuer, R. F., Smith, G. E.: J. Amer. Chem. Soc. *82*, 4797 (1960)
231. Vannikov, A. V.: Uspekhi khimii *44*, 1931 (1975)
232. Avaca, L. A., Bewick, A.: J. Appl. Electrochem. *2*, 349 (1972)
233. Kryukova, T. A., Sinyakova, S. I., Aref'eva, T. V.: Polarograficheskii analiz, Moscow, Goskhimizdat 1959
234. Demortier, A., Bard, A. J.: J. Amer. Chem. Soc. *95*, 3495 (1973)
235. Dewald, R. R., Jones, S. R., Schwartz, B. S.: J. Phys. Chem. *84*, 3272 (1980)
236. Sternberg, W., Markby, R., Wender, I.: J. Electrochem. Soc. *110*, 425 (1963)
237. Dubois, J. E., Dobin, G.: Tetrahedron Letters 2325 (1969)
238. Asahara, S., Senoo, M.: Pat. Japan 7240786. (1972); C. A. *78*, 3805 (1973)

239. Schindewolf, U., Neumann, B.: J. Phys. Chem. *83*, 423 (1979)
240. Birch, A. J., Smith, H.: Q. Rev. (Chem. Soc.) *12*, 17 (1958)
241. Krapcho, A. P., Bothner-By, A. A.: J. Amer. Chem. Soc. *81*, 3658 (1959)
242. Eastham, J. F., Keenan, C. W., Secor, H. V.: J. Amer. Chem. Soc. *81*, 6523 (1959)
243. Hatayama, T., Hamano, Y., Yamamoto, T. US Pat. 3700572 (1972); C. A. *78*, 37265 (1973)
244. Wagenknecht, J. H., Coleman, J. P.: in: Extended Abstracts, Electrochem. Soc. 80-1, p. 1075 (1980)
245. Misra, R. A., Yadav, A. R.: Bull. Chem. Soc. Japan *55*, 347 (1982)
246. Benkeser, R., Kaiser, E.: J. Amer. Chem. Soc. *85*, 2858 (1963)
247. Benkeser, R., Kaiser, E., Lambert, R.: J. Amer. Chem. Soc. *86*, 5272 (1964)
248. Misono, A., Osa, T., Yamagishi, T., Kadama, T.: J. Electrochem. Soc. *115*, 266 (1968)
249. Moritake, M., Fujii, M., Okazaki, T.: Japan Kokai 7379843 (1973); C. A. *80*, 59636 (1974)
250. Fujii, M., Moritake, M., Okazaki, T.: Japan Kokai 7379844 (1973); C. A. *80*, 59635 (1974)
251. Sternberg, H. W., Markby, R. E., Wender, J., Mohilner, D. M.: J. Amer. Chem. Soc. *89*, 186 (1967)
252. Lund, H.: in: Organic Electrochemistry, (eds.) Baizer, M. M., Lund, H., p. 873, New York, M. Dekker 1983 [2]
253. White, G. F., Morrison, A. B., Anderson, E. G. E.: J. Amer. Chem. Soc. *46*, 961 (1924)
254. Bolton, G. L., Robinson, M. G., Freeman, G. R.: Can. J. Chem. *54*, 1177 (1976)
255. Dewald, R. R., Dye, J. L., Eigen, M., Le Mayer, L.: J. Chem. Phys. *39*, 2388 (1963)
256. Dewald, R. R., Bezirjian, O. H.: J. Phys. Chem. *74*, 4155 (1970)
257. Zabusova, S. E., Tomilov, A. P., Filimonova, L. F., Alpatova, N. M.: Elektrokhimiya *16*, 970 (1980)
258. Coleman, J. P., Kobylecky, R. J., Utley, J. H. P.: Chem. Communs. 104 (1971)
259. Itaya, K., Kawai, M., Toshima, S.: Chem. Phys. Letters *42*, 179 (1976)
260. Itaya, K., Toshima, S.: Rev. Polarography *22*, 40 (1976)
261. Doblhofer, K., Gerischer, H.: Electrochim. Acta *20*, 215 (1975)
262. Itaya, K., Bard, A. J.: J. Phys. Chem *85*, 1358 (1981)

Author Index Volumes 101–138

Contents of Vols. 50–100 see Vol. 100
Author and Subject Index Vols. 26–50 see Vol. 50

The volume numbers are printed in italics

Alekseev, N. V., see Tandura, St. N.: *131*, 99–189 (1985).
Alpatova, N. M., Krishtalik, L. I., Pleskov, Y. V.: Electrochemistry of Solvated Electrons, *138*, 149–220 (1986).
Anders, A.: Laser Spectroscopy of Biomolecules, *126*, 23–49 (1984).
Asami, M., see Mukaiyama, T.: *127*, 133–167 (1985).
Ashe, III, A. J.: The Group 5 Heterobenzenes Arsabenzene, Stibabenzene and Bismabenzene. *105*, 125–156 (1982).
Austel, V.: Features and Problems of Practical Drug Design, *114*, 7–19 (1983).

Badertscher, M., Welti, M., Portmann, P., and Pretsch, E.: Calculation of Interaction Energies in Host-Guest Systems. *136*, 17–80 (1986).
Balaban, A. T., Motoc, I., Bonchev, D., and Mekenyan, O.: Topilogical Indices for Structure-Activity Correlations, *114*, 21–55 (1983).
Baldwin, J. E., and Perlmutter, P.: Bridged, Capped and Fenced Porphyrins. *121*, 181–220 (1984).
Barkhash, V. A.: Contemporary Problems in Carbonium Ion Chemistry I. *116/117*, 1–265 (1984).
Barthel, J., Gores, H.-J., Schmeer, G., and Wachter, R.: Non-Aqueous Electrolyte Solutions in Chemistry and Modern Technology. *11*, 33–144 (1983).
Barron, L. D., and Vrbancich, J.: Natural Vibrational Raman Optical Activity. *123*, 151–182 (1984).
Beckhaus, H.-D., see Rüchardt, Ch., *130*, 1–22 (1985).
Bestmann, H. J., Vostrowsky, O.: Selected Topics of the Wittig Reaction in the Synthesis of Natural Products. *109*, 85–163 (1983).
Beyer, A., Karpfen, A., and Schuster, P.: Energy Surfaces of Hydrogen-Bonded Complexes in the Vapor Phase. *120*, 1–40 (1984).
Binger, P., and Büch, H. M.: Cyclopropenes and Methylenecyclopropanes as Multifunctional Reagents in Transition Metal Catalyzed Reactions. *135*, 77–151 (1986).
Böhrer, I. M.: Evaluation Systems in Quantitative Thin-Layer Chromatography, *126*, 95–188 (1984).
Boekelheide, V.: Syntheses and Properties of the [2_n] Cyclophanes, *113*, 87–143 (1983).
Bonchev, D., see Balaban, A. T., *114*, 21–55 (1983).
Borgstedt, H. U.: Chemical Reactions in Alkali Metals *134*, 125–156 (1986).
Bourdin, E., see Fauchais, P.: *107*, 59–183 (1983).
Büch, H. M., see Binger, P.: *135*, 77–151 (1986).

Cammann, K.: Ion-Selective Bulk Membranes as Models. *128*, 219–258 (1985).
Charton, M., and Motoc, I.: Introduction, *114*, 1–6 (1983).
Charton, M.: The Upsilon Steric Parameter Definition and Determination, *114*, 57–91 (1983).
Charton, M.: Volume and Bulk Parameters, *114*, 107–118 (1983).
Chivers, T., and Oakley, R. T.: Sulfur-Nitrogen Anions and Related Compounds. *102*, 117–147 (1982).
Christoph, B., see Gasteiger, J.: *137*, 19–73 (1986).

Collard-Motte, J., and Janousek, Z.: Synthesis of Ynamines, *130*, 89–131 (1985).
Consiglio, G., and Pino, P.: Asymmetrie Hydroformylation. *105*, 77–124 (1982).

Coudert, J. F., see Fauchais, P.: *107*, 59–183 (1983).
Cox, G. S., see Turro, N. J.: *129*, 57–97 (1985).
Czochralska, B., Wrona, M., and Shugar, D.: Electrochemically Reduced Photoreversible Products of Pyrimidine and Purine Analogues. *130*, 133–181 (1985).

Dhillon, R. S., see Suzuki, A.: *130*, 23–88 (1985).
Dimroth, K.: Arylated Phenols, Aroxyl Radicals and Aryloxenium Ions Syntheses and Properties. *129*, 99–172 (1985).
Dyke, Th. R.: Microwave and Radiofrequency Spectra of Hydrogen Bonded Complexes in the Vapor Phase. *120*, 85–113 (1984).

Ebel, S.: Evaluation and Calibration in Quantitative Thin-Layer Chromatography. *126*, 71–94 (1984).
Ebert, T.: Solvation and Ordered Structure in Colloidal Systems. *128*, 1–36 (1985).
Edmondson, D. E., and Tollin, G.: Semiquinone Formation in Flavo- and Metalloflavoproteins. *108*, 109–138 (1983).
Eliel, E. L.: Prostereoisomerism (Prochirality). *105*, 1–76 (1982).
Emmel, H. W., see Melcher, R. G.: *134*, 59–123 (1986).
Endo, T.: The Role of Molecular Shape Similarity in Spezific Molecular Recognition. *128*, 91–111 (1985).

Fauchais, P., Bordin, E., Coudert, F., and MacPherson, R.: High Pressure Plasmas and Their Application to Ceramic Technology. *107*, 59–183 (1983).
Franke, J., and Vögtle, F.: Complexation of Organic Molecules in Water Solution. *132*, 135–170 (1986).
Fujita, T., and Iwamura, H.: Applications of Various Steric Constants to Quantitative Analysis of Structure-Activity Relationship. *114*, 119–157 (1983).
Fujita, T., see Nishioka, T.: *128*, 61–89 (1985).

Gann, L.: see Gasteiger, J.: *137*, 19–73 (1986).
Gasteiger, J., Hutchings, M. G., Christoph, B., Gann, L., Hiller, C., Löw, P., Marsili, M., Saller, H., Yuki, K.: A New Treatment of Chemical Reactivity: Development of EROS, an System for Reaction Prediction and Synthesis Design, *137*, 19–73 (1986).
Gärtner, A., and Weser, U.: Molecular and Functional Aspects of Superoxide Dismutases. *132*, 1–61 (1986).
Gerson, F.: Radical Ions of Phases as Studied by ESR and ENDOR Spectroscopy. *115*, 57–105 (1983).
Gielen, M.: Chirality, Static and Dynamic Stereochemistry of Organotin Compounds. *104*, 57–105 (1982).
Ginsburg, D.: Of Propellanes — and Of Spirans, *137*, 1–17 (1986).
Gores, H.-J., see Barthel, J.: *111*, 33–144 (1983).
Green, R. B.: Laser-Enhanced Ionization Spectroscopy. *126*, 1–22 (1984).
Groeseneken, D. R., see Lontie, D. R.: *108*, 1–33 (1983).
Gurel, O., and Gurel, D.: Types of Oscillations in Chemical Reactions. *118*, 1–73 (1983).
Gurel, D., and Gurel, O.: Recent Developments in Chemical Oscillations. *118*, 75–117 (1983).
Gutsche, C. D.: The Calixarenes. *123*, 1–47 (1984).

Heilbronner, E., and Yang, Z.: The Electronic Structure of Cyclophanes as Suggested by their Photoelectron Spectra. *115*, 1–55 (1983).
Heller, G.: A Survey of Structural Types of Borates and Polyborates. *131*, 39–98 (1985).
Hellwinkel, D.: Penta- and Hexaorganyl Derivatives of the Main Group Elements. *109*, 1–63 (1983).
Hess, P.: Resonant Photoacoustic Spectroscopy. *111*, 1–32 (1983).
Heumann, K. G.: Isotopic Separation in Systems with Crown Ethers and Cryptands. *127*, 77–132 (1985).
Hilgenfeld, R., and Saenger, W.: Structural Chemistry of Natural and Synthetic Ionophores and their Complexes with Cations. *101*, 3–82 (1982).
Hiller, C.: see Gasteiger, J., *137*, 19–73 (1986).

Holloway, J. H., see Selig, H.: *124*, 33–90 (1984).
Hutchings, M. G.: see Gasteiger, J., 19–73 (1986).

Iwamura, H., see Fujita, T.: *114*, 119–157 (1983).

Janousek, Z., see Collard-Motte, J.: *130*, 89–131 (1985).
Jørgensen, Ch. K.: The Problems for the Two-electron Bond in Inorganic Compounds. *124*, 1–31 (1984).
Jurczak, J., and Pietraszkiewicz, M.: High-Pressure Synthesis of Cryptands and Complexing Behaviour of Chiral Cryptands. *130*, 183–204 (1985).

Kaden, Th. A.: Syntheses and Metal Complexes of Aza-Macrocycles with Pendant Arms having Additional Ligating Groups. *121*, 157–179 (1984).
Kanaoka, Y., see Tanizawa, K.: *136*, 81–107 (1986).
Karpfen, A., see Beyer, A.: *120*, 1–40 (1984).
Káš, J., Rauch, P.: Labeled Proteins, Their Preparation and Application. *112*, 163–230 (1983).
Keat, R.: Phosphorus(III)-Nitrogen Ring Compounds. *102*, 89–116 (1982).
Keller, H. J., and Soos, Z. G.: Solid Charge-Transfer Complexes of Phenazines. *127*, 169–216 (1985).
Kellogg, R. M.: Bioorganic Modelling – Stereoselective Reactions with Chiral Neutral Ligand Complexes as Model Systems for Enzyme Catalysis. *101*, 111–145 (1982).
Kimura, E.: Macrocyclic Polyamines as Biological Cation and Anion Complexones – An Application to Calculi Dissolution. *128*, 113–141 (1985).
Kniep, R., and Rabenau, A.: Subhalides of Tellurium. *111*, 145–192 (1983).
Kobayashi, Y., and Kumadaki, I.: Valence-Bond Isomer of Aromatic Compounds. *123*, 103–150 (1984).
Koglin, E., and Séquaris, J.-M.: Surface Enhanced Raman Scattering of Biomolecules. *134*, 1–57 (1986).
Koptyug, V. A.: Contemporary Problems in Carbonium Ion Chemistry III Arenuim Ions – Strucfure and Reactivity. *122*, 1–245 (1984).
Kosower, E. M.: Stable Pyridinyl Radicals. *112*, 117–162 (1983).
Krebs, S., Wilke, J.: Angle Strained Cycloalkynes. *109*, 189–233 (1983).
Krief, A.: Synthesis and Synthetic Applications of 1-Metallo-1-Selenocyclopropanes and -cyclobutanes and Related 1-Metallo-1-silyl-cyclopropanes. *135*, 1–75 (1986).
Krishtalik, L. I.: see Alpatova, N. M.: *138*, 149–220 (1986).
Kumadaki, I., see Kobayashi, Y.: *123*, 103–150 (1984).

Laarhoven, W. H., and Prinsen, W. J. C.: Carbohelicenes and Heterohelicenes. *125*, 63–129 (1984).
Labarre, J.-F.: Up to-date Improvements in Inorganic Ring Systems as Anticancer Agents. *102*, 1–87 (1982).
Labarre, J.-F.: Natural Polyamines-Linked Cyclophosphazenes. Attempts at the Production of More Selective Antitumorals. *129*, 173–260 (1985).
Laitinen, R., see Steudel, R.: *102*, 177–197 (1982).
Landini, S., see Montanari, F.: *101*, 111–145 (1982).
Lau, K.-L., see Wong, N. C.: *133*, 83—157 (1986).
Lavrent'yev, V. I., see Voronkov, M. G.: *102*, 199–236 (1982).
Lontie, R. A., and Groeseneken, D. R.: Recent Developments with Copper Proteins. *108*, 1–33 (1983).
Löw, P.: see Gasteiger, J., *137*, 19–73 (1986).
Lynch, R. E.: The Metabolism of Superoxide Anion and Its Progeny in Blood Cells. *108*, 35–70 (1983).

Maas, G.: Transition-metal Catalyzed Decomposition of Aliphatic Diazo Compounds — New Results and Applications in Organic Synthesis, *137*, 75–253 (1986).
McPherson, R., see Fauchais, P.: *107*, 59–183 (1983).
Maercker, A., Theis, M.: Some Aspects of the Chemistry of Polylithiated Aliphatic Hydrocarbons, *138*, 1–61 (1986).
Majestic, V. K., see Newkome, G. R.: *106*, 79–118 (1982).
Mali, R. S.: see Narasimhan, N. S.: *138*, 63–147 (1986).

Manabe, O., see Shinkai, S.: *121*, 67–104 (1984).
Margaretha, P.: Preparative Organic Photochemistry. *103*, 1–89 (1982).
Marsili, M.: see Gasteiger, J., *137*, 19–73 (1986).
Martens, J.: Asymmetric Syntheses with Amino Acids. *125*, 165–246 (1984).
Matsui, Y., Nishioka, T., and Fujita, T.: Quantitative Structure-Reactivity Analysis of the Inclusion Mechanism by Cyclodextrins. *128*, 61–89 (1985).
Matzanke, B. F., see Raymond, K. N.: *123*, 49–102 (1984).
Mekenyan, O., see Balaban, A. T.: *114*, 21–55 (1983).
Melcher, R. G., Peter, Th. L., and Emmel, H. W.: Sampling and Sample Preparation of Environmental Material. *134*, 59–123 (1986).
Menger, F. M.: Chemistry of Multi-Armed Organic Compounds. *136*, 1–15 (1986).
Meurer, K. P., and Vögtle, F.: Helical Molecules in Organic Chemistry. *127*, 1–76 (1985).
Montanari, F., Landini, D., and Rolla, F.: Phase-Transfer Catalyzed Reactions. *101*, 149–200 (1982).
Motoc, I., see Charton, M.: *114*, 1–6 (1983).
Motoc, I., see Balaban, A. T.: *114*, 21–55 (1983).
Motoc, I.: Molecular Shape Descriptors. *114*, 93–105 (1983).
Müller, F.: The Flavin Redox-System and Its Biological Function. *108*, 71–107 (1983).
Müller, G., see Raymond, K. N.: *123*, 49–102 (1984).
Müller, W. H., see Vögtle, F.: *125*, 131–164 (1984).
Mukaiyama, T., and Asami, A.: Chiral Pyrrolidine Diamines as Efficient Ligands in Asymmetric Synthesis. *127*, 133–167 (1985).
Murakami, Y.: Functionalited Cyclophanes as Catalysts and Enzyme Models. *115*, 103–151 (1983).
Mutter, M., and Pillai, V. N. R.: New Perspectives in Polymer-Supported Peptide Synthesis. *106*, 119–175 (1982).

Naemura, K., see Nakazaki, M.: *125*, 1–25 (1984).
Nakatsuji, Y., see Okahara, M.: *128*, 37–59 (1985).
Nakazaki, M., Yamamoto, K., and Naemura, K.: Stereochemistry of Twisted Double Bond Systems, *125*, 1–25 (1984).
Narasimhan, N. S., Mali, R. S.: Heteroatom Directed Aromatic Lithiation Reactions for the Synthesis of Condensed Heterocyclic Compounds, *138*, 63–147 (1986).
Newkome, G. R., and Majestic, V. K.: Pyridinophanes, Pyridinocrowns, and Pyridinycryptands. *106*, 79–118 (1982).
Niedenzu, K., and Trofimenko, S.: Pyrazole Derivatives of Boron. *131*, 1–37 (1985).
Nishide, H., see Tsuchida, E.: *132*, 63–99 (1986).
Nishioka, T., see Matsui, Y.: *128*, 61–89 (1985).

Oakley, R. T., see Chivers, T.: *102*, 117–147 (1982).
Ogino, K., see Tagaki, W.: *128*, 143–174 (1985).
Okahara, M., and Nakatsuji, Y.: Active Transport of Ions Using Synthetic Ionosphores Derived from Cyclic and Noncyclic Polyoxyethylene Compounds. *128*, 37–59 (1985).

Paczkowski, M. A., see Turro, N. J.: *129*, 57–97 (1985).
Painter, R., and Pressman, B. C.: Dynamics Aspects of Ionophore Mediated Membrane Transport. *101*, 84–110 (1982).
Paquette, L. A.: Recent Synthetic Developments in Polyquinane Chemistry. *119*, 1–158 (1984).
Peters, Th. L., see Melcher, R. G.: *134*, 59–123 (1986).
Perlmutter, P., see Baldwin, J. E.: *121*, 181–220 (1984).
Pietraszkiewicz, M., see Jurczak, J.: *130*, 183–204 (1985).
Pillai, V. N. R., see Mutter, M.: *106*, 119–175 (1982).
Pino, P., see Consiglio, G.: *105*, 77–124 (1982).
Pleskov, Y. V.: see Alpatova, N. M.: *138*, 149–220 (1986).
Pommer, H., Thieme, P. C.: Industrial Applications of the Wittig Reaction. *109*, 165–188 (1983).
Portmann, P., see Badertscher, M.: *136*, 17–80 (1986).
Pressman, B. C., see Painter, R.: *101*, 84–110 (1982).
Pretsch, E., see Badertscher, M.: *136*, 17–80 (1986).
Prinsen, W. J. C., see Laarhoven, W. H.: *125*, 63–129 (1984).

Rabenau, A., see Kniep, R.: *111*, 145–192 (1983).

Rauch, P., see Káš, J.: *112*, 163–230 (1983).

Raymond, K. N., Müller, G., and Matzanke, B. F.: Complexation of Iron by Siderophores A Review of Their Solution and Structural Chemistry and Biological Function. *123*, 49–102 (1984).

Recktenwald, O., see Veith, M.: *104*, 1–55 (1982).

Reetz, M. T.: Organotitanium Reagents in Organic Synthesis. A Simple Means to Adjust Reactivity and Selectivity of Carbanions. *106*, 1–53 (1982).

Rolla, R., see Montanari, F.: *101*, 111–145 (1982).

Rossa, L., Vögtle, F.: Synthesis of Medio- and Macrocyclic Compounds by High Dilution Principle Techniques. *113*, 1–86 (1983).

Rubin, M. B.: Recent Photochemistry of α-Diketones. *129*, 1–56 (1985).

Rüchardt, Ch., and Beckhaus, H.-D.: Steric and Electronic Substituent Effects on the Carbon-Carbon Bond. *130*, 1–22 (1985).

Rzaev, Z. M. O.: Coordination Effects in Formation and Cross-Linking Reactions of Organotin Macromolecules. *104*, 107–136 (1982).

Saenger, W., see Hilgenfeld, R.: *101*, 3–82 (1982).

Saller, H.: see Gasteiger, J., *137*, 19–73 (1986).

Sandorfy, C.: Vibrational Spectra of Hydrogen Bonded Systems in the Gas Phase. *120*, 41–84 (1984).

Schlögl, K.: Planar Chiral Molecural Structures. *125*, 27–62 (1984).

Schmeer, G., see Barthel, J.: *111*, 33–144 (1983).

Schmidt, G.: Recent Developments in the Field of Biologically Active Peptides. *136*, 109–159 (1986).

Schmidtchen, F. P.: Molecular Catalysis by Polyammonium Receptors. *132*, 101–133 (1986).

Schöllkopf, U.: Enantioselective Synthesis of Nonproteinogenic Amino Acids. *109*, 65–84 (1983).

Schuster, P., see Beyer, A., see *120*, 1–40 (1984).

Schwochau, K.: Extraction of Metals from Sea Water. *124*, 91–133 (1984).

Shugar, D., see Czochralska, B.: *130*, 133–181 (1985).

Selig, H., and Holloway, J. H.: Cationic and Anionic Complexes of the Noble Gases. *124*, 33–90 (1984).

Séquaris, J.-M., see Koglin, E.: *134*, 1–57 (1986).

Shibata, M.: Modern Syntheses of Cobalt(III) Complexes. *110*, 1–120 (1983).

Shinkai, S., and Manabe, O.: Photocontrol of Ion Extraction and Ion Transport by Photofunctional Crown Ethers. *121*, 67–104 (1984).

Shubin, V. G. Contemporary Problemsn Carbonium Ion Chemistry II. *116/117*, 267–341 (1984).

Siegel, H.: Lithium Halocarbenoids Carbanions of High Synthetic Versatility. *106*, 55–78 (1982).

Sinta, R., see Smid, J.: *121*, 105–156 (1984).

Smid, J., and Sinta, R.: Macroheterocyclic Ligands on Polymers. *121*, 105–156 (1984).

Soos, Z. G., see Keller, H. J.: *127*, 169–216 (1985).

Steudel, R.: Homocyclic Sulfur Molecules. *102*, 149–176 (1982).

Steudel, R., and Laitinen, R.: Cyclic Selenium Sulfides. *102*, 177–197 (1982).

Suzuki, A.: Some Aspects of Organic Synthesis Using Organoboranes. *112*, 67–115 (1983).

Suzuki, A., and Dhillon, R. S.: Selective Hydroboration and Synthetic Utility of Organoboranes thus Obtained. *130*, 23–88 (1985).

Szele, J., Zollinger, H.: Azo Coupling Reactions Structures and Mechanisms. *112*, 1–66 (1983).

Tabushi, I., Yamamura, K.: Water Soluble Cyclophanes as Hosts and Catalysts. *113*, 145–182 (1983).

Takagi, M., and Ueno, K.: Crown Compounds as Alkali and Alkaline Earth Metal Ion Selective Chromogenic Reagents. *121*, 39–65 (1984).

Tagaki, W., and Ogino, K.: Micellar Models of Zinc Enzymes. *128*, 143–174 (1985).

Takeda, Y.: The Solvent Extraction of Metal Ions by Grown Compounds. *121*, 1–38 (1984).

Tam, K.-F., see Wong, N. C.: *133*, 83—157 (1986).

Tandura, St., N., Alekseev, N. V., and Voronkov, M. G.: Molecular and Electronic Structure of Penta- and Hexacoordinate Silicon Compounds. *131*, 99–189 (1985).

Tanizawa, K., and Kanaoka, Y.: Design of Biospecific Compounds which Simulate Enzyme-Substrate Interaction. *136*, 81–107 (1986).

Theis, M.: see Maercker, A.: *138*, 1–61 (1986).

Thieme, P. C., see Pommer, H.: *109*, 165–188 (1983).
Tollin, G., see Edmondson, D. E.: *108*, 109–138 (1983).
Trofimenko, S., see Niedenzu, K.: *131*, 1–37 (1985).
Trost, B. M.: Strain and Reactivity: Partners for Delective Synthesis. *133*, 3–82 (1986).
Tsuchida, E., and Nishide, H.: Hemoglobin Model – Artificial Oxygen Carrier Composed of Porphinatoiron Complexes. *132*, 63–99 (1986).
Turro, N. J., Cox, G. S., and Paczkowski, M. A.: Photochemistry in Micelles. *129*, 57–97 (1985).

Ueno, K., see Tagaki, M.: *121*, 39–65 (1984).
Urry, D. W.: Chemical Basis of Ion Transport Specificity in Biological Membranes. *128*, 175–218 (1985).

Veith, M., and Recktenwald, O.: Structure and Reactivity of Monomeric, Molecular Tin(II) Compounds. *104*, 1–55 (1982).
Venugopalan, M., and Vepřek, S.: Kinetics and Catalysis in Plasma Chemistry. *107*, 1–58 (1982).
Vepřek, S., see Venugopalan, M.: *107*, 1–58 (1983).
Vögtle, F., see Rossa, L.: *113*, 1–86 (1983).
Vögtle, F.: Concluding Remarks. *115*, 153–155 (1983).
Vögtle, F., Müller, W. M., and Watson, W. H.: Stereochemistry of the Complexes of Neutral Guests with Neutral Crown Molecules. *125*, 131–164 (1984).
Vögtle, F., see Meurer, K. P.: *127*, 1–76 (1985).
Vögtle, F., see Franke, J.: *132*, 135–170 (1986).
Volkmann, D. G.: Ion Pair Chromatography on Reversed-Phase Layers. *126*, 51–69 (1984).
Vostrowsky, O., see Bestmann, H. J.: *109*, 85–163 (1983).
Voronkov, M. G., and Lavrent'yev, V. I.: Polyhedral Oligosilsequioxanes and Their Homo Derivatives. *102*, 199–236 (1982).
Voronkov, M. G., see Tandura, St. N.: *131*, 99–189 (1985).
Vrbancich, J., see Barron, L. D.: *123*, 151–182 (1984).

Wachter, R., see Barthel, J.: *111*, 33–144 (1983).
Watson, W. H., see Vögtle, F.: *125*, 131–164 (1984).
Welti, M., see Badertscher, M.: *136*, 17–80 (1986).
Weser, U., see Gärtner, A.: *132*, 1–61 (1986).
Wilke, J., see Krebs, S.: *109*, 189–233 (1983).
Wong, N. C., Lau, K.-L., and Tam, K.-F.: The Application of Cyclobutane Derivatives in Organic Synthesis. *133*, 83–157 (1986).
Wrona, M., see Czochralska, B.: *130*, 133–181 (1985).

Yamamoto, K., see Nakazaki, M.: *125*, 1–25 (1984).
Yamamura, K., see Tabushi, I.: *113*, 145–182 (1983).
Yang, Z., see Heilbronner, E.: *115*, 1–55 (1983).
Yuki, K.: see Gasteiger, J., *137*, 19–73 (1986).

Zollinger, H., see Szele, I.: *112*, 1–66 (1983).

Organolithium Compounds/ Solvated Electrons

Some Aspects of the Chemistry of Polylithiated Aliphatic Hydrocarbons

A. Maercker and M. Theis

Heteroatom Directed Aromatic Lithiation Reactions for the Synthesis of Condensed Heterocyclic Compounds

N. S. Narasimhan and R. S. Mali

Electrochemistry of Solvated Electrons

N. M. Alpatova, L. I. Krishtalik and Yu. V. Pleskov

AKADEMIE-VERLAG BERLIN

www.ingramcontent.com/pod-product-compliance
Lightning Source LLC
LaVergne TN
LVHW060623300926
849162LV00008B/1700